BIRKHÄUSER

Progress in Mathematical Physics
Volume 58

Joan Girbau
Lluís Bruna

Stability by Linearization of Einstein's Field Equation

Birkhäuser

Authors:

Joan Girbau
Lluís Bruna
Departament de Matemàtiques
Universitat Autònoma de Barcelona
Edifici C
08193 Bellaterra (Cerdanyola del Vallès) – Barcelona
Spain
e-mail: girbau@mat.uab.cat
lbruna@xtec.cat

2000 Mathematics Subject Classification: 83C05, 83C35, 58J35, 58J45, 58J05, 53C80

Library of Congress Control Number: 2010920204

Bibliographic information published by Die Deutsche Bibliothek.
Die Deutsche Bibliothek lists this publication in the Deutsche Nationalbibliografie; detailed bibliographic data is available in the Internet at http://dnb.ddb.de

ISBN 978-3-0346-0303-4

P.O. Box 133, CH-4010 Basel, Switzerland
Part of Springer Science+Business Media
Printed on acid-free paper produced from chlorine-free pulp. TCF ∞
Printed in Germany

ISBN 978-3-0346-0303-4
9 8 7 6 5 4 3 2 1

e-ISBN 978-3-0346-0304-1
www.birkhauser.ch

A les nostres famílies

Contents

Introduction

The fundamental aim of this monograph is to present in a coherent and self-contained manner the main results on stability by linearization of Einstein's equation, both in the case of the vacuum and in the case of matter. Before describing the contents of each chapter in detail, a brief explanation of the concept of stability is in order.

In classical mechanics, the potential V of the gravitational field created by a distribution of matter of density ρ fulfils Poisson's equation

$$\Delta V = 4\pi K\rho ,$$

where Δ is the Laplace operator in $\mathbb{R}^3$, $\Delta = \sum_i \partial^2/\partial x_i^2$, and K is the gravitational constant. However, in general relativity, the equation analogous to the foregoing is Einstein's equation, which relates the Lorentzian metric $\tilde{g}$ (which is the concept that replaces the gravitational potential V of classical mechanics) and the stress-energy tensor T of the matter responsible for the gravitational field. Einstein's equation is written

$$G(\tilde{g}) = \chi T ,$$

where χ is a universal constant and $G(\tilde{g})$ is Einstein's tensor of the metric $\tilde{g}$ (see Chapter II for the precise definitions).

Since Einstein's equation, written in coordinates, is equivalent to a nonlinear system of 10 second-order partial differential equations, it is no surprise that, except for problems with a high degree of symmetry, in most situations it is impossible to obtain exact solutions to this equation. It is for this reason that, since the introduction of the theory, the usual practice in many situations has been to replace the true equation with its linearization, in the belief that solutions to the latter depart very little from solutions to the true equation. However, it was not until the 1970s that certain researchers began to ask if this way of proceeding was in fact correct. Thus Y. Choquet-Bruhat and S. Deser [22] proved that in the case of metrics close to Minkowski's metric $\tilde{\eta}$ of $\mathbb{R}^4$, it made sense to linearize Einstein's equation at the initial metric $\tilde{\eta}$. Let us state here that when solutions of the linearized equation with respect to an initial metric differ little from solutions to the true equation (and, therefore, the linearization makes sense), Einstein's equation is said to be stable by linearization at the initial metric. This would be the case, for example, for Minkowski's initial metric.

The foregoing case of the linearization of Einstein's equation at Minkowski's initial metric (which was addressed by Einstein) is particularly interesting from many points of view, and it is at the origin of the concept of the gravitational wave, as will be explained later. When there is no matter to create a gravitational field, we are dealing with the case of special relativity, and the corresponding metric is Minkowski's metric $\tilde{\eta}$, which fulfils Einstein's equation in the vacuum $G(\tilde{\eta}) = 0$. When we then address the case of a gravitational field created by matter with a small stress-energy tensor T, the Lorentzian metric $\tilde{g}$ associated with the gravitational field is assumed to be a small perturbation of Minkowski's metric; that is, $\tilde{g} = \tilde{\eta} + \tilde{h}$, with $\tilde{h}$ being small. Then the linearization of Einstein's equation at the initial metric $\tilde{\eta}$ can be written with appropriate coordinates (see Chapter III) in the form

$$\frac{1}{2}\left(\frac{1}{c^2}\frac{\partial^2}{\partial t^2} - \frac{\partial^2}{\partial x_1^2} - \frac{\partial^2}{\partial x_2^2} - \frac{\partial^2}{\partial x_3^2}\right)\tilde{h}_{\alpha\beta} = \chi S_{\alpha\beta} \ ,$$

where S is the tensor defined in terms of the stress-energy tensor T by

$$S = T - \frac{1}{2}(\mathrm{tr}_{\tilde{\eta}} T)\tilde{\eta} \ .$$

In regions where there is no matter (in the vacuum) we have $T = S = 0$, and then the linearization of Einstein's equation in this region is none other than the wave equation

$$\left(\frac{1}{c^2}\frac{\partial^2}{\partial t^2} - \frac{\partial^2}{\partial x_1^2} - \frac{\partial^2}{\partial x_2^2} - \frac{\partial^2}{\partial x_3^2}\right)\tilde{h}_{\alpha\beta} = 0 \ .$$

Therefore, if it makes sense to substitute Einstein's equation by its linearization, the metric $\tilde{g}$ corresponding to the gravitational field created is approximately the sum of Minkowski's metric and a small perturbation $\tilde{h}$ whose components $\tilde{h}_{\alpha\beta}$ are propagated in the vacuum as if they were waves (since they fulfil the wave equation).

In the 1970s, stability by linearization of Einstein's equation in the vacuum was extensively studied in the literature, and results of great beauty were obtained which we believe should be far better known than they are, both by mathematicians and by physicists. In addition to the already cited work by Y. Choquet-Bruhat and S. Deser, there is a series of important works in this regard, among which we may mention [23], [32], [34], [52], [2], [49]. An extensive bibliography can be found on the subject, for example, in [50]. For recent results we should cite [8], [26] and [27].

All the above-mentioned works refer to Einstein's equation in the vacuum. However, very little literature exists on stability by linearization of Einstein's equation with matter, a case that is particularly interesting from the point of view of physics. For instance, if we interpret the universe as a Robertson-Walker model of metric $\tilde{g}$ and stress-energy tensor T, the explosion of a supernova in a distant galaxy will produce a perturbation of the initial stress-energy tensor T, which will give rise to a perturbation of the initial metric $\tilde{g}$. The perturbed universe will cease to be of the Robertson-Walker type. In order to study the effects of such an explosion, is it admissible or not to work with the linearized Einstein equation? Note that this is not a question of linearizing Einstein's equation in the vacuum, but rather of linearizing Einstein's equation with matter (since Robertson-Walker models describe universes with matter). The authors of this monograph studied precisely this case in [14], [15] and [16].

We are now ready to describe the contents of the book in detail. The first two chapters cover preliminary material. Chapter I consists of a presentation (very concise and schematic) of the concepts on pseudo-Riemannian manifolds that are necessary for an understanding of the rest of this monograph. Its purpose is to clarify from the outset both the notation used and the definitions of basic concepts. While the theory is set out in an ordered manner in this chapter, together with references to most of the proofs, its content implicitly presupposes certain prior (very basic) knowledge on the part of the reader as regards Riemannian manifolds. Chapter II consists of an introduction to the theory of relativity, both special and general. Unlike the preceding chapter, the theory is set out in a much more intuitive way and is addressed to readers having no prior knowledge of the subject. Thus this chapter may be of use for mathematicians who have no background in physics.

Since the main aim of the book, as already stated, is to present the results on stability by linearization of Einstein's equation, before dealing fully with the subject we consider it appropriate to present in Chapter III the oldest, most basic and most paradigmatic example of linearization of this equation (studied by Einstein himself in his works [30] and [31])): the linearization corresponding to the case where the initial metric is Minkowski's metric and which relates Einstein's equation with the wave equation.

The techniques normally employed in the study of stability by linearization of Einstein's equation do not use the solutions to the equation directly, but rather Cauchy's data of these solutions in a 3-dimensional hypersurface. This makes the Cauchy problem for Einstein's equation an essential tool in all the results on stability by linearization. Nevertheless, Cauchy's problem for Einstein's equation with matter constitutes an enormous field of research that has given rise to a vast amount of literature. An extensive, up-to-date bibliography on the subject can be found, for example, in the article by H. Friedrich and A. Rendall [37]. Given such a plethora of literature, it is a far from easy task to choose a path to follow in order to make a self-contained, intelligible presentation which will be sufficient for the results on stability by linearization both in the vacuum and with matter. This is what we have attempted in Chapter IV of this monograph, which we trust will be attractive for a broad range of readers (and not just those interested in stability by linearization of Einstein's equation). The approach we adopt for proving the fundamental theorem in Chapter IV is inspired by Y. Choquet-Bruhat's [21] paper, published in 1952, although in that work only Cauchy's problem for Einstein's equation in the *vacuum* is considered.

In Chapters V, VI, VII and VIII we deal with the body of the subject itself which gives its name to this monograph.

Chapter V covers the basic concepts of stability by linearization of Einstein's equation: stability in the vacuum, used in most of the works published in the 1970s and early 1980s (refs: [22], [23], [32], [34], [2], [49], [52]), and the new concept of stability with matter ([14], [15] and [16]).

Chapter VI contains the main results on the linearization of Einstein's equation in the case where the initial submanifold M on which the Cauchy data are given is compact (results by A. Fischer, J. E. Marsden and V. Moncrief, [34], [52]).

Chapter VII is basically devoted to the proof of stability in the case where the initial metric is Minkowski's metric. The result by Y. Choquet-Bruhat and S. Deser (in the vacuum), [22], is adapted to the case of weak gravitational fields created by a distribution of matter from the point of view of the new concept of stability with matter introduced in Chapter V. For the proof of this result, it is necessary to use a theorem by M. Cantor ([17], [18]), which establishes that the Euclidean Laplacian in $\mathbb{R}^n$ gives an isomorphism between certain Sobolev spaces with weights. The proof of this result provided in this monograph is different from that in M. Cantor's original papers in order to underline the reason why Sobolev spaces with weights are necessary here rather than ordinary Sobolev spaces.

Y. Choquet-Bruhat, A. Fischer and J. E. Marsden, [23], extended the above-mentioned result on stability at the initial Minkowski metric to the case where the initial metric (instead of Minkowski's metric) converges to Minkowski's metric at infinity in any space-like direction. Section VII.5 in Chapter VII is devoted to this result, which we adapt to the case where matter is present. The version we give of it (Theorem VII.8) generalizes the result by the cited authors even in the case of the vacuum, which is the case they considered. The proof of Theorem VII.8 requires strong results on Fredholm operators in $\mathbb{R}^n$ due to L. A. Bagirov, V. A. Kondrat'ev, R. B. Lockhart and R. C. McOwen, [6], [45], [46], [47], [48]. We have attempted to summarize all the results by these authors that are required for the proof of our result in a single theorem (Theorem VII.14), in such a way that the reader may follow our proof simply on the basis of its statement.

Finally, Chapter VIII is devoted to the study of stability by linearization of Einstein's equation when the initial metric and the initial stress-energy tensor correspond to a Robertson-Walker cosmological model ([14], [15] and [16]). The main theorems in this chapter require certain technical results related with the Laplacian in hyperbolic space ([13], [16] and [12]), for some of which a proof different from that in the original papers is also given. We should also cite [1], [38] and [42] for analogous results on asymptotically hyperbolic manifolds involving weighted Sobolev spaces (while the results in [13], [16] and [12] concern usual Sobolev spaces).

Acknowledgements

The authors wish to express their gratitude for the invaluable help provided by different people in the writing of this monograph. First of all, to Joaquim Bruna (the second author's brother) for providing us with some notes of his own which form the basis for Section VII.3, Chapter VII, and Subsection VIII.3.1, Chapter VIII.

We also wish to thank Xavier Cabré and Robert Lockhart for the bibliographic information they provided for us regarding elliptic operators and Fredholm operators in $\mathbb{R}^n$. We should also mention the many conversations that the first author held with Josep Maria Burguès concerning different subjects related with the monograph, especially those on Sobolev spaces and elliptic operators.

Chapter I

Pseudo-Riemannian Manifolds

This chapter is a brief summary of the main concepts and results about pseudo-Riemannian and Lorentzian manifolds which will be widely used in the rest of the book. While the theory is presented from the beginning in a systematic way, some proofs have been omitted and others are simply hinted at. That is why the reading of this chapter assumes previous knowledge of the subject. We refer to [53] for further treatment.

I.1 Connections

Let M be a C^∞ differentiable manifold of dimension n. Let $\mathcal{X}(M)$ be the set of C^∞ vector fields on M and $\mathcal{F}(M)$ the set of C^∞ functions. $\mathcal{X}(M)$ is an $\mathcal{F}(M)$-module. A *connection* or a *covariant derivative* on M is an operator

$$\begin{array}{rccc} \nabla : & \mathcal{X}(M) \times \mathcal{X}(M) & \longrightarrow & \mathcal{X}(M) \\ & (X, Y) & \rightarrow & \nabla_X Y \end{array}$$

satisfying the following properties:

a) $\nabla_{X_1+X_2} Y = \nabla_{X_1} Y + \nabla_{X_2} Y$.

b) $\nabla_X (Y_1 + Y_2) = \nabla_X Y_1 + \nabla_X Y_2$.

c) $\nabla_{fX} Y = f \nabla_X Y$, where $f \in \mathcal{F}(M)$.

d) $\nabla_X (fY) = X(f) Y + f \nabla_X Y$, where $f \in \mathcal{F}(M)$.

We say that a connection ∇ is without *torsion* if it satisfies the following:

$$\nabla_X Y - \nabla_Y X = [X, Y] \text{ for each couple } X, Y \in \mathcal{X}(M) \ .$$

From the definition we can easily deduce the *localization principle* for connections:

Theorem I.1. *Let X, Y, X', Y' be C^∞ vector fields on M. Let U be an open set in M. If $X = X'$ on U and if $Y = Y'$ on U, then $\nabla_X Y = \nabla_{X'} Y'$ on U.*

Proof. We first prove that $\nabla_X Y = \nabla_{X'} Y$ on U and then (in a similar way) that $\nabla_{X'} Y = \nabla_{X'} Y'$ on U. We will just write the first proof. Let $Z = X - X'$, which vanishes on U. If we prove that $\nabla_Z Y$ vanishes on U the equality will be proved. It is sufficient to see that if $p \in U$ then $(\nabla_Z Y)_p = 0$. For a fixed p, let f be a C^∞ function on M equal to 0 at p and equal to 1 outside a neighborhood V of p with compact closure such that $\bar{V} \subset U$. Then $fZ = Z$ on M. Therefore, $\nabla_Z Y = \nabla_{fZ} Y = f \nabla_Z Y$. Obviously, we have $(\nabla_Z Y)_p = 0$. □

The consequence of the localization principle is that all connections ∇ on M induce a connection ∇_U on every open set U in M in the following way. If X, Y are fields on U and if $p \in U$, then $((\nabla_U)_X Y)_p$ is defined as the value at p of $\nabla_{\tilde{X}} \tilde{Y}$, where $\tilde{X}$ and $\tilde{Y}$ are C^∞ vector fields on M which match X and Y respectively in a neighborhood V of p with compact closure $\bar{V} \subset U$. The localization principle ensures that the definition of $((\nabla_U)_X Y)_p$ does not depend on the neighborhood V, nor on the extensions $\tilde{X}$, $\tilde{Y}$ of X and Y.

From now on, we will omit U and we will write only ∇ to denote the connection on M as well as the one induced on each open set U. If U is an open set in M and $e_1 \dots e_n$ are vector fields on U giving a basis of the tangent space $T_x(M)$ for each point $x \in U$, as $\nabla_{e_i} e_j$ is a vector field on U, at each point it will be a linear combination of the vectors of the basis. That is, $\nabla_{e_i} e_j = \sum_k \Gamma_{ij}^k e_k$. We will now adopt Einstein's convention, considering that every super-index and sub-index designed by the same letter will be added if the contrary is not explicitly indicated. Thus, we can write $\nabla_{e_i} e_j = \Gamma_{ij}^k e_k$. The coefficients Γ_{ij}^k are called *Christoffel symbols* of the connection relative to the basis $\{e_1 \dots e_n\}$. In a local chart $(U, x^1 \dots x^n)$ we usually consider the basis $\{\frac{\partial}{\partial x^1} \dots \frac{\partial}{\partial x^n}\}$. In this basis, if the connection is without torsion, we have $\Gamma_{ij}^k = \Gamma_{ji}^k$ (as $[\frac{\partial}{\partial x^i}, \frac{\partial}{\partial x^j}] = 0$).

In such a local chart, if X and Y are vector fields on U they can be written as

$$X = X^i \frac{\partial}{\partial x^i} \; , \; Y = Y^j \frac{\partial}{\partial x^j}$$

(we have suppressed the summations according to our adopted Einstein's convention), and by virtue of the covariant derivative properties we have

$$\begin{aligned} \nabla_X Y = \nabla_X \left(Y^j \frac{\partial}{\partial x^j} \right) &= X(Y^j) \frac{\partial}{\partial x^j} + Y^j \nabla_X \frac{\partial}{\partial x^j} \\ &= X(Y^j) \frac{\partial}{\partial x^j} + Y^j X^i \Gamma_{ij}^k \frac{\partial}{\partial x^k} \; . \end{aligned}$$

Therefore,

$$\nabla_X Y = (X(Y^k) + X^i Y^j \Gamma_{ij}^k) \frac{\partial}{\partial x^k} \; . \tag{I.1}$$

This formula allows us to state the following proposition:

Proposition I.2. *Let ∇ be a connection on a differentiable manifold M. Let $p \in M$ and let X and Y be vector fields in a small neighborhood of p. To know the value $(\nabla_X Y)_p$ we just need the value of X at the point p and of Y on a small curve $x(t)$ such that $x(0) = p$ and $\dot{x}(0) = X_p$. In other words, if X' and Y' are two other vector fields on a neighborhood of p such that $X_p = X'_p$ and $Y = Y'$ on a curve $x(t)$ satisfying the former conditions, then $(\nabla_X Y)_p = (\nabla_{X'} Y')_p$.*

Proof. By definition, the Christoffel symbols are determined in any local chart by the connection. Consequently, the terms $X^i Y^j \Gamma^k_{ij}$ of (I.1) are determined at p only by the values of X and Y at p, and the terms $X(Y^k)$ (as they are directional derivatives at p) are determined by the value of X at p and the value of Y on a small curve $x(t)$ such that $x(0) = p$ and $\dot{x}(0) = X_p$. □

Whenever we have a curve $x(t)$ on M and a vector field Y defined only on this curve (this is usually called a vector field with support on the curve) then, by the latter proposition, we are allowed to consider $\nabla_{\dot{x}(t)} Y$ (even if not $\dot{x}(t)$ nor Y are defined in a neighborhood of every point of the curve). In particular, whenever $x(t)$ is a curve on M, for each t the expression $\nabla_{\dot{x}(t)}\dot{x}(t)$ makes sense. In a local chart $(U, x^1 \dots x^n)$ we have

$$\nabla_{\dot{x}}\dot{x} = \left(\frac{d^2x^k}{dt^2} + \Gamma^k_{ij} \frac{dx^i}{dt} \frac{dx^j}{dt} \right) \frac{\partial}{\partial x^k}$$

The curves $x(t)$ that satisfy $\nabla_{\dot{x}}\dot{x} = 0$ are called *geodesic curves.*

Let us introduce a notation that will be used from now on. If Y is a vector field in a local chart $(U, x^1 \dots x^n)$, we will denote by $\nabla_i Y^j$ the components of $\nabla_{\frac{\partial}{\partial x^i}} Y$ in the basis $\{\frac{\partial}{\partial x^1} \cdots \frac{\partial}{\partial x^n}\}$. Thus, owing to (I.1), we have:

$$\nabla_i Y^k = \partial_i Y^k + Y^j \Gamma^k_{ij} \ ,$$

where ∂_i indicates $\partial/\partial x^i$.

Let $\mathcal{I}^r_s(M)$ be the space of C^∞ r times contravariant and s times covariant tensor fields on M. Denote by $\mathcal{I}^0_0(M) = \mathcal{F}(M)$ the space of C^∞ functions on M. Let $\mathcal{I}(M)$ be the direct sum of all $\mathcal{I}^r_s(M)$ for r, $s = 0, 1, 2 \dots$. A derivation of $\mathcal{I}(M)$ is a mapping $D : \mathcal{I}(M) \to \mathcal{I}(M)$ which is $\mathbb{R}$-linear and satisfies the following properties:

1) D applies $\mathcal{I}^r_s(M)$ into $\mathcal{I}^r_s(M)$ (it preserves the type of the tensor).
2) $D(A \otimes B) = DA \otimes B + A \otimes DB$.
3) If C is a tensor contraction, $C : \mathcal{I}^r_s(M) \to \mathcal{I}^{r-1}_{s-1}(M)$, and if $K \in \mathcal{I}^r_s(M)$, then $D(CK) = C(DK)$ (that is, D commutes with the contractions).

In analogy with Theorem I.1 we can say that if D is a derivation of $\mathcal{I}(M)$ and if K and K' are elements of $\mathcal{I}^r_s(M)$ that match in an open set U, then DK and

DK' also match on U (localization principle). This implies (just as in the case of a covariant derivative) that any derivation D of $\mathcal{I}(M)$ induces a derivation D_U of $\mathcal{I}(U)$ for any open set U in M (we will denote this derivation by the same letter D, omitting the sub-index U).

The following theorem shows that every connection can act in a natural way on any tensor field and not only on vector fields:

Theorem I.3. *For a given connection ∇ on M and a vector field X on M there exists a unique derivation D of $\mathcal{I}(M)$ that satisfies*

1) *$D(f) = X(f)$ for $f \in \mathcal{F}(M)$.*
2) *$DY = \nabla_X Y$ for $Y \in \mathcal{X}(M)$.*

We will denote this derivation by ∇_X. Thus, $\nabla_X K$ will make sense for $K \in \mathcal{I}^r_s(M)$.

Proof. If $\omega \in \mathcal{I}^0_1(M)$ and if $X \in \mathcal{X}(M)$, as D must commute with the contractions we have $(D\omega)(Y) = C(D\omega \otimes Y) = C(D(\omega \otimes Y) - \omega \otimes DY) = D(C(\omega \otimes Y)) - C(\omega \otimes DY) = D(\omega(Y)) - \omega(DY) = X(\omega(Y)) - \omega(\nabla_X Y)$. This defines D on the elements of $\mathcal{I}^0_1(M)$.

If K is an element of $\mathcal{I}^r_s(M)$ and $(U, x^1 \dots x^n)$ is a local chart, K is expressed in this chart as:

$$K = K^{i_1 \dots i_r}_{j_1 \dots j_s} \frac{\partial}{\partial x^{i_1}} \otimes \dots \otimes \frac{\partial}{\partial x^{i_r}} \otimes dx^{j_1} \otimes \dots \otimes dx^{j_s} \; . \tag{I.2}$$

Since D can be localized in U, to find the value of DK at any point of U it will suffice to apply D to the former expression of K. Due to property 2) of the definition of derivation, finding D on functions $K^{i_1 \dots i_r}_{j_1 \dots j_s}$, on the vector fields $\partial/\partial x^i$ and on the 1-forms dx^j will suffice. However, we have already seen that D is uniquely determined on all these elements. □

If K is a tensor which in a local chart has the expression (I.2), we will denote the components of $\nabla_{\partial/\partial x^i} K$ by $\nabla_i K^{i_1 \dots i_r}_{j_1 \dots j_s}$. That is to say:

$$\nabla_{\partial/\partial x^i} K = \nabla_i K^{i_1 \dots i_r}_{j_1 \dots j_s} \frac{\partial}{\partial x^{i_1}} \otimes \dots \otimes \frac{\partial}{\partial x^{i_r}} \otimes dx^{j_1} \otimes \dots \otimes dx^{j_s} \; .$$

From the proof of Theorem I.3 it is easily deduced that in a local chart

$$\nabla_{\partial/\partial x^i} dx^k = -\Gamma^k_{ij} dx^j \; ,$$

and hence, if ω is a 1-form,

$$\nabla_i \omega_j = \partial_i \omega_j - \Gamma^k_{ij} \omega_k \; . \tag{I.3}$$

I.2 First results on pseudo-Riemannian manifolds

A C^∞ *pseudo-Riemannian manifold* is a couple (M, g), where M is a C^∞ manifold of finite dimension n, and g is a C^∞, symmetric, 2-covariant tensor field on M which is non-singular at each point $p \in M$ and whose *index* ν is constant. Let us remember that if $p \in M$, the index of g_p is the number of $-$ signs that appear in every diagonalization of g_p. The tensor g will be called the *metric tensor*. If $(U, x^1 \dots x^n)$ is a local chart of M we will denote it by $g_{ij} = g(\frac{\partial}{\partial x^i}, \frac{\partial}{\partial x^j})$. In a local chart, the metric g can be written as

$$g = g_{ij} dx^i \otimes dx^j \ .$$

The fact that g is C^∞ and non-singular at each p implies that the index ν_p of g_p is constant on each connected component of M. However, we have imposed that the value of ν should be the same for all connected components.

When the index of the metric is 0 that metric is named Riemannian and the couple (M, g) is called a *Riemannian manifold*. When $n = 4$ and $\nu = 1$ we have a Lorentzian metric and the couple (M, g) is called a *Lorentzian manifold*.

I.2.1 Associate connection

In a pseudo-Riemannian manifold (M, g) there exists a natural connection ∇ owing to the following Riemann's theorem:

Theorem I.4. *Let (M, g) be a pseudo-Riemannian manifold. There exists on M a unique connection fulfilling the two following properties:*

1) *It has no torsion, that is, it satisfies $[X, Y] = \nabla_X Y - \nabla_Y X$ for any couple of vector fields X, Y.*

2) *It satisfies $X(g(Y, Z) = g(\nabla_X Y, Z) + g(Y, \nabla_X Z)$ for every vector field X, Y, Z (connections with this property are called compatible with the metric).*

The unique connection whose existence is shown by this theorem is called the *Levi-Civita connection* or *Riemannian connection.*

Proof. We write the expression

$$X(g(Y, Z) = g(\nabla_X Y, Z) + g(Y, \nabla_X Z)$$

and the analogous ones obtained by cyclically permuting the vector fields X, Y, Z, which are

$$Y(g(Z, X) = g(\nabla_Y Z, X) + g(Z, \nabla_Y X) \ ,$$
$$Z(g(X, Y) = g(\nabla_Z X, Y) + g(X, \nabla_Z Y) \ .$$

We then add the two first and subtract the third one. Considering the property $[X,Y] = \nabla_X Y - \nabla_Y X$ we can easily obtain

$$g(\nabla_X Y, Z) = \tfrac{1}{2}\{X(g(Y,Z)) + Y(g(Z,X)) - Z(g(X,Y) \\ - g(X,[Y,Z]) - g(Y,[X,Z]) - g(Z,[Y,X])\} \ . \tag{I.4}$$

This equality provides the value of $\nabla_X Y$ uniquely at each point p (since we know the inner product of $(\nabla_X Y)_p$ with any Z_p). Then it can be proved that the connection ∇ defined at each p by (I.4) satisfies all the required properties. □

In a local chart $(U, x^1 \dots x^n)$ the formula (I.4) is equivalent to the following expression:

$$\Gamma^r_{ij} = \frac{1}{2} g^{rk}(\partial_i g_{jk} + \partial_j g_{ki} - \partial_k g_{ij}) \ , \tag{I.5}$$

where (g^{rk}) stands for the inverse matrix of (g_{rk}) and ∂_i for $\partial/\partial x^i$.

Since in a local chart the equations of geodesics are

$$\frac{d^2x^k}{dt^2} + \Gamma^k_{ij}\frac{dx^i}{dt}\frac{dx^j}{dt} = 0 \ ,$$

given any point p and any vector $v \in T_p(M)$, only a unique geodesic $x(t)$ exists, defined by t in a certain interval $-\varepsilon < t < \varepsilon$, such that $x(0) = p$ and $\dot{x}(0) = v$. If any geodesic $x(t)$ defined for t in a small interval can be enlarged for each $t \in \mathbb{R}$, then the pseudo-Riemannian manifold is called *complete*.

I.2.2 Curvature

We define the *curvature tensor* of a pseudo-Riemannian manifold (M, g) as the mapping

$$\begin{array}{ccc} \mathcal{X}(M) \times \mathcal{X}(M) \times \mathcal{X}(M) & \longrightarrow & \mathcal{X}(M) \\ (X,Y,Z) & \rightarrow & R(X,Y)Z \end{array} \ ,$$

where $R(X,Y)Z = \nabla_X\nabla_Y Z - \nabla_Y\nabla_X Z - \nabla_{[X,Y]}Z$, where ∇ is the Riemannian connection. From the latter definition, from the properties of the covariant derivative and from $[fX,Y] = f[X,Y] - Y(f)X$, when f is a function, we can deduce the following properties:

1) $R(X,Y)Z = -R(Y,X)Z$.

2) $R(X_1 + X_2, Y)Z = R(X_1,Y)Z + R(X_2,Y)Z$. A similar property is also true when Y or Z split into the sum of two vector fields.

3) $R(fX,Y)Z = R(X,fY)Z = R(X,Y)fZ = fR(X,Y)Z$, when f is a function.

In a local chart $R(\partial/\partial x^j, \partial/\partial x^k)\partial/\partial x^i$ is a linear combination of $\{\partial/\partial x^i\}$. Hence,

$$R\left(\frac{\partial}{\partial x^j}, \frac{\partial}{\partial x^k}\right)\frac{\partial}{\partial x^i} = R^r_{ijk}\frac{\partial}{\partial x^r} .$$

The coefficients R^r_{ijk} are the components of the curvature tensor on this basis. Since

$$R\left(\frac{\partial}{\partial x^j}, \frac{\partial}{\partial x^k}\right)\frac{\partial}{\partial x^i} = \nabla_{\partial/\partial x^j}\nabla_{\partial/\partial x^k}\frac{\partial}{\partial x^i} - \nabla_{\partial/\partial x^k}\nabla_{\partial/\partial x^j}\frac{\partial}{\partial x^i} ,$$

we obtain the following expression for the components of the curvature tensor:

$$R^r_{ijk} = \partial_j\Gamma^r_{ki} - \partial_k\Gamma^r_{ji} + (\Gamma^s_{ki}\Gamma^r_{js} - \Gamma^s_{ji}\Gamma^r_{ks}) .$$

Owing to properties 2) and 3), in a local chart we have

$$\begin{aligned} R(X,Y)Z &= R(X^i\partial/\partial x^i\,,\, Y^j\partial/\partial x^j)Z^k\partial/\partial x^k \\ &= X^iY^jZ^kR(\partial/\partial x^i\,,\,\partial/\partial x^j)\partial/\partial x^k = X^iY^jZ^kR^s_{kij}\partial/\partial x^s , \end{aligned}$$

an expression which proves that the value of $R(X,Y)Z$ at each point of this local chart only depends on the values of X , Y, Z at that point. Therefore, at any point $p \in M$ the curvature tensor provides a mapping

$$T_p(M) \times T_p(M) \times T_p(M) \longrightarrow T_p(M) .$$

Consequently, we have a three-times covariant and one-time contravariant tensor at this point. That justifies the use of the word tensor in the expression *curvature tensor.*

The curvature tensor, apart from its already mentioned elemental properties, also satisfies the following ones:

- $g(R(X,Y)Z,W) = -g(R(X,Y)W,Z)$.
- $R(X,Y)Z + R(Y,Z)X + R(Z,X)Y = 0$ *(first Bianchi identity).*
- $(\nabla_X R)(Y,Z,W) + (\nabla_Y R)(Z,X,W) + (\nabla_Z R)(X,Y,W) = 0$ *(second Bianchi identity).*

These will not be proved in this book.

Let us now talk about the *Ricci tensor* and the *scalar curvature.* The curvature tensor $\mathcal{R}$, as a one-time contravariant and three-times covariant tensor, is written as follows in any local chart:

$$\mathcal{R} = R^r_{kij}\frac{\partial}{\partial x^r} \otimes dx^i \otimes dx^j \otimes dx^k .$$

The two-times covariant tensor obtained from the former tensor by contraction between the contravariant factor and the first covariant factor is called the Ricci tensor. It will be expressed as *Ric*. That is,

$$Ric = R_{kj} dx^j \otimes dx^k \ ,$$

with $R_{kj} = R^i_{kij}$. If we consider a two-times covariant tensor as a mapping that maps every pair of vector fields into a function, we will have

$$Ric(X,Y) = dx^i(R(\frac{\partial}{\partial x^i}, X)Y) \ . \tag{I.6}$$

This expression does not depend on the coordinates $x^1 \dots x^n$. We can show (using the first Bianchi identity and the property of the cuvature tensor asserting that $g(R(X,Y)Z,W) = -g(R(X,Y)W,Z))$ that the Ricci tensor is symmetric, that is $R_{ij} = R_{ji}$.

The *scalar curvature*, which we will denote by R, is defined as the contraction between the two factors of the Ricci tensor, considering the first factor as a contravariant one through the metric:

$$R = g^{ij} R_{ij} \ .$$

We will finally discuss the *sectional curvature.*

Proposition I.5. *Let $x \in M$. Let E_x be a vector subspace of dimension 2 of $T_x(M)$. Let $\{u_1, u_2\}$ be a basis of E_x. The number*

$$\frac{g(R(u_1,u_2)u_2,u_1)}{g(u_1,u_1)g(u_2,u_2) - g(u_1,u_2)^2} \tag{I.7}$$

does not depend on the chosen particular basis of E_x.

Proof. Let $\{v_1, v_2\}$ be another basis of E_x. Let A be the matrix of the change from $\{u_i\}$ to $\{v_i\}$. Using the properties of the curvature tensor we have

$$g(R(v_1,v_2)v_2,v_1) = (\det A) g(R(u_1,u_2)v_2,v_1) = (\det A)^2 g(R(u_1,u_2)u_2,u_1) \ .$$

Furthermore, as the denominator of (I.7) is the determinant of the matrix of the restriction of g to E_x, in the basis $\{u_1, u_2\}$, we have

$$g(v_1,v_1)g(v_2,v_2) - g(v_1,v_2)^2 = (\det A)^2 (g(u_1,u_1)g(u_2,u_2) - g(u_1,u_2)^2) \ .$$

The proposition is proved by these two equalities. □

The number (I.7) is called the *sectional curvature* of (M,g) at point x, corresponding to the plane E_x. We will denote it by $K(x, E_x)$. We say that (M,g) has constant sectional curvature at point x if $K(x, E_x)$ does not depend on the chosen subspace $E_x \subset T_x(M)$ of dimension 2. We say that (M,g) has constant sectional curvature if a certain number $K \in \mathbb{R}$ exists, such that $K = K(x, E_x)$ for each $x \in M$ and for all subspace E_x of dimension 2 of $T_x(M)$. We can easily show that in a manifold (M, g) of constant sectional curvature K, the curvature tensor has the following expression:

$$R(X,Y)Z = K(g(Y,Z)X - g(X,Z)Y) \ .$$

I.2.3 Covariant differentiation and divergence

We will introduce the concepts of *covariant differentiation* and *divergence* of a tensor field. Let $K \in \mathcal{I}_s^r(M)$. Let $x \in M$ and $v \in T_x(M)$. Then $(\nabla_v K)_x$ makes sense, owing to Proposition I.2. Let us write, to make it shorter, $E_x = \otimes^r T_x(M) \otimes (\otimes^s T_x(M)^*)$. The map

$$\begin{array}{ccc} T_x(M) & \longrightarrow & E_x \\ v & \to & \nabla_v K \end{array}$$

is linear and, consequently, it can be understood as an element of $E_x \otimes T_x(M)^*$. The element of $\otimes^r T_x(M) \otimes (\otimes^{s+1} T_x(M)^*)$ obtained in this way will be denoted by $(\nabla K)_x$. The tensor field that associates $(\nabla K)_x$ to each point x will be simply denoted by ∇K and will be called a *covariant differentiation* of K. In coordinates, if the expression of K is (I.2), then the expression of ∇K will be:

$$\nabla K = \nabla_i K^{i_1 \dots i_r}_{j_1 \dots j_s} \frac{\partial}{\partial x^{i_1}} \otimes \dots \otimes \frac{\partial}{\partial x^{i_r}} \otimes dx^{j_1} \otimes \dots \otimes dx^{j_s} \otimes dx^i \ .$$

We can easily see that the property $X(g(Y, Z) = g(\nabla_X Y, Z) + g(Y, \nabla_X Z)$ of the covariant derivative is equivalent to saying that the covariant differentiation of the metric tensor vanishes: $\nabla g = 0$.

When $K \in \mathcal{I}_s^r(M)$, with $r \neq 0$, we define the *divergence* of K as the element of $\mathcal{I}_s^{r-1}(M)$ obtained from ∇K by contracting the first contravariant factor and the last covariant one. In a local chart,

$$(\text{div } K)^{i_2 \dots i_r}_{j_1 \dots j_s} = \nabla_i K^{i i_2 \dots i_r}_{j_1 \dots j_s} \ .$$

Since in a pseudo-Riemannian manifold every tangent space can be canonically identified with its dual by means of the metric, we can always identify $\mathcal{I}_s^0(M)$ with $\mathcal{I}_{s-1}^1(M)$, and the hypothesis $r \neq 0$ that we assumed to define divergence is not necessary any more. As the covariant differentiation of the metric tensor vanishes, to switch a covariant factor to contravariant (or vice versa) and then apply the covariant differentiation is exactly the same as first applying the covariant differentiation and then switching the same factor from covariant to contravariant (or vice versa).

I.2.4 Divergence of the Ricci tensor

In relativity it is very important to know the divergence of the Ricci tensor, given by the following proposition.

Proposition I.6. *We have*

$$\operatorname{div}(Ric) = \frac{1}{2} dR \ ,$$

where dR indicates the exterior derivative of the scalar curvature.

Proof. We will do the computation in a local chart. Owing to the second Bianchi identity, we have:

$$\nabla_r R^i_{jkh} + \nabla_h R^i_{jrk} + \nabla_k R^i_{jhr} = 0 \ .$$

The last term is equal to $-\nabla_k R^i_{jrh}$. Considering $i = r$ in the former identity and adding with respect to i we obtain

$$\nabla_i R^i_{jkh} + \nabla_h R^i_{jik} - \nabla_k R^i_{jih} = 0 \ .$$

From this we can deduce

$$g^{jk}\nabla_i R^i_{jkh} + g^{jk}\nabla_h R^i_{jik} - g^{jk}\nabla_k R^i_{jih} = 0 \ . \tag{I.8}$$

Now we need to transform the first term of the latter identity in a suitable way. We would like to see:

$$g^{jk}\nabla_i R^i_{jkh} = -g^{ir}\nabla_i R^k_{rkh} \ . \tag{I.9}$$

Indeed,

$$\begin{aligned} R^i_{jkh} &= (g^{ir} g_{rm}) R^m_{jkh} = g^{ir} g\left(\frac{\partial}{\partial x^i}, R\left(\frac{\partial}{\partial x^k}, \frac{\partial}{\partial x^h}\right)\frac{\partial}{\partial x^j}\right) \\ &= -g^{ir} g\left(\frac{\partial}{\partial x^j}, R\left(\frac{\partial}{\partial x^k}, \frac{\partial}{\partial x^h}\right)\frac{\partial}{\partial x^i}\right) = -g^{ir} g_{jm} R^m_{rkh} \ . \end{aligned}$$

Hence, $g^{jk} R^i_{jkh} = -g^{ir} R^k_{rkh}$. Using the property that the covariant differentiation of g vanishes, we obtain (I.9). Consequently, (I.8) can be expressed as

$$-g^{ir}\nabla_i R^k_{rkh} + g^{jk}\nabla_h R^i_{jik} - g^{jk}\nabla_k R^i_{jih} = 0 \ .$$

In other words,

$$-g^{ir}\nabla_i R_{rh} + g^{jk}\nabla_h R_{jk} - g^{jk}\nabla_k R_{jh} = 0 \ .$$

The first and last terms are identical, so this can be written as

$$-2(\operatorname{div}\,\mathrm{Ric})_h + \nabla_h R = 0 \ ,$$

which proves our proposition. □

I.2.5 Lie derivative and infinitesimal isometries

The *Lie derivative* $L_X K$ of a tensor field K with respect to a vector field X is defined as follows: Let $\{\varphi_t\}$ be the local one-parameter group of transformations corresponding to the vector field X. $L_X K$ is the tensor field of the same type as K, which at every $x \in M$ has the value

$$(L_X K)_x = \lim_{t\to 0} \frac{1}{t}\left((d\varphi_t)^{-1}(K_{\varphi_t(x)}) - K_x\right) .$$

We can prove that if Y is a vector field then $L_X Y = [X,Y]$, and that if f is a function then $L_X(f) = X(f)$. It can also be proved that L_X is a derivation of $\mathcal{I}(M)$ which commutes with contractions. Therefore, the expression of L_X on any tensor can be calculated using the expressions of L_X on the functions and on vector fields (in analogy with the procedure used for the covariant derivative). Later on, we will be interested in the following expression of the Lie derivative of a 2-covariant tensor α:

$$(L_X\alpha)(Y,Z) = X(\alpha(Y,Z)) - \alpha([X,Y],Z) - \alpha(Y,[X,Z]) . \tag{I.10}$$

In the case of a pseudo-Riemannian manifold this expression can be written in a different way, as the following proposition ensures:

Proposition I.7. *If α is a two-times covariant tensor in a pseudo-Riemannian manifold, we have*

$$(L_X\alpha)(Y,Z) = (\nabla_X\alpha)(Y,Z) + \alpha(\nabla_Y X, Z) + \alpha(Y, \nabla_Z X) .$$

Proof. The covariant derivative of α has the following expression (similar to (I.10)):

$$(\nabla_X\alpha)(Y,Z) = X(\alpha(Y,Z) - \alpha(\nabla_X Y, Z) - \alpha(Y, \nabla_X Z) .$$

Replacing here $\nabla_X Y$ with $\nabla_Y X + [X,Y]$ and $\nabla_X Z$ with $\nabla_Z X + [Z,X]$, we obtain

$$\begin{aligned} X(\alpha(Y,Z)) = (\nabla_X\alpha)(Y,Z) + \alpha(\nabla_Y X, Z) + \alpha(Y, \nabla_Z X) \\ + \alpha([X,Y],Z) + \alpha(Y,[X,Z]) . \end{aligned}$$

Replacing this with (I.10) we find the desired expression. □

Corollary I.8. *In a pseudo-Riemannian manifold (M,g) we have*

$$(L_X g)(Y,Z) = g(\nabla_Y X, Z) + g(Y, \nabla_Z X) .$$

Proof. It suffices to apply the former proposition to the metric tensor, considering that $\nabla_X g = 0$ for any vector field X. □

A vector field X is called an *infinitesimal isometry* or a *Killing vector field* if its corresponding local one-parameter group of transformations $\{\varphi_t\}$ consists of isometries. This is equivalent to saying that for each $x \in M$ and each v, $w \in T_x(M)$ the following is fulfilled: $g_x(v,w) = g_{\varphi_t(x)}(\varphi_t'(x)v, \varphi_t'(x)w)$ for each t belonging to the definition interval of $\varphi_t(x)$ (here $\varphi_t'(x)$ stands for the derivative or tangent linear map of φ_t at point x). The latter equation is equivalent to $L_X g = 0$.

I.3 Laplacians

Let (M, g) be a pseudo-Riemannian manifold. Let α and β be elements of $\mathcal{I}_r^0(M)$ and let $p \in M$. Let us choose coordinates in a neighborhood of p. The inner product $(\alpha, \beta)(p)$ of α and β at point p is defined as:

$$(\alpha, \beta)(p) = g^{i_1 j_1}(p) \dots g^{i_r j_r}(p) \alpha_{i_1 \dots i_r}(p) \beta_{j_1 \dots j_r}(p) ,$$

which does not depend on the chosen coordinates. If we consider β to be a contravariant tensor through the metric, we can write the former expression in a shorter way:

$$(\alpha, \beta)(p) = \alpha_{i_1 \dots i_r}(p) \beta^{i_1 \dots i_r}(p) .$$

Consequently, (α, β) will be a function on M.

Let us suppose now, to simplify, that the manifold M is orientable (this condition is not essential, we just use it in order to make the exposition more simple). Let ν be the index of the metric g (the number of $-$ signs that appear in any diagonalization of g at any point). Let us consider an atlas of M with coordinate changes of positive jacobian. Then the volume element $d\mu$ is the n-differential form that in any chart of this atlas with coordinate functions $x^1 \dots x^n$ is given by

$$d\mu = \sqrt{(-1)^\nu \det\, (g_{ij}) dx^1 \wedge \dots \wedge dx^n} ,$$

an expression which is invariant under coordinate changes of positive jacobian (note, besides, that $(-1)^\nu \det\, (g_{ij}) > 0$). The global inner product of α and β is defined by

$$\langle \alpha, \beta \rangle = \int_M (\alpha, \beta) d\mu \tag{I.11}$$

as long as the integral makes sense. When the manifold is non-orientable, the volume element must be defined as a measure (and not as an n-form).

We denote by ∇^* the adjoint of the covariant differentiation with respect to the former inner product. If $\beta \in \mathcal{I}_{r+1}^0(M)$, $\nabla^* \beta$ will be the element of $\mathcal{I}_r^0(M)$ such that for each $\alpha \in \mathcal{I}_r^0(M)$ with compact support fulfils

$$\langle \alpha, \nabla^* \beta \rangle = \langle \nabla \alpha, \beta \rangle .$$

It is a well-known fact (readily proven) that the expression of $\nabla^* \beta$ in components with respect to a local chart is

$$(\nabla^* \beta)_{i_1 \dots i_r} = -g^{ij} \nabla_i \beta_{j i_1 \dots i_r} = -\nabla^j \beta_{j i_1 \dots i_r} .$$

The operator $\Delta_R = \nabla^* \nabla$, called the *rough Laplacian*, maps r-tensors into r-tensors and has the following expression in coordinates:

$$(\Delta_R \alpha)_{i_1 \dots i_r} = -\nabla^j \nabla_j \alpha_{i_1 \dots i_r} .$$

If α and β are r-tensors with compact support, we then have

$$\langle \Delta_R \alpha, \beta \rangle = \langle \nabla^* \nabla \alpha, \beta \rangle = \langle \nabla \alpha, \nabla \beta \rangle \ .$$

The operators ∇, ∇^* and Δ_R act on every r-tensor. We introduce now the operators δ and Δ which act solely on antisymmetric r-tensors, that is, on r-forms. The *codifferentiation* δ is defined as the adjoint of the exterior derivative d with respect to the inner product (I.11). That is to say, if η is an $(r+1)$-form, $\delta\eta$ is the r-form such that

$$\langle d\omega, \eta \rangle = \langle \omega, \delta\eta \rangle \ ,$$

for each r-form ω with compact support. It is a well-known fact (readily proven) that the expression of $\delta\eta$ in components with respect to a local chart is

$$(\delta\eta)_{i_1\dots i_r} = -g^{ij}\nabla_i \eta_{ji_1\dots i_r} = -\nabla^j \eta_{ji_1\dots i_r} \ .$$

The operator $\Delta = d\delta + \delta d$ is called the *de Rham Laplacian* and it maps r-forms into r-forms. From the definitions it follows that

$$\langle \Delta\omega, \eta \rangle = \langle \delta\omega, \delta\eta \rangle + \langle d\omega, d\eta \rangle = \langle \omega, \Delta\eta \rangle \ ,$$

if ω and η have compact support.

$\Delta\omega$ has the following expression in coordinates:

$$(\Delta\omega)_{i_1\dots i_r} = -\nabla^j \nabla_j \omega_{i_1\dots i_r} + \sum_{k=1}^{r} (-1)^k (\nabla_{i_k} \nabla^j - \nabla^j \nabla_{i_k}) \omega_{ji_1\dots\widehat{i_k}\dots i_r} \ , \tag{I.12}$$

where the notation $\widehat{i_k}$ indicates that the index i_k is suppressed. For further references on this expression of the Laplacian we refer to [54]. To calculate the last term of expression (I.12) we may use the following identity, known as *Ricci's identity*:

$$\begin{aligned} &((\nabla_X \nabla_Y - \nabla_Y \nabla_X - \nabla_{[X,Y]})\omega)(X_1 \dots X_r) \\ &= -\sum_{k=1}^{r} \omega(X_1 \dots R(X,Y)X_k \dots X_r) \ . \end{aligned}$$

We can express the former equation in coordinates as follows:

$$(\nabla_i \nabla_j - \nabla_j \nabla_i)\omega_{i_1\dots i_r} = -\sum_{k=1}^{r} R^s_{i_k ij} \omega_{i_1\dots(s)_k\dots i_r} \ , \tag{I.13}$$

where $(s)_k$ indicates that index s is placed in position k. Replacing Ricci's identity in (I.12) we obtain:

$$\begin{aligned} (\Delta\omega)_{i_1\dots i_r} = &-\nabla^j \nabla_j \omega_{i_1\dots i_r} + \sum_{k=1}^{r} g^{ab} R_{i_k a} \omega_{i_1\dots(b)_k\dots i_r} \\ &- \sum_{k\neq l} g_{ai_k} g^{bc} g^{de} R^a_{bi_l e} \omega_{i_1\dots(c)_k\dots(d)_l\dots i_r} \ . \end{aligned} \tag{I.14}$$

I.4 Sobolev spaces of tensors on Riemannian manifolds

We will assume throughout this chapter that the reader is familiar with the properties of the functional spaces L^p in $\mathbb{R}^n$. Let (M,g) be a Riemannian manifold. Therefore, g will be positive definite. We will identify contravariant tensors with covariant ones through the metric g. For a given r-tensor T (which we can assume to be covariant, for instance), we denote by $|T|$ the function $\sqrt{(T,T)}$, where $(\ ,\ )$ is the local inner product introduced in the previous section. If p is a real number >1, we can say that an r-tensor T is in L^p if it has measurable coefficients on any local chart and $|T|$ (which is a function on M) belongs to $L^p(d\mu)$ (let us remember that $d\mu$ denotes the volume measure of (M,g)). We will denote by $L^p(M)_r$ the space of r-tensors that are in L^p.

If $T \in L^p(M)_r$, we will denote by $\|T\|_p$ the number

$$\|T\|_p = \left(\int_M |T(x)|^p d\mu(x)\right)^{1/p} .$$

Let $\mathcal{D}(M)_r$ be the subspace consisting of all C^∞ r-tensors with compact support. For a given positive integer s we denote by $\mathcal{W}^{p,s}(M)_r$ the completion of $\mathcal{D}(M)_r$ by the norm

$$||T||_{p,s} = \sum_{k=0}^{s} ||\nabla^{(k)}T||_p \ , \tag{I.15}$$

where $\nabla^{(k)}$ stands for the k times iteration of the covariant differentiation ∇. The space $\mathcal{W}^{p,s}(M)_r$ is called the L^p-Sobolev space of r-tensors of regularity degree s.

To describe $\mathcal{W}^{p,s}(M)_r$ we need to use *generalized covariant derivatives.* We say that $T \in L^p(M)_r$ has a generalized (or weak) k-th order covariant differentiation operator, $\nabla^{(k)}T$, in L^p if for every C^∞ $(r+k)$-covariant tensor α with compact support, we have

$$|\langle T, (\nabla^*)^{(k)}\alpha\rangle| \le C||\alpha||_q \ ,$$

for some constant C independent of α, with q being the conjugate exponent of p, defined by $1/p+1/q=1$. Then, the map $\alpha \to \langle T, (\nabla^*)^{(k)}\alpha\rangle$ is continuous on $L^q(M)_r$ and an $f \in L^p(M)_r$ exists such that

$$\langle T, (\nabla^*)^{(k)}\alpha\rangle = \langle f, \alpha\rangle \ .$$

Thus, we define $\nabla^{(k)}T = f$.

Let $L^{p,s}(M)_r$ be the space of r-tensors T in L^p which have weak covariant differentiation operators $\nabla^{(k)}T$ in L^p for $k \le s$, endowed with the norm $||\ \ ||_{p,s}$. As $L^{p,s}(M)_r$ is a complete space, $\mathcal{W}^{p,s}(M)_r$ is nothing but the closure of $\mathcal{D}(M)_r$ inside $L^{p,s}(M)_r$. From the definition, it follows that relations such as

$$\langle \nabla^{(k)}T, S\rangle = \langle T, \nabla^{*(k)}S\rangle$$

are true for $T \in \mathcal{W}^{p,s}(M)_r$ and $S \in \mathcal{W}^{q,s}(M)_r$, with $1/p + 1/q = 1$.

Any differential operator of order m defined intrinsically on M can act (in the sense of weak derivatives) on the elements of $\mathcal{W}^{p,s}(M)_r$ when $s \geq m$. For instance, $\Delta_R = \nabla^* \nabla$ acting on $T \in \mathcal{W}^{p,s}(M)_r$ is the r-tensor f in L^p such that $\langle T, \Delta_R \alpha \rangle = \langle f, \alpha \rangle$ for each r-covariant differentiable tensor with compact support α.

It is a well-known fact that, when (M, g) is complete, $\mathcal{W}^{p,s}(M)_r = L^{p,s}(M)_r$ for $s = 0$, 1 (chapter 2 of [4]). The former equality stands for all $s \geq 2$ if the manifold (M, g), besides being complete, fulfills certain conditions on derivatives of the curvature tensor (see [5]).

The well-known *Sobolev lemma* states, among other conditions, that when $M = \mathbb{R}^n$ and $s > \frac{n}{p} + k$, each element $T \in \mathcal{W}^{p,s}(\mathbb{R}^n)_r$ is of class C^k. This result is still true for any compact manifold and also for complete manifolds which fulfill quite general conditions (positive injectivity radius and bounded curvature), [4]. Owing to this, for a big enough s we can consider the elements of $\mathcal{W}^{p,s}(M)_r$ as true vector fields, r times covariant with a certain degree of differentiability $k < s - \frac{n}{p}$.

When M is a compact manifold it can be easily proved that the norms $\| \quad \|_{p,s}$ corresponding to different Riemannian metrics are equivalents and, therefore, the Sobolev spaces $\mathcal{W}^{p,s}(M)_r$ do not depend on the Riemannian metric used. Bearing in mind that any paracompact manifold admits a Riemannian metric, in the case of compact manifolds the Sobolev spaces $\mathcal{W}^{p,s}(M)_r$ are associated to the manifold M (without the necessity of specifying any concrete Riemannian metric on M). In such a case it is not possible to canonically identify covariant tensors and contravariant ones, and $\mathcal{W}^{p,s}(M)_r$ would be the corresponding Sobolev space for r-covariant tensors.

When $p = 2$, its conjugate exponent q is also 2 and the space $\mathcal{W}^{2,s}(M)_r$ happens to be a Hilbert space with inner product $\langle \ \rangle_s$ defined by

$$\langle T, S \rangle_s = \sum_{k \leq s} \langle \nabla^{(k)} T, \nabla^{(k)} S \rangle ,$$

which induces a norm equivalent to the one introduced before: $\| \quad \|_{2,s}$. In the case of $p = 2$ we shall write $\mathcal{W}^s(M)_r$ instead of $\mathcal{W}^{2,s}(M)_r$. In other words, if we do not specify p, we will understand that $p = 2$. The space $\mathcal{W}^s(M)_r$ is sometimes denoted by $H^s(M)_r$, even though we will never use this notation in this book, because in the case of manifolds $H^s(M)$ it traditionally denotes the s-th degree space of cohomology.

We must still introduce another kind of spaces called the *local Sobolev spaces*. Let (M, g) be a Riemannian manifold. We provide the space $\mathcal{D}(M)_r$ (space of the C^∞ r-covariant tensors on M with compact support) with the well-known topology (analogous to that of spaces of functions) with respect to which the notion of convergence to zero is the following. A sequence $\{\psi_j\}$ of elements of $\mathcal{D}(M)_r$ converges to zero if and only if: 1) The supports of all the ψ_j are contained in a fixed compact K, and 2) In any local chart $(U, x^1 \dots x^n)$, with U of compact closure, the coefficients of ψ in this chart and all their derivatives converge to zero.

We denote by $\mathcal{D}'(M)_r$ the topological dual of space $\mathcal{D}(M)_r$. The C^∞ r-covariant tensors on M can be considered elements of $\mathcal{D}'(M)_r$ owing to the following construction. If T is a C^∞ r-covariant tensor on M, we may think that T acts on any $\psi \in \mathcal{D}(M)_r$ by means of

$$T[\psi] = \int_M (T, \psi) d\mu \ .$$

Through this action of T on the $\psi \in \mathcal{D}(M)_r$, T is an element of $\mathcal{D}'(M)_r$. In general, the space $\mathcal{D}'(M)_r$ can be interpreted as the space of the r-covariant tensors T which in any local chart $(U, x^1 \dots x^n)$ are expressed as

$$T = T_{i_1 \dots i_r}(x) dx^{i_1} \otimes \dots \otimes dx^{i_r} \ ,$$

with local coefficients $T_{i_1 \dots i_r}$ that are distributions on the open set U.

The *local Sobolev space* $\mathcal{W}^{p,s}_{loc}(M)_r$ is defined as the space of those $T \in \mathcal{D}'(M)_r$ such that $\varphi T \in \mathcal{W}^{p,s}(M)_r$ for every C^∞ function φ with compact support. Despite the fact that M is not assumed to be compact, imposing that for every φ with compact support we have $\varphi T \in \mathcal{W}^{p,s}(M)_r$ implies, as before, that the space $\mathcal{W}^{p,s}_{loc}(M)_r$ does not depend on the Riemannian metric used and, consequently, it can be associated to any differentiable paracompact manifold. In the case of $p = 2$ we shall write $\mathcal{W}^{s}_{loc}(M)_r$ instead of $\mathcal{W}^{2,s}_{loc}(M)_r$. In these local Sobolev spaces, the Sobolev lemma also works without any additional condition on M.

I.5 Lorentzian manifolds

I.5.1 Definitions

A *Lorentzian manifold* is a pseudo-Riemannian manifold (V, g), where V is a differentiable manifold of dimension 4 and the index ν of g is 1. Consequently, g is a covariant tensor field of order 2 (that is, an inner product g_x on each tangent space $T_x(V)$, which depends smoothly on x), such that for each point $x \in V$ one can find a basis of $T_x(V)$ with respect to which, the matrix of the inner product g_x is

$$g_x = \begin{pmatrix} -1 & & & \\ & 1 & & \\ & & 1 & \\ & & & 1 \end{pmatrix} . \tag{I.16}$$

At each point $x \in V$ the concepts of *time-like vector*, *space-like vector* and *light-like vector* can be defined. A vector $v \in T_x(V)$ is time-like if $g_x(v, v) < 0$. It is space-like if $g_x(v, v) > 0$. Finally, $v \in T_x(V)$ is a light-like vector if $g_x(v, v) = 0$.

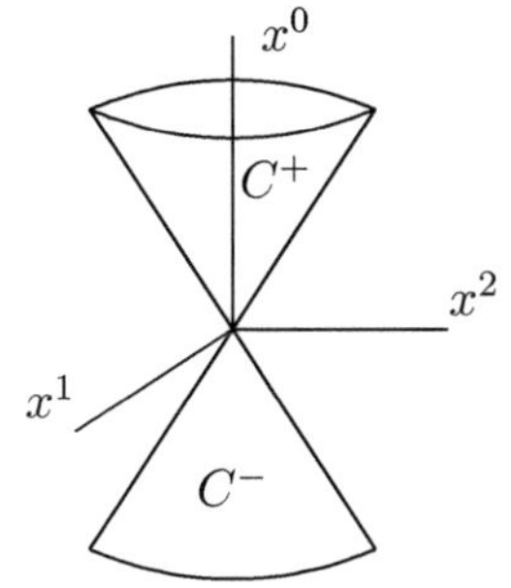

Let $\{e_0, e_1, e_2, e_3\}$ be a basis of $T_x(V)$ for which the matrix of g is as in (I.16). The condition that a certain vector $v \in T_x(V)$ of components $v^0 \dots v^3$ in this basis is time-like is written as

$$-(v^0)^2 + (v^1)^2 + (v^2)^2 + (v^3)^2 < 0 \ .$$

Let C_x be the set of time-like vectors of $T_x(V)$. It is obvious that C_x has two connected components, C_x^+ and C_x^-.

(V, g) is called time-orientable if for each $x \in V$ a connected component C_x^+ can be chosen so that the following coherence property is fulfilled: a differentiable vector field X can be found such that $X_x \in C_x^+$ for each $x \in V$.

By means of a standard argument on partitions of unity it can be easily proved that a Lorentzian manifold, (V, g), is time-orientable if and only if there exists a C^∞ vector field X on V such that at each $x \in V$, X_x is time like (that is, $g(X, X) < 0$ at each point).

A *space-time* is a Lorentzian manifold oriented with respect to time. In a space-time each $v \in T_x(V)$ such that $v \in C_x^+$ is said to *point to the future.*

I.5.2 Specific notation for Lorentzian manifolds

In the context of Lorentzian manifolds, every time that we use greek letters we will assume that they take values from 0 to 3, while the latin letters can be assumed to take values between 1 and 3. Consequently, a coordinate system (x^0, x^1, x^2, x^3) can also be written as (x^α), or (x^0, x^i). Additionally, as we very often work simultaneously with a Lorentzian metric and with the Riemannian metric induced by this one in some submanifold of dimension 3, we distinguish the former with a tilde. Thus, from now on, we will denote a Lorentzian manifold by $(V, \tilde{g})$, whereas $\tilde{\nabla}$ will be the covariant derivative induced by $\tilde{g}$, $\widetilde{Ric}$ its Ricci tensor, $\tilde{R}$ will be its scalar curvature, etc.

Chapter II

Introduction to Relativity

This chapter presents basic concepts of relativity, both special and general, aimed at readers who do not have previous knowledge of the subject. Of course, there is a plethora of literature on the subject. Good references for main results are [57], [40] and [55]. Since having a clear concept of the stress-energy tensor appearing in Einstein's equation of general relativity requires certain elementary concepts in classical fluid mechanics, the chapter begins with a brief synopsis of this part of mechanics.

II.1 Classical fluid mechanics

II.1.1 A lemma on derivation of integrals

A one-parameter family of diffeomorphisms in an open set U of $\mathbb{R}^3$ is a differentiable map

$$\begin{array}{ccccc} U & \times & I & \xrightarrow{\varphi} & \mathbb{R}^3 \\ (x & , & t) & \longrightarrow & \varphi_t(x) \ , \end{array}$$

where I is an open interval of $\mathbb{R}$ containing the origin, such that the following two properties hold:

(1) For each t in I, φ_t is a diffeomorphism of U on $\varphi_t(U)$.

(2) $\varphi_0 =$id.

If φ_t is a family satisfying the latter, we write $\vec{v}(x) = (d\varphi_t(x)/dt)_{t=0}$. Then $\vec{v}$ is a vector field on U called the *velocity field* of the family φ_t at the instant $t = 0$.

Lemma II.1. *If $F(x,t)$ is a differentiable function on $U \times I$ and V is an arbitrary open set in U such that $\varphi_t(V) \subset U \ \forall t$, then one has:*

$$\left[\frac{d}{dt}\int_{\varphi_t(V)} F(x,t)dx\right]_{t=0} = \int_V \left(\frac{\partial F}{\partial t} + div(F\vec{v})\right)_{t=0} dx \ .$$

A sketch of the proof: By the theorem on the change of variables, we have

$$\int_{\varphi_t(V)} F(x,t)dx = \int_V F(\varphi_t(x),t)J_t(x)dx \ ,$$

where $J_t(x)$ stands for the Jacobian of φ_t. Since φ_0 is the identity, it is clear that $(dJ_t(x)/dt)_{t=0} = \operatorname{div}\vec{v}$. Deriving the latter equality gives:

$$\left[\frac{d}{dt}\int_{\varphi_t(V)} F(x,t)dx\right]_{t=0} = \int_V \left(\sum \frac{\partial F}{\partial x^i}v^i + \frac{\partial F}{\partial t} + F(x,t)\operatorname{div}\vec{v}\right)_{t=0} dx \ .$$

Then it is immediate to see that the right side of the latter formula is equal to the right side of the required equality.

II.1.2 Mass of a fluid. Continuity equation

In order to describe the motion of a fluid, let us first introduce the concept of *mass density.* For describing the mass of any region in continuous media, we will suppose that there exists a differentiable function $\rho(x,t)$ such that for every region V the integral $\int_V \rho(x,t)dx$ gives the mass contained in this region at time t. The function ρ is called the *mass density* of the medium (or simply, density).

Assume that the motion of a fluid is described by a differentiable map $\varphi : U \times I \times I \longrightarrow \mathbb{R}^3$, where U is an open set of $\mathbb{R}^3$ and I an open interval of $\mathbb{R}$, in the way that $\varphi(x,t_0,t_1)$ gives the position at time t_1 of the fluid particle whose position at time t_0 was x. Then $\varphi(x,t,t) = x$. Given an initial instant t_0, it is supposed that $\varphi_t(x) = \varphi(x,t_0,t_0+t)$ is a one-parameter family of diffeomorphisms in the sense of the above subsection. Therefore its velocity field $\vec{v}(x,t_0) = (d\varphi_t(x)/dt)_{t=0}$ will depend on the initial time t_0. We will assume in the sequel that an initial instant $t_0 = 0$ is fixed. If V is any region contained in U, the region $\varphi_t(V)$ contains at time t all fluid particles contained in the region V at the initial time $t = 0$. Then, by the conservation of mass, $m(\varphi_t(V),t) = m(V,0)$. Deriving this formula yields $dm(\varphi_t(V),t)/dt = 0$. That is,

$$\left[\frac{d}{dt}\int_{\varphi_t(V)} \rho(x,t)dx\right]_{t=0} = 0 \ .$$

Apply the lemma and take into account that the latter equality should hold for all regions V to obtain

$$\left[\frac{\partial\rho}{\partial t} + \operatorname{div}(\rho\vec{v})\right]_{t=0} = 0 \ .$$

In general, for all t,

$$\frac{\partial\rho}{\partial t} + \operatorname{div}(\rho\vec{v}) = 0 \ .$$

This equation is known as the continuity equation.

II.1.3 Total force

Let us define the *momentum* $\vec{\mathcal{P}}(V,t)$ of the fluid contained in the region V at time t as $\int_V \rho(x,t)\vec{v}(x,t)dx$. We could also define (similar to the case of n particles) the *total force* acting on a fluid region V at time $t = 0$ as $\vec{F}(V,0) = \left(d\vec{\mathcal{P}}(\varphi_t(V),t)/dt\right)_{t=0}$. Then we will have for the i-th component of this force

$$\begin{aligned} F^i(V,0) &= \left[\frac{d}{dt}\int_{\varphi_t(V)} \rho(x,t)v^i(x,t)dx\right]_{t=0} \\ &= \int_V \left(\frac{\partial(\rho v^i)}{\partial t} + \sum \frac{\partial(\rho v^i v^k)}{\partial x^k}\right)_{t=0} dx\ . \end{aligned}$$

If we have a physical law telling us what the total force is at each instant, then the previous definition will enable us to calculate the fluid motion (by calculating the velocity field and by further integrating to obtain trajectories $\varphi_t(x)$ of the fluid particles). However, if we do not know the total force a priori, the latter definition enables us to calculate a posteriori the total force acting on each region V at a given time $t = 0$ by experimental observation of the trajectories.

Experience shows that the motion of a fluid is a result of *external forces* (e.g. gravity if the fluid is in a free fall) and of some *internal forces* or *forces of constraint*. If we have a system of n particles with an external force acting upon each of them ($\vec{f}_i$ is the force acting upon the i-th particle), the total external force acting upon the system is the sum of these forces ($\sum_i \vec{f}_i$). When continuous media are concerned, describing the external force acting upon each region requires us to assume that there exists a differentiable (vector-valued) function $\vec{f}(x,t)$ such that for every region V the integral $\int_V \vec{f}(x,t)dx$ gives the total external force acting upon the medium contained in the region at time t. The function $\vec{f}(x,t)$ is called the *density of external force* acting upon the medium.

If we suppose the density of external force to be known, the internal force acting upon a region V of the fluid at time t, $\vec{F}_{\text{int}}(V,t)$, is defined as the difference of the total force and the external force acting upon the medium, that is, $\vec{F}_{\text{int}}(V,t) = \vec{F}(V,t) - \int_V \vec{f}(x,t)dx$. In this way, the motion equation states $\int_V \vec{f}(x,t)dx + \vec{F}_{\text{int}}(V,t) = \vec{F}(V,t)$. In order to a priori calculate the fluid motion, physical laws are required that should a priori say what the external and internal forces are. We already dispose of physical laws describing external forces in some important cases (for example, when the fluid evolves by the action of gravity or when there are no external forces, $\vec{f} = 0$). We are now going to give a physical law (Cauchy principle) describing internal forces.

II.1.4 Cauchy principle

Cauchy stated that there exists a differentiable map

$$\begin{array}{ccc} U\times I\times \mathbb{R}^3 & \stackrel{\vec{T}}{\longrightarrow} & \mathbb{R}^3 \\ (x,t,\vec{v}) & \longrightarrow & \vec{T}(x,t,\vec{v}) \end{array} \quad ,$$

linear with respect to $\vec{v}$ for all fixed x and t, giving the internal force $\vec{F}_{\text{int}}(V,t)$ for every region V contained in U by the formula

$$\vec{F}_{\text{int}}(V,t) = \int_{x\in\partial V} \vec{T}(x,t,\vec{n}(x))\, ds \ .$$

Here, $\vec{n}(x)$ stands for the unitary exterior normal vector to the boundary ∂V of V (it is assumed that ∂V is a differentiable manifold) and ds is the area element of ∂V. For every fixed x and t, the map $\vec{T}$ is a linear map from $\mathbb{R}^3$ to $\mathbb{R}^3$, that is, a tensor of type (1, 1). Therefore for any fixed t the field $\vec{T}$ is a tensor field of type (1, 1) on U called the *stress tensor.*

The Cauchy principle is based on the following intuitive fact: since the force exerted by the rest of the fluid upon the fluid contained in the region V is transmitted to V through the boundary ∂V, it seems logical that this force would be expressed by an integral extended to ∂V. Less intuitive is the part of the Cauchy principle requiring that $\vec{T}$ depends *linearly* on the normal vector $\vec{n}(x)$. In fact, Cauchy proved this linear dependence using more intuitive assumptions.

The stress tensor depends on the nature of the fluid. There are particular cases of fluids for which the stress tensor takes the form $\vec{T}(x,t,\vec{v}) = -p(x,t)\vec{v}$. These fluids are called *perfect.* In this type of fluids, the internal force acting upon a region V (the force exerted by the rest of the fluid upon V) is

$$\vec{F}_{\text{int}}(V,t) = -\int_{x\in\partial V} p(x,t)\vec{n}(x) ds \ .$$

When a fluid is perfect the integral contains a normal force on the boundary of V only (there are no tangential components at the boundary). The function $p(x,t)$ appearing in the definition of perfect fluids is called *pressure.* The negative sign refers to the fact that the force exerted by the rest of the fluid upon the region V has direction opposite to that of the exterior normal vector.

II.1.5 Differential expression of the motion equations

Let $\vec{T}$ be the stress tensor of our fluid. The vector $\vec{T}(x,t,\vec{v})$ will have three components, (T^1,T^2,T^3) where $T^j(x,t,\vec{v}) = \sum_i T_i^j(x,t)v^i$ (since $\vec{T}$ depends on $\vec{v}$

linearly). Write $\vec{Y}^j = (T_1^j, T_2^j, T_3^j)$. With this notation, we have $T^j(x,t,\vec{v}) = \sum T_i^j v^i = g(\vec{Y}^j, \vec{v})$, where g denotes the usual inner product in $\mathbb{R}^3$. By the divergence theorem,

$$\int_{x\in\partial V} T^j(x,t,\vec{n}(x))\, ds = \int_V \operatorname{div} \vec{Y}^j(x,t)\, dx = \int_V \left(\sum_k \frac{\partial T_k^j}{\partial x^k}\right) dx \ .$$

Since $\operatorname{div}\vec{T}$ is the vector whose components are $\sum_k \partial T_k^j/\partial x^k$ and taking into account the expression for the external force obtained in Section II.1.3, the motion equations are written as

$$\int_V \left(\frac{\partial(\rho\vec{v})}{\partial t} + \operatorname{div}\,(\rho\vec{v}\otimes\vec{v})\right) dx = \int_V \vec{f}(x,t) dx + \int_V \operatorname{div} T\, dx \ .$$

This should hold for any region V. Therefore

$$\frac{\partial(\rho\vec{v})}{\partial t} + \operatorname{div}\,(\rho\vec{v}\otimes\vec{v}) = \vec{f} + \operatorname{div} T \ .$$

II.2 Kinematics of special relativity

II.2.1 Inertial systems

A *physical reference system* is constituted by:

1. Three perpendicular coordinate axes moving in ordinary space.
2. At each point (x, y, z) at rest with respect to the above mentioned axes, an observer provided with a clock in a way that different clocks of different observers are "synchronized". The sense of the term "synchronized" will be explained in the sequel.

Let us explain the meaning of the assumption that the clocks of different observers are synchronized. Consider two observers situated at two points A and B of the reference system (A and B move together with the system and are therefore at rest with respect to each other). Assume that the observer situated at A emits a ray of light in the direction of B at time t_0 according to his clock. Suppose that this ray arrives at B at time t_0' according to B's clock and, reflected by a mirror, it comes back to A at time t_1 according to A's clock. If at any repetition of this experience the formula $t_0' - t_0 = t_1 - t_0'$ holds, we will say that the clocks of A and B are synchronized.

An *inertial system* is a physical reference system where no force acts upon each of the observers of the system. This means that if one of them releases any object without giving it initial velocity, this object will always remain at the place where it was released by the observer, at rest with respect to him.

II.2.2 Postulates of special relativity

Special relativity is based on the following postulates:

1. There exist inertial systems.
2. If S and S' are physical reference systems in uniform motion of parallel translation with respect to each other, then the fact that one of these systems is inertial implies that the other one is also inertial. Reciprocally, two arbitrary inertial systems move rectilinearly and uniformly with respect to each other (with speed that may be zero).
3. All inertial systems play the same role in the theory.
4. Two events occurring at a point at rest with respect to an inertial system S at two different times t_1 and t_2 are seen from any other inertial system S' in the same temporal order as seen by an observer in S.
5. Invariance of the velocity of light in vacuum: If an observer of an inertial system S contemplates propagation of a ray of light in the vacuum emitted by any luminous source (this luminous source could move or remain at rest with respect to the observer), he will see the ray as propagated rectilinearly and uniformly. Moreover, the module of the velocity vector of the above mentioned ray with respect to the inertial system S is a constant that does not depend on the inertial system S. That is, an observer from another inertial system S' who would measure the module of the velocity vector of this ray of light from S' would find the same value.

Let us make some comments on these postulates. Inertial systems as the ones described above do not exist in reality, since the action of gravitational fields makes it impossible to find a system (infinite in all directions) where no force would act upon the observers. However, special relativity pretends to model physical phenomena not affected by gravitational fields and therefore it postulates that inertial systems exist. As far as the second postulate is concerned, we should say that we will see in what follows that no inertial system could move with respect to another inertial system at a speed greater than that of light. So we provisionally admit the second postulate as stated, although we will have to modify it later on.

The third postulate is quite ambiguous as stated above. Though we could have written it in a more concise way, we do not do it in order not to make this chapter too long: we will clarify the sense of this postulate when we apply it to particular cases.

Classical mechanics satisfies all of the above stated postulates, except for that of invariance of the velocity of light. This postulate was proved in a well-known experiment by Michelson and Morley. In contrast, in classical electromagnetism (Maxwell-Lorentz laws) the third postulate fails to hold (a privileged reference is required: the ether).

II.2.3 Lorentz transformations

Suppose that S and S' are two inertial systems. Any event seen by an observer in S will be determined by the coordinates (x^1, x^2, x^3) of the point of space where it has occurred (with respect to the three axes in S) and by the time t (according to synchronized clocks in S) when it has occurred. Thus, any event contemplated from S corresponds to a point (t, x^1, x^2, x^3) in $\mathbb{R}^4$. The same event contemplated from S' will correspond to another point (t', x'^1, x'^2, x'^3) in $\mathbb{R}^4$. Therefore, given two inertial systems S and S', we have a map $f : \mathbb{R}^4 \longrightarrow \mathbb{R}^4$ assigning to the coordinates (t, x^1, x^2, x^3) of an event contemplated from S the coordinates (t', x'^1, x'^2, x'^3) of the same event contemplated from S'. This map f is called the *Lorentz transformation* associated to the change of inertial system from S to S' (the order of writing S and S' is important). The map f is invertible and the inverse is exactly the map associated to the change from S' to S.

The second postulate implies that each Lorentz transformation is an affinity. Indeed, assume that f is a Lorentz transformation associated to the change of two inertial systems S and S'. Let r be a straight line in $\mathbb{R}^4$ that is not contained in a hyperplane of the form $t =$const. Then one can parametrize r as follows:

$$\begin{cases} t = t \\ x^i(t) = v^i t + b^i \ , \ i = 1, 2, 3. \end{cases}$$

For an observer in S, this straight line in $\mathbb{R}^4$ represents the motion of a particle $P(t)$ of coordinates $(x^1(t), x^2(t), x^3(t))$ in a uniform motion of parallel translation with respect to S of velocity $\vec{v} = (v^1, v^2, v^3)$. Then there exists an inertial system S'' whose origin is in this particle. The system S'' moves rectilinearly and uniformly with respect to S' (second postulate). Therefore observers in S' will describe the motion of the origin of S'' by means of a straight line in their $\mathbb{R}^4$ which is not contained in a hyperplane $t' =$const. By the definition of f, this straight line should be $f(r)$. This proves that f transforms any straight line that is not contained in a hyperplane $t =$const into a straight line that is not contained in a hyperplane $t' =$const. If f transformed any straight line of $\mathbb{R}^4$ into a straight line, a well-known result in geometry would imply that f is an affinity. It is not the moment now to explain in detail how to avoid the fact that we only know that what is transformed in straight lines are those straight lines that are not contained in a hyperplane $t =$const: we will learn later on how to expound this fact under some more restrictive assumptions (see Remark II.2).

We admit that any Lorentz transformation is an affinity and aim at explicitly establishing the form of such a transformation. Consider a Lorentz transformation f associated to the change between two inertial systems S and S'. Without any loss of generality, assume that we have chosen origins of space and time in S and S' in a way that at time $t = t' = 0$ the space origins in S and S' coincide. Assume moreover that the x^1 axis in S has the direction of the velocity vector of the uniform motion of S' as seen from S. This implies that any point of S' situated on the x^1 axis in S at a certain time always belongs to this axis. Let P' and Q' be two points at rest with respect to S' that at some time belong to the x^1 axis in S. Take the x'^1 axis as the axis determined by P' and Q'. In this way, the x'^1 axis always coincides with the x^1 axis.

Since f is an affinity transforming (by virtue of the above selections) origin into origin at time $t = t' = 0$ and transforming each point having coordinates $x^2 = x^3 = 0$ to a point having coordinates $x'^2 = x'^3 = 0$, f should be as follows:

$$\left\{\begin{array}{lllll} t' & = & a_0^0 t & +a_1^0 x^1 & +a_2^0 x^2 & +a_3^0 x^3 \\ x'^1 & = & a_0^1 t & +a_1^1 x^1 & +a_2^1 x^2 & +a_3^1 x^3 \\ x'^2 & = & & & a_2^2 x^2 & +a_3^2 x^3 \\ x'^3 & = & & & a_2^3 x^2 & +a_3^3 x^3 \end{array}\right. . \tag{II.1}$$

The fourth postulate implies that $a_0^0 > 0$. Write $v = -a_0^1/a_0^0$ (a physical interpretation of v will be given later on). Suppose that a light ray is emitted from the origin of S at time $t = 0$ in all directions. For an observer in S', this light ray also comes out from his origin at time $t' = 0$. Consider now a guy, named John, travelling in space (there is no need for him to be at either of two inertial systems). Take the following event: “John starts seeing the light ray”. For an observer in S, this event has coordinates (t, x^1, x^2, x^3) satisfying $-c^2t^2 + (x^1)^2 + (x^2)^2 + (x^3)^2 = 0$, where c is the velocity of light in the vacuum. For an observer in S', the same event has coordinates (t', x'^1, x'^2, x'^3) such that $-c^2t'^2 + (x'^1)^2 + (x'^2)^2 + (x'^3)^2 = 0$. Therefore f should map the points (t, x^1, x^2, x^3) belonging to the cone in $\mathbb{R}^4$ given by $-c^2t^2+(x^1)^2+(x^2)^2+(x^3)^2 = 0$ to the points (t', x'^1, x'^2, x'^3) belonging to the cone $-c^2t'^2 + (x'^1)^2 + (x'^2)^2 + (x'^3)^2 = 0$. In order for a linear map to transform the first cone into the second one, we should have

$$\begin{aligned} &-c^2t'^2 + (x'^1)^2 + (x'^2)^2 + (x'^3)^2 \\ &= k\left\{-c^2t^2 + (x^1)^2 + (x^2)^2 + (x^3)^2\right\} , \end{aligned} \tag{II.2}$$

where (t', x'^i) denotes the transformed point $f(t, x^i)$ and k is a constant $\neq 0$.

Let us first study the transformation induced by f on the $\mathbb{R}^4$ plane satisfying $x^2 = x^3 = 0$ (which is transformed into $x'^2 = x'^3 = 0$ by (II.1)). This transformation is as follows:

$$\left\{\begin{array}{lll} t' & = & a_0^0 t + a_1^0 x^1 \\ x'^1 & = & -v a_0^0 t + a_1^1 x^1 \end{array}\right. \tag{II.3}$$

(recall that we have defined $v = -a_0^1/a_0^0$). Equations (II.2) and (II.3) imply (in the $\mathbb{R}^4$ plane satisfying $x^2 = x^3 = 0$) the following three equalities:

$$-c^2(a_1^0)^2 + (a_1^1)^2 = k \text{ (terms in } (x^1)^2 \text{)} , \tag{II.4}$$

$$-c^2(a_0^0)^2 + v^2(a_0^0)^2 = -c^2k \text{ (terms in } t^2 \text{)} , \tag{II.5}$$

$$-c^2 a_0^0 a_1^0 - v a_0^0 a_1^1 = 0 \text{ (terms in } tx^1 \text{)} .$$

By the latter equality, we have $a_1^0 = -va_1^1/c^2$. By substituting in (II.4), we obtain

$$k = (a_1^1)^2 \left(1 - \frac{v^2}{c^2}\right) .$$

We should assume $v^2 \neq c^2$ since k has to be $\neq 0$. It follows from (II.5) that

$$k = (a_0^0)^2 \left(1 - \frac{v^2}{c^2}\right) . \tag{II.6}$$

By comparing the two values obtained for k, we have $(a_1^1)^2 = (a_0^0)^2$. If necessary, we can change orientation of the x'^1 axis in order to have $a_1^1 > 0$. Therefore $a_1^1 = a_0^0$. With this argument, the transformation matrix (II.3) takes the form

$$\begin{pmatrix} a_0^0 & -va_0^0/c^2 \\ -va_0^0 & a_0^0 \end{pmatrix} .$$

By calculating the inverse matrix, one can see that the inverse transformation of (II.3) is

$$\begin{cases} t &= \dfrac{t'}{a_0^0(1-v^2/c^2)} &+ \dfrac{vx'^1}{c^2 a_0^0(1-v^2/c^2)} \\ x^1 &= \dfrac{vt'}{a_0^0(1-v^2/c^2)} &+ \dfrac{x'^1}{a_0^0(1-v^2/c^2)} \end{cases} . \tag{II.7}$$

Let us now use the third postulate. Consider two events occurring at the origin of S at times $t = 0$ and $t = 1$. The first event is seen from S' in its origin of coordinates and the second event (by virtue of (II.3)) is seen at the point $x'^1 = -a_0^0 v$ of the x'^1-axis. The distance between these two points in S' is $a_0^0|v|$ (the distance is always positive). If we repeat the same experiment with S and S' interchanged, we will obtain $|v|/a_0^0(1 - v^2/c^2)$ (by (II.7)). The third postulate claims that these two values should be equal, therefore we have

$$a_0^0 = \frac{1}{a_0^0(1 - v^2/c^2)} .$$

Since $a_0^0 > 0$, the factor $1 - v^2/c^2$ should be > 0. Hence we should assume that $|v| < c$. In addition from the latter equality and (II.6) we have

$$a_0^0 = 1/\sqrt{1 - v^2/c^2} \ , \ k = 1 \ .$$

With this in mind, equations (II.1) are written as follows:

$$\begin{cases} t' &= (1/\alpha)t & -(v/c^2\alpha)x^1 & +a_2^0 x^2 & +a_3^0 x^3 \\ x'^1 &= -(v/\alpha)t & +(1/\alpha)x^1 & +a_2^1 x^2 & +a_3^1 x^3 \\ x'^2 &= & & a_2^2 x^2 & +a_3^2 x^3 \\ x'^3 &= & & a_2^3 x^2 & +a_3^3 x^3 \end{cases} , \tag{II.8}$$

where α stands for $\sqrt{1-v^2/c^2}$. Moreover, since we know that $k=1$, by equating the terms containing x^1x^2 and x^2t of the two sides of equality (II.2), we obtain

$$(v/\alpha)a_2^0 + (1/\alpha)a_2^1 = 0 \; .$$
$$-c^2a_2^0/\alpha - a_2^1(v/\alpha) = 0 \; .$$

The first of these equalities implies that $a_2^1 = -va_2^0$. By substituting into the second equality, we obtain $a_2^0 = 0$. Therefore $a_2^1 = -va_2^0 = 0$. The same argument applied to the terms containing x^1x^3 and x^1t in (II.2) gives $a_3^1 = a_3^0 = 0$. Then (II.8) is written as follows:

$$\left\{ \begin{array}{ccccc} t' & = & (1/\alpha)t & -(v/c^2\alpha)x^1 & & \\ x'^1 & = & -(v/\alpha)t & +(1/\alpha)x^1 & & \\ x'^2 & = & & & a_2^2x^2 & +a_3^2x^3 \\ x'^3 & = & & & a_2^3x^2 & +a_3^3x^3 \end{array} \right. . \tag{II.9}$$

By a direct calculation, this expression implies $(x'^1)^2 - c^2t'^2 = (x^1)^2 - c^2t^2$. Since (II.2) should also hold for $k=1$, we have $(x'^2)^2 + (x'^3)^2 = (x^2)^2 + (x^3)^2$. Therefore the transformation given by the two last formulas in (II.9) is a Euclidean isometry. We can isometrically change the axes x^2, x^3 in S (or in S') in order to obtain $x'^2 = x^2$ and $x'^3 = x^3$ in the new coordinates. Then formulas (II.9) with this modification explicitly give the Lorentz transformation switching from the system S to S'.

By what we have already said, the inverse transformation of the first two equations in (II.9) will be

$$\left\{ \begin{array}{cccc} t & = & (1/\alpha)t' & +(v/c^2\alpha)x'^1 \\ x^1 & = & (v/\alpha)t' & +(1/\alpha)x'^1 \end{array} \right. . \tag{II.10}$$

Let us now give a physical interpretation of the quantity v appearing in the above equations. At the origin of S' ($x'^1 = x'^2 = x'^3 = 0$), the relation between time t and time t' will be $t = (1/\alpha)t'$. Therefore the origin of S' seen from S will be $x^1 = (v/\alpha)t' = vt$, $x^2 = 0$, $x^3 = 0$. This is why the observers in S see that the origin of S' moves uniformly at velocity v. Therefore v is the velocity of S' with respect to S.

Let us endow $\mathbb{R}^4$ with the metric

$$\tilde{\eta} = -c^2dt^2 + (dx^1)^2 + (dx^2)^2 + (dx^3)^2 \; .$$

This is the so-called Minkowski's metric which has the particular feature that Lorentz transformations are isometries with respect to this metric (this follows from (II.2) with $k=1$).

Remark II.2. In the above reasonings on the explicit form of Lorentz transformations, we have had to suppose that $|v| < c$, which is not directly implied by the above stated postulates. In order to deduce this from the postulates, we should slightly modify the second postulate, restricting it to inertial systems that move with respect to others at a velocity $< c$. Then the argument applied for proving that if f is a Lorentz transformation and r is a straight line in $\mathbb{R}^4$ that is not contained at a hyperplane of the form t =constant, then $f(r)$ is also a straight line in $\mathbb{R}^4$ that is not contained in a hyperplane t' =const, in the new situation of the second postulate modified would only prove that, if r is a straight line in $\mathbb{R}^4$ having time-like vector (with respect to the Minowski metric), then $f(r)$ is a straight line in $\mathbb{R}^4$ having time-like vector. One cannot conclude from this that f is an affinity. In order to claim this, we would need a result stating that if f is a bijective map from $\mathbb{R}^4$ to $\mathbb{R}^4$ transforming straight lines of time-like vector (with respect to the Minkowski metric) into straight lines having time-like vector, then f is an affinity. This result is proven in the Appendix concluding this chapter.

II.2.4 Inertial systems and the Minkowski space

The space $\mathbb{R}^4$ endowed with the metric $\tilde{\eta}$ is called *Minkowski space.* We will say that a basis $\{e_0, e_1, e_2, e_3\}$ of $\mathbb{R}^4$ is *orthonormal with respect to the Minkowski metric* $\tilde{\eta}$ if $\tilde{\eta}(e_i, e_j) = \delta_{ij}$ for $i, j = 1, 2, 3$ (where δ_{ij} are the Kronecker symbols), $\tilde{\eta}(e_0, e_i) = 0$ for $i = 1, 2, 3$, and $\tilde{\eta}(e_0, e_0) = -c^2$. We will say that it is *positive orthonormal* if, in addition to the above conditions, we have $e_0 \in C^+$ (see the notation introduced in Section I.5.1).

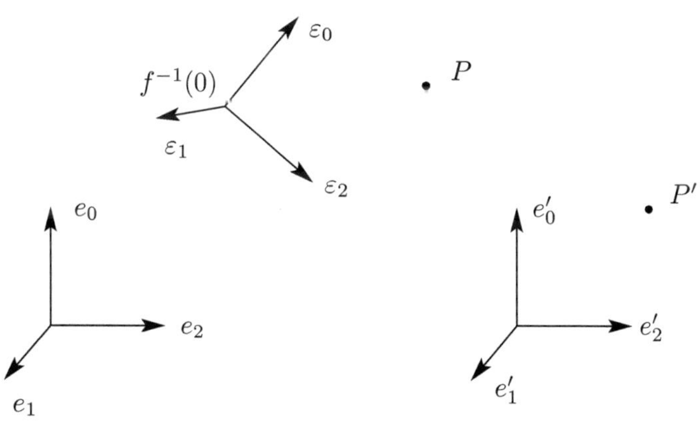

Up to now, given any inertial system we have represented events as points belonging to $\mathbb{R}^4$. If we have another inertial system S', we represent its events as points belonging to "another" $\mathbb{R}^4$. The map f assigning to a point P corresponding to some event in the "first" $\mathbb{R}^4$ a point P' belonging to the second $\mathbb{R}^4$ and corresponding to the same event, is the Lorentz transformation associated to the change from S to S'. A Lorentz transformation (being an affinity) has the form $f(\vec{x}) = F(\vec{x}) + \vec{a}$, with linear F. Let $\{e'_0, e'_1, e'_2, e'_3\}$ be the canonical basis of the second $\mathbb{R}^4$. Write $\varepsilon_\alpha = F^{-1}(e'_\alpha)$, where $\alpha = 0, 1, 2, 3$ (see the notation introduced in Section I.5.2). Consider the reference of the first $\mathbb{R}^4$ formed by the origin $f^{-1}(0)$ and the vectors $\{\varepsilon_\alpha\}_{\alpha=0\ldots3}$. It is easy to see that the point P' belonging to the second $\mathbb{R}^4$ has the same coordinates (with respect to the canonical basis) as those of the point $P = f^{-1}(P')$ belonging to the first $\mathbb{R}^4$ in the reference $\{f^{-1}(0); \varepsilon_\alpha\}$.

This simple remark reveals that the following two points of view are completely equivalent:

1. Each inertial system uses a different $\mathbb{R}^4$ in order to represent events. Lorentz transformations relate points corresponding to the same event in these different $\mathbb{R}^4$.

2. All observers, whatever inertial system they are in, employ the same $\mathbb{R}^4$ for representing events. That is, an event always is a point in $\mathbb{R}^4$ (the same point, independent of the inertial system of the observer). However, the observers in a certain inertial system use a concrete reference $\{O; \varepsilon_0, \varepsilon_1, \varepsilon_2, \varepsilon_3\}$ for measuring the coordinates of a point P (with the basis $\{\varepsilon_\alpha\}$ being orthonormal and positive). The reference $\{O; \varepsilon_0, \varepsilon_1, \varepsilon_2, \varepsilon_3\}$ depends on the particular inertial system. In this way, different inertial systems assign different coordinates to the same point P.

(Recall that an inertial system is formed by three axes in ordinary space and infinite persons having clocks (one person at each point) and satisfying certain properties, while a reference in the Minkowski space is constituted by an origin and a basis in this space.) In the sequel, we will always follow the second point of view enabling us to interpret any physical phenomenon within the same space-time $(\mathbb{R}^4, \tilde{\eta})$. From this point of view, one should never work with Lorentz transformations. Any problem related to measuring lengths or times in inertial systems becomes a geometric problem in the Minkowski space. In order to illustrate this fact, we are going to adopt this point of view for representing contraction of lengths and dilation of times upon a change of an inertial system.

II.2.5 Contraction of lengths

Assume that S and S' are two inertial systems. Suppose that the second system moves with respect to the first one at velocity v and that the motion is made in the direction of the x^1 and the x'^1 axes of the two systems. Suppose also that the origins of the two systems coincide for $t = t' = 0$. Consider a rigid rod at rest with respect to the system S lying along the x^1 axis of this system, one of the extremes being at the origin and the other one at $x^1 = l$. For observers in S the rod has length l. Discarding the coordinates x^2 and x^3, the observers in S will represent the motion of the origin of S' by the straight line in the Minkowski space satisfying the equation $x^1 = vt$ (see the figure), that is, $t = x^1/v$, which will be the t' axis for observers in S'. The x'^1 axis will be the straight line perpendicular (according to the Minkowski metric) to the line $t = x^1/v$ and passing through the origin.

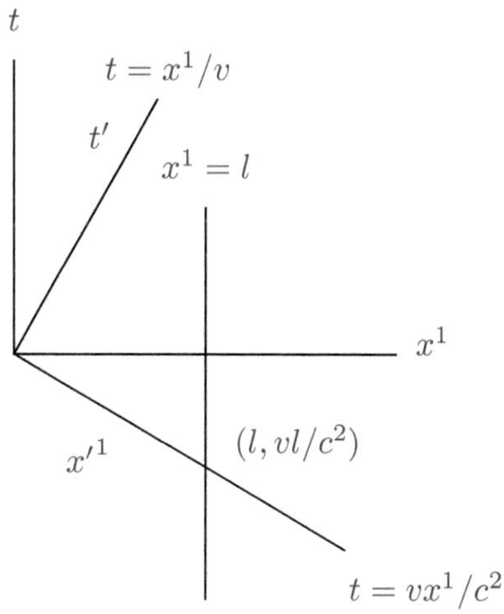

This straight line is $t = vx^1/c^2$. The two extremes of the rod will be represented at any time (in the Minkowski space) by the parallel lines $x^1 = 0$ and $x^1 = l$. The x'^1 axis (given by $t = vx^1/c^2$) intersects the above lines at the points $(0,0)$ and $(l, vl/c^2)$ (here the first coordinate of each point refers to space and the second one refers to time). The distance d of the rod measured by observers in S' will be the distance in the Minkowski space between these two points, that is,

$$d = \sqrt{1 - v^2/c^2}\, l \ .$$

For observers in S', the length of the rod which moves from their point of view, is smaller than the length l of this rod in the system S where it has been remaining at rest.

II.2.6 Proper time of a particle

Suppose that you are at the origin of coordinates of three axes of a "physical" inertial system S. Assume that you say "yah" at time t_0 according to your clock and say again "yah" at time t_1 (suppose that $t_1 > t_0$). There is a positive orthonormal reference of the Minkowski space $\{O; \varepsilon_0, \varepsilon_1, \varepsilon_2, \varepsilon_3\}$ that corresponds to your inertial system S, and both events we have mentioned (you say "yah" twice) correspond to certain points A and B of the Minkowski space that have coordinates $A = (t_0, 0, 0, 0)$ and $B = (t_1, 0, 0, 0)$ in the previous reference. Then the point A will be $A = O + t_0\varepsilon_0$ and B will be $B = O + t_1\varepsilon_0$. The interval of time elapsed on your clock between these two events is $t_1 - t_0$. Let us now calculate the distance between A and B in the Minkowski space. We have $d(A, B) = \sqrt{\tilde{\eta}(B - A, B - A)} = \sqrt{\tilde{\eta}((t_1 - t_0)\varepsilon_0, (t_1 - t_0)\varepsilon_0)} = \sqrt{-(t_1 - t_0)^2 c^2} = ic(t_1 - t_0)$ where $i = \sqrt{-1}$. Therefore the time elapsed between the two events measured with your clock is $d(A, B)/ic$. Since the distance is an intrinsic concept, any observer at any inertial system by calculating $d(A,B)/ic$ in his coordinates will know what the interval $t_1 - t_0$ on your clock is. On the other hand, the segment of the straight line in the Minkowski space that connects A and B represents your life during the time interval $[t_0, t_1]$. It is clear that the vector $B - A$ of this segment is a time-like vector which, moreover, belongs to C^+ (is directed to the future).

Let us now imagine a slightly more complex situation. Assume that we have an inertial system S and that one of its clocks starts to uniformly move with respect to S, describing a straight segment $\overline{PQ}$ of the ordinary space. Suppose, however, that arriving at Q it suddenly changes direction and switches to describing a segment $\overline{QR}$, again in uniform motion. The points P, Q and R are assumed to be at rest with respect to the inertial system S. Let t_P, t_Q, t_R be the times showed by the clock at P, Q and R, respectively. Consider three events: "the clock is at P", "the clock is at Q," "the clock is at R." There are three points $\mathcal{P}$, $\mathcal{Q}$, $\mathcal{R}$ in the Minkowski space that correspond to these three events. When the clock moves from P to Q we can consider that this happens in an inertial system, so we can calculate the time interval $t_Q - t_P$ using the above stated procedure in such a way that the time elapsed during the travel from P to Q, according to the clock itself, will be $t_Q - t_P = (1/ic)d(\mathcal{P}, \mathcal{Q})$. The same argument can be applied to calculate the time interval $t_R - t_Q$. Let us postulate that no jump happens at the clock time when it suddenly changes direction at the point Q. The total travel time elapsed will be, according to the travelling clock, $t_R - t_P = (t_R - t_Q) + (t_Q - t_P) = (1/ic)(d(\mathcal{P}, \mathcal{Q}) + d(\mathcal{Q}, \mathcal{R}))$. On the other hand, the polygonal of the Minkowski space connecting $\mathcal{P}$, $\mathcal{Q}$ and $\mathcal{R}$ represents the life of the travelling clock during the time interval $[t_P, t_R]$ of its time. Observe that each of these segments has time-like vector. Moreover, the vectors $\mathcal{Q} - \mathcal{P}$ and $\mathcal{R} - \mathcal{Q}$ belong to C^+.

Assume now another situation which is still more complex. Suppose that we have an inertial system S and that one of its clocks, r, starts to move with respect to S with an arbitrary differentiable motion (not uniform) until it stops after some time. We would like to measure the elapsed time according to the clock itself (the difference between the clock time at the end of the motion and at the beginning). The travel will be represented by an arc of a differentiable curve γ belonging to the Minkowski space. This arc can be approximated by a polygonal of segments $\overline{\mathcal{P}_0\mathcal{P}_1}$, …, $\overline{\mathcal{P}_{n-1}\mathcal{P}_n}$. This polygonal will correspond to the representation in the Minkowski space of the travel of a particle moving along a trajectory that is close to that described by the clock r. If the clock moved along this polygonal (previous situation), it would show travel time equal to $(1/ic)(d(\mathcal{P}_0, \mathcal{P}_1) + \ldots + d(\mathcal{P}_{n-1}, \mathcal{P}_n))$. Taking the limit (which is nothing else than the definition of length of an arc of a curve) we conclude that the elapsed travel time according to the clock r itself is

$$(1/ic) \text{ length of } \gamma \text{ from } \mathcal{P}_0 \text{ to } \mathcal{P}_n .$$

On the other hand, each one of the segments $\overline{\mathcal{P}_i\mathcal{P}_{i+1}}$ has a time-like vector oriented towards the future. When taking the limit, we will have to suppose that γ has at each point a time-like tangent vector oriented towards the future.

Summing up: *the motion of any particle is represented by a curve of the Minkowski space of time-like tangent vector at each point. The length of an arc of this curve between two points multiplied by the factor* $1/ic$ *represents the proper time of the particle between the time instants corresponding to two extremes of this arc. Therefore the proper time* τ *of a particle is the arc length divided by* ic.

II.2.7 Time dilation

Let us come back to the situation described in Section II.2.5. Let S and S' be two inertial systems. Suppose that the second system moves with respect to the first one at velocity v and that the motion is carried out in the direction of the x^1 and the x'^1 axes of two systems. Assume also that the origins of two systems coincide for $t = t' = 0$. Observers in S represent the events in the Minkowski space using some axes (x^1, t) (let us drop the coordinates x^2 and x^3). Observers in S' represent the events using other axes (x'^1, t'). As we have already seen above in Section II.2.5, the t' axis has equation $t = x^1/v$ and the x'^1 axis has equation $t = vx^1/c^2$. Assume now that an observer called Maria situated at the origin of S says "now" at time $t = 0$ of her clock and says again "now" at time T of her clock. For observers in S, the elapsed time between these two events is T. Let us see what time will be measured from S' between the same events. First, we clarify in what way observers in S' would measure this time. When Maria says "now" for the first time there is an observer in S', called Jaume, at the same point where Maria is, looking at his watch and writing down the time. When Maria says "now" for the second time there is another observer in S', called Marta, at the same point where Maria is, writing down the time. The time measure from S' will be the difference between times written down by Marta and by Jaume. Since the time written down by Jaume is $t' = 0$, the difference of times will be nothing else than Marta's time when she takes her note. Maria's life is represented by the t axis in the Minkowski space. Jaume's life is represented by the t' axis. The event described as "Maria says *now* for the second time" is represented by the point P of the Minkowski space having coordinates $(0, T)$ with respect to the axes (x^1, t). Marta's life is represented by the straight line passing through the point P parallel to the t' axis. Let A' be the intersection point of the x'^1 axis with the straight line passing through P parallel to the t' axis (see the figure). Let B' be the intersection point of the t' axis with the straight line passing through P parallel to the x'^1 axis. Marta's proper time when she takes the note is $T' = d(P, A')/ic = d(B', O)/ic$.

An immediate calculation shows that the point B' has coordinates

$$B' = \left(\frac{vT}{1 - v^2/c^2}, \frac{T}{1 - v^2/c^2} \right)$$

with respect to the axes (x^1, t).

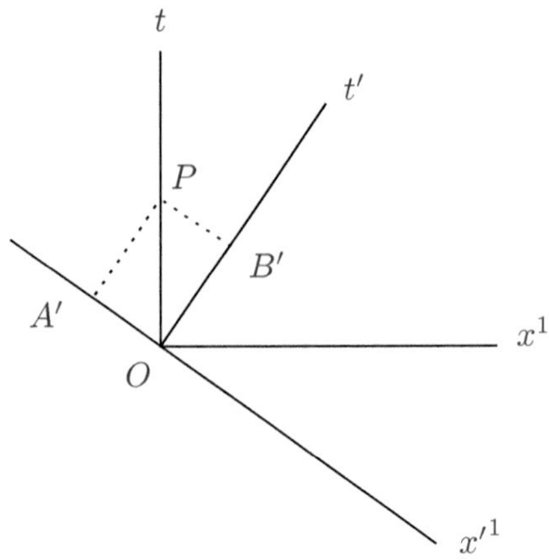

Therefore

$$T' = \frac{d(B', O)}{ic} = \frac{T}{\sqrt{1 - v^2/c^2}} .$$

We see that time T' is greater than time T. For observers in S', Maria moves and the time in S' is greater than Maria's proper time. In fact, Maria is younger with respect to the age attributed to her by observers in S'.

II.3 Dynamics of special relativity

II.3.1 Mass and momentum

Using a procedure based on the theory of elastic collisions (we do not describe this procedure here for the sake of brevity in our exposition), a real positive number m is assigned to each material particle. This number is called the *mass* of the particle (let us emphasize that the mass m we are speaking about is the rest-mass).

Recall that the motion of any particle is described by a curve γ in the Minkowski space with time-like tangent vector oriented towards the future. This curve can be parametrized by the proper time τ which is an intrinsic parameter (does not depend on any reference system). The vector $\vec{u} = (d\gamma/d\tau)_{\tau_0}$ is called *Minkowski velocity* of this particle at instant τ_0. If m is the mass of such a particle, the vector $\vec{\mathcal{P}} = (md\gamma(\tau)/d\tau)_{\tau_0}$ will be called the *Minkowski momentum* of this particle at an instant τ_0.

Let us now choose an inertial system S. This system provides a reference $\{O, \varepsilon_0, \varepsilon_1, \varepsilon_2, \varepsilon_3\}$ in the Minkowski space. In the coordinates (t, x^1, x^2, x^3) of this reference, the curve γ (parametrised by t) is written as follows: $x^0 = t\,, x^1 = \gamma^1(t)\,, x^2 = \gamma^2(t)\,, x^3 = \gamma^3(t)$. If we now express t as a function of the proper time

τ, we will obtain the following parametrisation with respect to τ: $x^0 = t(\tau)\,, x^1 = \gamma^1(\tau)\,, x^2 = \gamma^2(\tau)\,, x^3 = \gamma^3(\tau)$. We have:

$$\begin{aligned}\vec{\mathcal{P}} = m\frac{d\vec{\gamma}}{d\tau} &= \left(m\frac{dt}{d\tau}, m\frac{d\gamma^1}{d\tau}, m\frac{d\gamma^2}{d\tau}, m\frac{d\gamma^3}{d\tau}\right) \\ &= \frac{dt}{d\tau}\left(m, m\frac{d\gamma^1}{dt}, m\frac{d\gamma^2}{dt}, m\frac{d\gamma^3}{dt}\right) = \frac{dt}{d\tau}(m, mv^1, mv^2, mv^3)\,,\end{aligned}$$

where $\vec{v} = (v^1, v^2, v^3)$ is the usual velocity of the particle with respect to the inertial system S. By definition of proper time, we obtain

$$\tau(t) = (1/ic)\int_0^t \sqrt{\tilde{\eta}(d\vec{\gamma}/d\xi, d\vec{\gamma}/d\xi)}d\xi\ .$$

Therefore

$$\begin{aligned}d\tau/dt &= (1/ic)\sqrt{\tilde{\eta}(d\vec{\gamma}/dt, d\vec{\gamma}/dt)} = (1/ic)\sqrt{\tilde{\eta}((1, v^1, v^2, v^3), (1, v^1, v^2, v^3))} \\ &= (1/c)\sqrt{c^2 - \|\vec{v}\|^2} = \sqrt{1 - \|\vec{v}\|^2/c^2}\ ,\end{aligned}$$

where $\|\vec{v}\| = \sqrt{(v^1)^2 + (v^2)^2 + (v^3)^2}$. Hence the Minkowski momentum, with respect to the coordinates corresponding to an inertial system, is expressed as follows:

$$\vec{\mathcal{P}} = \frac{1}{\sqrt{1 - \|\vec{v}\|^2/c^2}}(m, mv^1, mv^2, mv^3)\ .$$

(Observe that the latter formula implicitly implies that the module $\|\vec{v}\|$ of the velocity vector $\vec{v}$ of the particle with respect to an inertial system should be smaller than c.) By the Taylor expansion $(1/\sqrt{1-x^2}) = 1 + x^2/2 + \ldots$, we have:

$$\vec{\mathcal{P}} = (m + \frac{m\|\vec{v}\|^2}{2c^2}, mv^1, mv^2, mv^3) + \ldots$$

where we have only taken into account the velocity terms of order not greater than 2. Therefore we see that for small velocities with respect to an inertial system, the three last components of $\vec{\mathcal{P}}$ approximately correspond to the classical momentum, while the zero component (time component) approximately corresponds to the mass incremented by the kinetic energy divided by c^2. The total *energy* of the particle with respect to an inertial system is defined as the time component of $\vec{\mathcal{P}}$ multiplied by c^2. Therefore the energy is not an intrinsic concept since it depends on the inertial system.

II.3.2 Collision laws. Equivalence of mass and energy

One of the collision laws for two particles in classical mechanics is that of conservation of momentum: the momentum of the system formed by two particles remains constant before and after the impact. This law governs any type of collisions, both elastic and non-elastic. In classical mechanics, an elastic collision of two particles is defined as a collision where the kinetic energy is conserved, that is, where the kinetic energy of the system before and after the impact remains unaltered. Completely non-elastic collisions are defined in classical mechanics as those where two particles stay together after the impact. There are many intermediate classes of collisions between those that are elastic and completely non-elastic. In non-elastic collisions there are losses of kinetic energy by definition; this energy transforms into heat.

How should collision laws be stated in relativity? By virtue of Postulate 3 of special relativity, the law we would like to state should be intrinsic, that is, independent of any inertial system. Since the unique concept in relativity which is similar to the classical momentum is the Minkowski momentum, it seems reasonable to take conservation of the Minkowski momentum of the system (formed by two particles) before and after the impact to be the relativity collision law for two particles. Physical laws should not be proved. The criteria we will use to admit this law are to be intrinsic (it is!) and the coincidence at small velocities with the classical law. At small velocities, conservation of the Minkowski momentum implies that of classical momentum (three space components) and that of energy (time component). Therefore, at small velocities, conservation of the Minkowski momentum gives us the classical law for elastic collisions. But how are non-elastic collisions contemplated within the framework of relativity theory?

In order to give a response to the last question, let us study a collision of two particles with the same mass m and moving with respect to an inertial system at velocities having equal modules but directed in opposite directions. Choose the axes in this system in order to situate the two particles over the x axis. Suppose that the velocities of the particles are $\vec{v} = (v, 0, 0)$ and $-\vec{v} = (-v, 0, 0)$. The Minkowski momentum of the system prior to the collision will be

$$\left(\frac{m}{\sqrt{1-v^2/c^2}}, \frac{mv}{\sqrt{1-v^2/c^2}}, 0, 0, \right) + \left(\frac{m}{\sqrt{1-v^2/c^2}}, -\frac{mv}{\sqrt{1-v^2/c^2}}, 0, 0\right).$$

That is

$$\left(\frac{2m}{\sqrt{1-v^2/c^2}}, 0, 0, 0\right).$$

If one managed to maintain the two particles together after the impact at rest with respect to the inertial system S (as it happens in a completely non-elastic collision in classical mechanics), the Minkowski momentum posterior to the impact would be $(m', 0, 0, 0)$, where m' is the sum of the masses of two particles. By equating this vector to that prior to the impact, we see that if these two momentums have to be equal, then the equality $m' = 2m/\sqrt{1-v^2/c^2}$ should hold. Therefore we would have to assume that the mass of the system posterior to the impact is greater than the mass $2m$ of the system prior to the impact. The difference $m' - 2m$ of masses after and before the collision would be $2m((1/\sqrt{1-v^2/c^2})-1) \cong mv^2/c^2$. In this way, we should admit that the kinetic energy is inverted into a mass increment. Therefore if the particles stay together after the collision, then either the system mass is augmented by a quantity approximately equal to the classical kinetic energy divided by c^2 or this energy is transformed into heat (another form of energy). In any case, we should admit that any form of energy (kinetic, caloric, etc.) is equivalent to the mass multiplied by c^2.

II.3.3 Minkowski force

Consider a moving particle of mass m. Let $\vec{\mathcal{P}} = md\gamma/d\tau$ be the Minkowski momentum of this particle. Define the *Minkowski force* of the particle as $\vec{F} = d\vec{\mathcal{P}}/d\tau$. It is clear that the Minkowski force is an intrinsic concept associated to the moving particle. Now, given an inertial system S, the vector $\vec{F}$ will have four components as referred to the reference in the Minkowski space associated to S: $\vec{F} = (F^0, F^1, F^2, F^3)$. The three space components are given by

$$F^i = \frac{d}{dt}\left(\frac{mv^i}{\sqrt{1-\|\vec{v}\|^2/c^2}}\right)\frac{dt}{d\tau},$$

where (v^1, v^2, v^3) stands for the (ordinary) velocity vector of the particle with respect to the inertial system. Then, with the notation of $\alpha = \sqrt{1-\|\vec{v}\|^2/c^2}$, we obtain by a direct calculation that

$$\frac{d}{dt}\left(\frac{mv^i}{\alpha}\right) = \frac{1}{\alpha}\frac{mdv^i}{dt} + \alpha^{-3}\frac{mv^i}{c^2}\frac{d\|\vec{v}\|}{dt}.$$

If at instant τ_0 of the proper time of the particle we choose an inertial system such that the particle is at rest with respect to it at that instant, then at this time and with respect to this system the three space components of $\vec{F}$ with respect to the reference $\{O; \varepsilon_0, \varepsilon_1, \varepsilon_2, \varepsilon_3\}$ associated to the inertial system will be $F^i = mdv^i/dt$ (the classical Newton's force). The time component F^0 in this reference equals zero as we will see now. Indeed, since $dt/d\tau = 1/\alpha$, we have

$$F^0 = \frac{d}{dt}\left(\frac{m}{\alpha}\right)\frac{dt}{d\tau} = \frac{d}{dt}\left(\frac{m}{\alpha}\right)\frac{1}{\alpha} = \frac{m\alpha^{-4}\|\vec{v}\|}{c^2}\frac{d\|\vec{v}\|}{dt},$$

which vanishes if $\vec{v} = 0$.

Summarizing the above argument, if at an instant τ_0 of the proper time of the particle we take an inertial system such that the particle does not move with respect to this system at that moment, then the components of the Minkowski force are $(0,\ mdv^1/dt,\ mdv^2/dt,\ mdv^3/dt)$. This enables us to introduce the definition of *proper inertial system* of the particle (at each instant of its proper time) as an inertial system such that the particle does not move with respect to it at that moment. The three space components of the Minkowski force with respect to the reference of the Minkowski space associated to such a system coincide with the Newton's force in this inertial system, with the time component equal to zero.

II.3.4 Relativistic fluid dynamics

The variables appearing in the equations of motion of a fluid in classical mechanics (see Subsection II.1.5) are the following: mass density ρ, stress tensor $\mathcal{T}$, density of the external force $\vec{f}$, velocity field of the fluid $\vec{v}$. It is clear that all these variables depend on the the inertial system we choose. With minimum possible changes in the classical equations, we would like to find other *intrinsic* equations in the relativity framework in such a way that these new equations will be similar to classical ones in any particular inertial system with respect to which the fluid moves at small velocities. For writing intrinsic equations in the relativity framework, we should start from intrinsic concepts that would substitute the classical ones such as mass density, stress tensor, density of the external force, etc. This is what we are going to do in what follows.

We can think of a fluid as a set of moving particles. The life of each particle in this fluid will be represented in the Minkowski space-time by a curve having time-like tangent vector. Let us assume, as is done in classical mechanics, that two fluid particles never coincide at the same place. This implies that the curves representing particles' lives never intersect each other. Each of these curves is parametrized by the particle's proper time. The tangent vector at every point of these curves will be a time-like vector field oriented towards the future whose norm is ic. Therefore the life of the fluid is represented by a family of curves tangent to a time-like vector field $\vec{u}$, of norm ic and oriented towards the future, defined on some open set V of the Minkowski space $(\mathbb{R}^4, \tilde{\eta})$. The field $\vec{u}$ is called the (Minkowski) velocity field of the fluid.

Let P be a point in V. There is one and only one particle of the fluid whose life is a curve $\gamma(\tau)$ passing through P. Let τ_0 be the particle's proper time at which $\gamma(\tau_0) = P$. Consider an orthonormal reference of the Minkowski space of origin P having the first vector (time vector) equal to $\vec{u}_P$, that is, a reference of the form $\{P; \vec{u}_P, \vec{\varepsilon}_1, \vec{\varepsilon}_2, \vec{\varepsilon}_3,\}$. This reference corresponds to an inertial system S_P called the *proper inertial system* of the fluid particle whose life is represented by γ at instant τ_0 of its proper time (the instant τ_0 of the particle's proper time corresponds to the initial instant $t = 0$ of the inertial system S_P). The *proper mass-energy density* of the fluid is defined as the function ρ on V representing at each point P the classical mass density of the fluid measured from the inertial system S_P at its origin and at its initial instant. The *Minkowski density of the external force* (acting on the fluid) is defined as the vector field $\vec{f}$ on V which for each P in V is taken to be the vector whose components in the reference $\{P; \vec{u}_P, \vec{\varepsilon}_1, \vec{\varepsilon}_2, \vec{\varepsilon}_3\}$ are $(0, f^1, f^2, f^3)$, where f^1, f^2, f^3 are the three components of the density of the external force in classical mechanics measured from the inertial system S_P at its origin at time $t = 0$. The fluid *stress tensor* $\mathcal{T}$ is defined as the (1,1)-type tensor field on V which for each P in V is taken to be the tensor $\mathcal{T}_P$ whose components in the reference $\{P; \vec{u}_P, \vec{\varepsilon}_1, \vec{\varepsilon}_2, \vec{\varepsilon}_3\}$ are determined by means of the following conditions: $(\mathcal{T}_P)^i_0 = (\mathcal{T}_P)^0_i = 0$ for $i = 1, 2, 3$, and $(\mathcal{T}_P)^i_j$ (when $i, j = 1, 2, 3$) are the components of the fluid stress tensor according to classical mechanics, measured from the inertial system S_P at its origin at time $t = 0$. All these concepts are intrinsic.

In the Minkowski space, consider the equation $\operatorname{div}(\rho\vec{u} \otimes \vec{u} - \mathcal{T}) = \vec{f}$ where $\vec{u}$ is the Minkowski velocity field of the fluid, ρ is the proper density, $\mathcal{T}$ is the stress tensor in the Minkowski space, and $\vec{f}$ is the Minkowski density of the external force. Since all variables that appear in the last equation are intrinsically defined, this equation is intrinsic. But what is the relation of this equation with classical equations of motion of a fluid? Assume that we are situated in an inertial system S and that the fluid moves with respect to S at small velocities. At each point of the fluid and at each time instant, there will be a proper inertial system S' (such that a fluid particle is situated at that instant at the origin of S' and moves at zero speed with respect to S' at that moment). These systems are those we have already used for defining the concepts of proper density, Minkowski density of the external force, etc. If all measurements (those of proper density, of the Minkowski density of the external force, etc.) we have to carry out in each of systems S', are all made in S, what would happen? We would obtain almost the same values as those had we carried out our measurements in each of systems S' (since each S' moves with respect to S at a very small speed). Since we would obtain almost the same values, let us carry out all our measurements in S. In this way, we assume that the proper density coincides with the density measured in S, etc. The inertial system S will provide, as usual, a reference to the Minkowski space. Let us write the equation $\operatorname{div}(\rho\vec{u} \otimes \vec{u} - \mathcal{T}) = \vec{f}$ component-wise with respect to this reference. The equation will be

$$\sum_j \frac{\partial}{\partial x^j}(\rho u^\alpha u^j - \mathcal{T}^{\alpha j}) + \frac{\partial}{\partial t}(\rho u^\alpha u^0 - \mathcal{T}^{\alpha 0}) = f^\alpha ,$$

where, according to the convention adopted in Section I.5.2, the Latin indices range from 1 to 3 and the Greek ones run from 0 to 3. Since we suppose that all proper systems S' coincide with S, we have $\mathcal{T}^{\alpha 0} = 0$. On the other hand, the vector $\vec{u}$ for small velocities is approximately equal to $(1, v^1, v^2, v^3)$, where $\vec{v} = (v^1, v^2, v^3)$ is the field of ordinary velocities of the fluid with respect to the inertial system S. Therefore we see that three of the last equations (those corresponding to $\alpha = 1, 2, 3$) can be approximated by $\operatorname{div}(\rho\vec{v} \otimes \vec{v} - \mathcal{T}) + \partial(\rho\vec{v})/\partial t = \vec{f}$ (vector-valued equation on $\mathbb{R}^3$) and that the first equation (which corresponds to $\alpha = 0$) can be approximated by $\operatorname{div}(\rho\vec{v}) + \partial\rho/\partial t = 0$. Both equations are classical equations of the fluid motion. This argument suggests that we accept the equation $\operatorname{div}(\rho\vec{u} \otimes \vec{u} - \mathcal{T}) = \vec{f}$ as the equation that governs the fluid motion in relativity (since it is an intrinsic equation which is approximated by classical equations for small velocities).

II.3.5 Stress-energy tensor of a fluid

The *stress-energy tensor* of a fluid is defined as the contravariant 2-tensor field (in an open set of the Minkowski space) expressed as $T = (1/c^2)(\rho\vec{u} \otimes \vec{u} - \mathcal{T})$, using the notation introduced in the previous section. When it is convenient for us we can think, through the Minkowski metric, that T is one-time contravariant and one-time covariant, or two-times covariant. With this definition, the motion equation of a fluid would be written as

$$c^2 \text{div } T = \vec{f} \ . \tag{II.11}$$

If we think of T as a linear map, it is clear that T has only one time-like eigenvector $\vec{u}$ with norm ic and oriented towards the future. Then $-\rho$ is the eigenvalue that corresponds to this eigenvector. This tells us that the knowledge of T implies that of ρ, $\vec{u}$ and of the stress tensor.

Let us now see that for perfect fluids the stress-energy tensor is

$$T = \frac{1}{c^2}\left(\rho + p/c^2\right)\vec{u} \otimes \vec{u} + \frac{p}{c^2}\tilde{\eta} \ ,$$

where $\tilde{\eta}$ is the Minkowski metric interpreted as a contravariant tensor. In order to see this, denote by T the stress-energy tensor as we have already done before, $T = (1/c^2)(\rho\vec{u}\otimes\vec{u} - \mathcal{T})$, and by $\mathcal{A}$ the right side in the above formula. We would like to see that $T = \mathcal{A}$. At any point P in the Minkowski space, take an orthonormal reference, $\{P; \varepsilon_0, \varepsilon_1, \varepsilon_2, \varepsilon_3\}$ with $\varepsilon_0 = \vec{u}_P$. By the previous definition of T, we have $T^{ij} = -\mathcal{T}^{ij}/c^2$. Since $\mathcal{T}$ corresponds to a perfect fluid, $\mathcal{T}(x, \vec{v}, t) = -\rho(x,t)\vec{v}$. So $\mathcal{T}_{ij} = -\rho\delta_{ij} = p\tilde{\eta}_{ij}$ (where i, j range from 1 to 3). Thus $T^{ij} = -\mathcal{T}^{ij}/c^2 = p\tilde{\eta}^{ij}/c^2$. We have also $T^{i0} = -\mathcal{T}^{i0}/c^2 = 0$, $T^{00} = \rho/c^2$. It is clear now that the components of $\mathcal{A}$ in this reference coincide with those of T we have already calculated. Therefore $T = \mathcal{A}$.

In the case of a perfect fluid, there are six unknown scalar variables in equation (II.11). We would like to know these variables based on our knowledge of $\vec{f}$. These six unknowns are ρ, p and four components of $\vec{u}$. Further, ρ and p are normally related through a state equation, so that p is a known function of ρ. On the other hand, the norm of $\vec{u}$ is ic. These two arguments reduce the number of unknowns down to four. If external forces $\vec{f}$ are known, one can calculate from the vector equation $c^2\text{div } T = \vec{f}$ (in first-order partial derivatives) the four independent unknowns, if we know them on a hypersurface in the space (initial conditions). Therefore, if all variables of the fluid are known at an instant $t = 0$ with respect to an inertial system, we know the evolution of this fluid in time.

II.4 General relativity

II.4.1 Fundamentals

The objective of special relativity is to model physical phenomena occurring under the lack of gravity. General relativity pretends to introduce gravitational fields into relativity theory. It had taken a long time for Einstein to guess that gravitational forces did not exist in reality, that they rather appear as a result of the particular choice of a reference system of each observer. Indeed, an observer subject to the terrestrial gravitational field would not note the effects of this field if he was in a free fall. On the contrary, an observer situated inside a closed box under the lack of gravity will feel subject to a gravitational field if the box in which he is enclosed moves uniformly accelerated with respect to another inertial observer. In this case, if he freely releases a particle of mass m, he will see that this particle is subject to a constant relative acceleration (with respect to him) independent of m and will attribute this acceleration to a gravitational field (in gravitational fields the acceleration of a particle is independent of its mass). This argument made Einstein intuitively suppose that gravitational forces did not exist, but are a result of choosing a particular reference system. If under the lack of gravity the appropriate framework for presentation of relativity theory was the Minkowski space and any physical phenomenon had an adequate geometrical representation in this space, the framework that is adequate to the case of presence of gravity would be a space-time endowed with another metric. That is, the only thing gravitational fields would do is to deform the metric of the space-time which would no longer be that of Minkowski and would become another, more general, metric.

As we have already said, special relativity models physical phenomena under the lack of gravity. Let us briefly recall the way this modelling is carried out.

Any event is represented by a point in the Minkowski space. The life of any material particle is a curve belonging to this space whose tangent vector at any point is time-like and oriented towards the future. The *arc length* of this curve (which is imaginary) divided by ic represents the particle's proper time. If the particle is not subject to any force, the curve representing its life is a straight line (that is, a geodesic of the Minkowski space) whose vector is time-like and oriented towards the future. The life of a light ray is also a straight line in the Minkowski space whose vector $\vec{v}$ satisfies the following two conditions: 1) $\vec{v}$ belongs to the closure of the cone C^+; 2) $\tilde{\eta}(\vec{v},\vec{v}) = 0$, where $\tilde{\eta}$ is the Minkowski metric.

An inertial reference system (three perpendicular axes travelling in ordinary space together with observers provided with clocks, etc.) is represented in the Minkowski space by a reference $\{O;\varepsilon_0,\varepsilon_1,\varepsilon_2,\varepsilon_3\}$ having orthonormal basis $\{\varepsilon_\alpha\}$ and such that $\varepsilon_0 \in C^+$. To any material particle whose life is represented in the Minkowski space by the curve $\gamma(\tau)$, a proper inertial system is assigned at any instant τ of its proper time (this assignment is not unique) with respect to which the particle remains at rest at that instant. This inertial system is represented by a reference $\{\gamma(\tau);\varepsilon_\alpha(\tau)\}$ to the Minkowski space that constitutes a mobile reference along the curve $\gamma(\tau)$. An observer is nothing else than a moving material particle endowed at each moment of its life by a proper inertial system. Therefore we can think of such an observer to be a curve $\gamma(\tau)$ of the Minkowski space provided with a moving frame $\{\gamma(\tau);\varepsilon_\alpha(\tau)\}$ orthonormal for all values of τ, with the condition that $\varepsilon_0(\tau)$ is tangent to $\gamma(\tau)$ for all τ and $\varepsilon_0(\tau) \in C^+$.

To each material particle, a mass m is assigned (its rest-mass, or its mass measured in a proper inertial system). This mass can, in principle, be changed due to collisions or other physical phenomena. If the life of the particle is represented by the curve $\gamma(\tau)$ of the Minkowski space, the vectors $\vec{\mathcal{P}} = m\dot{\gamma}$ and $\vec{F} = m\tilde{\nabla}_{\dot{\gamma}}\dot{\gamma}$ at each instant τ are called the Minkowski momentum and the Minkowski force, respectively, at the instant τ (here, $\tilde{\nabla}$ stands for the covariant derivative associated to the Minkowski metric $\tilde{\eta}$ according to the notational conventions in Section I.5.2).

If the particle of mass m whose life is $\gamma(\tau)$ at a certain moment of its life comes across an observer represented by a moving frame $\{\gamma'(\tau');\varepsilon_\alpha(\tau')\}$ with $\gamma(\tau_0) = \gamma'(\tau_0')$, then the time component of the particle's momentum $\vec{\mathcal{P}}(\tau_0)$ with respect to the reference $\{\gamma'(\tau_0');\varepsilon_\alpha(\tau_0')\}$ multiplied by c^2 is the particle's mass-energy measured by the observer. Therefore energy is a concept that depends on the observer.

Einstein thought that the geometrical description of the previous paragraph, valid under the lack of gravitational fields, would still be valid when gravitational fields were present, with only the substitution of the Minkowski metric by another similar Lorentzian metric. Einstein was working locally on open sets of $\mathbb{R}^4$. If one wished to work globally, space-times $(V,\tilde{g})$ should be considered.

II.4.2 Einstein's field equation

The metric $\tilde{g}$ of the Lorentzian manifold $(V,\tilde{g})$ which should be considered, is to be determined in some way by the matter that creates the gravitational field. But, in what way?

First, even though gravitational forces do not exist in reality and are the result of a bad choice of coordinates, we should explain in what way these apparent forces appear for us. If we freely release a particle, this particle's life will be represented by a geodesic $\gamma(\tau)$ of the Lorentzian manifold $(V,\tilde{g})$. The condition that γ is geodesic is written as $\tilde{\nabla}_{\dot{\gamma}}\dot{\gamma} = 0$, where $\tilde{\nabla}$ denotes the covariant differential associated with the metric $\tilde{g}$. In a coordinate system (x^α), this equation becomes $d^2\gamma^\lambda/d\tau^2 = -\sum\tilde{\Gamma}^\lambda_{\mu\nu}(d\gamma^\mu/d\tau)(d\gamma^\nu/d\tau)$. The second derivatives $d^2\gamma^\lambda/d\tau^2$ will be responsible for the particle's apparent acceleration. Observe that this apparent acceleration depends on the Christoffel symbols $\tilde{\Gamma}^\lambda_{\mu\nu}$, with these latter depending on the coefficients $\tilde{g}_{\alpha\beta}$ of the metric and its derivatives. This leads us to the following question: In classical mechanics, does there exist any mathematical object whose first-order derivatives would appear in the expression describing acceleration of a freely released particle subject to the action of a gravitational field? The response is in the affirmative, and the mathematical object referred to in the question is the potential function of the gravitational field (whose gradient with the opposite sign gives the above mentioned acceleration). Therefore the coefficients $\tilde{g}_{\alpha\beta}$ of the metric $\tilde{g}$ play the same role in general relativity as that performed by the potential function of a gravitational field in classical mechanics. This is the reason to sometimes call the coefficients $\tilde{g}_{\alpha\beta}$ *gravitational potentials*.

How do we determine the potential function of a gravitational field in classical mechanics using the matter that originates the field? If the matter that originates the gravitational field is a continuous medium whose density function is ρ, the gravitational potential V originated by this matter satisfies the Poisson equation

$$\Delta V = 4\pi K \rho \ ,$$

where K is the Newtonian constant of gravitation and Δ is the Laplace operator $\partial^2/\partial x_1^2 + \partial^2/\partial x_2^2 + \partial^2/\partial x_3^2$. The Poisson equation is a second-order partial differential equation from which, given ρ, the function V can be recovered using some boundary conditions.

Einstein thought that a similar law should exist in general relativity which would enable us to calculate the coefficients $\tilde{g}_{\alpha\beta}$ of the metric $\tilde{g}$ under some knowledge of the matter causing the gravitational field. The fundamental Physical Laws (Maxwell's electromagnetic laws, Newton's gravitational law, etc.) are not proved. Simply, a physicist introduces them in order to explain the behavior of more complex phenomena. Einstein wanted to find a law to determine the metric $\tilde{g}$ based on knowledge of the matter causing the gravitational field. He started from the classical Poisson law and simply worked by similarity with the classical theory. In this classical theory there is only one function to be determined (the potential V) while the number of components of the relativistic $\tilde{g}$ (a symmetric 2-tensor on a four-dimensional manifold) is 10. He thought then that the Poisson equation should be substituted by 10 scalar equations or by a single tensor equation involving symmetric 2-tensors. Therefore the idea was that the classical Poisson equation should be replaced by a tensor equation having the form

$$G = \chi T \ , \tag{II.12}$$

where G was a symmetric 2-tensor whose components with respect to any local chart were functions of $\tilde{g}_{\alpha\beta}$ and of its derivatives up order 2 (since the Poisson equation involves derivatives of order not greater than 2 only), χ was a constant and T a symmetric 2-tensor containing all information on the matter. In relativistic fluid mechanics (in special relativity), the stress-energy tensor is a symmetric 2-tensor giving a complete description of the medium's matter. Einstein then thought that a natural candidate to be the tensor T in his equation was the stress-energy tensor. According to his philosophy, the step from special to general relativity should have been made by only replacing the Minkowski metric $\tilde{\eta}$ by another appropriate Lorentzian metric $\tilde{g}$ (which should have emerged as a solution of this equation). We have seen above in the particular case of perfect fluids what form the stress-energy tensor had in special relativity. According to Einstein's philosophy, in general relativity (and in the same case of perfect fluids) the same expression should hold with another metric.

Recall that the equation that governs the life of a fluid in special relativity is $c^2 \text{div } T = \vec{f}$, where $\vec{f}$ stands for the density of the Minkowski external force acting upon the fluid. Here, in general relativity, we suppose that the continuous medium (not necessarily a fluid) is the one that generates the gravitational field and also we assume that gravitational forces do not exist. Therefore the equality $\text{div } T = 0$ should hold in our case (where external forces do not exist).

Let us continue with the equation (II.12) we are looking for. In a Lorentzian manifold (and also in any pseudo-Riemannian manifold), what symmetric 2-tensors do we know whose components in any local chart would be functions of $\tilde{g}_{\alpha\beta}$ and of their derivatives up to order 2? The first tensor having these characteristics that comes to our mind is the Ricci tensor. In principle, it seems then that this Ricci tensor could be a candidate to be the tensor G on the left side of the equation $G = \chi T$. But since the right side in this equation is the tensor T whose divergence vanishes, the left side should also have divergence zero. What is the divergence of the Ricci tensor? By virtue of Proposition I.6, the divergence of the Ricci tensor is $(1/2)d\tilde{R}$, where $d\tilde{R}$ denotes the exterior derivative of the scalar curvature $\tilde{R}$. This immediately implies that the Ricci tensor has the same divergence as the tensor $(1/2)\tilde{R}\tilde{g}$, where $\tilde{g}$ denotes the metric tensor. Therefore if one would like to have a tensor having divergence zero, the Ricci tensor is of no avail, while the tensor $\widetilde{Ric} - (1/2)\tilde{R}\tilde{g}$ does help (recall that in Section I.5.2, we have denoted by $\widetilde{Ric}$ the Ricci tensor of $\tilde{g}$). Moreover, the tensor $\widetilde{Ric} - (1/2)\tilde{R}\tilde{g}$ satisfies the other requirements the left side in the equation $G = \chi T$ should have, more precisely: to be symmetric and to depend on the coefficients of the metric $\tilde{g}$ and on its derivatives up to order 2. This is why Einstein took as the left side G in his equations the tensor $\widetilde{Ric} - (1/2)\tilde{R}\tilde{g}$ (which is called an *Einstein tensor*).

If an extra term having the form $\Lambda\tilde{g}$, where Λ is an arbitrary constant, is added to the tensor $\widetilde{Ric} - (1/2)\tilde{R}\tilde{g}$, then the resulting tensor $\widetilde{Ric} - (1/2)\tilde{R}\tilde{g} + \Lambda\tilde{g}$ also satisfies all requirements imposed on the left side of the equation $G = \chi T$. Therefore, in principle, we could also take it as the left side in the equation. Poincaré proved that the only 2-tensors that satisfied all above mentioned requirements (be symmetric, have divergence zero, have components that would depend on the coefficients $\tilde{g}_{\alpha\beta}$ of the metric and on their derivatives up to order 2) were those having the form $\widetilde{Ric} - (1/2)\tilde{R}\tilde{g} + \Lambda\tilde{g}$. Therefore the equation Einstein was looking for (Einstein's equation) should be as follows:

$$\widetilde{Ric} - (1/2)\tilde{R}\tilde{g} + \Lambda\tilde{g} = \chi T \ . \tag{II.13}$$

In principle, if one would like the Minkowski metric $\tilde{\eta}$ to be a solution of the latter equation when there is no matter ($T = 0$), since for the Minkowski metric both the Ricci tensor and the scalar curvature vanish, the coefficient Λ in (II.13), called the *cosmological constant*, should be equal to zero. This is what Einstein admitted at a first instance. Nevertheless, later he was conflicted between the possibility that it were different from zero and that it were equal to zero. Here, we assume this term to be equal to zero.

The Poisson equation $\Delta V = 4\pi K\rho$ describing the classical gravitational potential V created by a matter with density ρ should be an approximation of Einstein's equation. We will see in Chapter III that this fact implies that χ is approximately equal to $8K\pi/c^4$, where K is Newton's constant of gravitation.

Another form of writing Einstein's equation is as follows:

$$\widetilde{Ric}(\tilde{g}) = \chi(T - \frac{1}{2}(\mathrm{tr}_{\tilde{g}}T)\tilde{g}) \,, \tag{II.14}$$

where $\mathrm{tr}_{\tilde{g}}T$ denotes the trace of T with respect to the metric $\tilde{g}$. It is very easy to see the equivalence of (II.13) with $\Lambda = 0$ and (II.14). Indeed, in coordinates (II.13) with $\Lambda = 0$ can be written

$$\tilde{R}_{\alpha\beta} - \frac{1}{2}\tilde{R}\tilde{g}_{\alpha\beta} = \chi T_{\alpha\beta} \,.$$

Let $(\tilde{g}^{\alpha\beta})$ be the inverse matrix of $(\tilde{g}_{\alpha\beta})$. Multiply the latter equality by $\tilde{g}^{\alpha\beta}$ and sum with respect to α, β to obtain

$$\tilde{R} - 2\tilde{R} = \chi \mathrm{tr}\; T \,.$$

Therefore $\tilde{R} = -\chi \mathrm{tr}\; T$. If we substitute this into equation (II.13) with $\Lambda = 0$, we obtain (II.14).

II.5 Cosmological models

Among a number of cosmological models proposed within the framework of general relativity starting from the moment when Einstein published in 1917 his first paper on cosmology, the most simple and plausible models are due to Robertson and Walker which will be our topic in this section.

Imagine the matter of the universe as a perfect fluid where each particle is a galaxy. There is no other force acting upon fluid particles (or galaxies) except for those exerted upon one another through gravity. The particles' motion is represented by geodesics in a space-time $(V, \tilde{g})$. The metric $\tilde{g}$ is related to the matter by means of Einstein's equation. If we admit certain "natural" assumptions we are going to describe in what follows, we will be able to give a more concrete expression for the metric $\tilde{g}$. The assumptions we admit are the following:

1. The motions of each fluid particle are represented by a geodesic of V with time-like tangent vector oriented towards the future. One and only one of these geodesics passes through each point $x \in V$. Denote by $\mathcal{F}$ the foliation of dimension 1 (flow) constituted by these geodesics, and by F_x the subspace of $T_x(V)$ tangent to the foliation $\mathcal{F}$.

2. The distribution assigning to each $x \in V$ the subspace $F_x^{\perp}$ orthogonal to F_x is integrable. Denote this foliation by $\mathcal{F}^{\perp}$. The leaves of $\mathcal{F}^{\perp}$ will be the "space" common for all galaxies at a certain instant.

3. With $x \in V$ fixed, for each pair (u, v) of unitary vectors of $F_x^{\perp}$ there exists an isometry Φ of $(V, \tilde{g})$ leaving x fixed, that preserves the foliation $\mathcal{F}$ and such that $(d\Phi)_x$ transforms u into v. This assumption is equivalent to saying that there are no spatial privileged directions (*spatial isotropy*) at each point.

   Based on these assumptions, the following result can be proved.

Proposition II.3. *Each point $p \in V$ has a neighbourhood U that is a domain of a local chart (U, t, x^1, x^2, x^3) where $\partial/\partial t$ is tangent to the flow $\mathcal{F}$, with $-\varepsilon < x^i < \varepsilon$, $-\varepsilon < t < \varepsilon$, $p = (0,0,0,0)$, and where the metric $\tilde{g}$ is as follows:*

$$\tilde{g} = -dt^2 + f(t)^2 \hat{g} \ , \tag{II.15}$$

where $f(t)$ is a function of t and $\hat{g}$ is a Riemannian metric having constant sectional curvature $\kappa = 0$, $\kappa = 1$ or $\kappa = -1$ over the hypersurface S_0 of U given by $t = 0$.

Though this proposition is local, it suggests that the most simple space-times $(V, \tilde{g})$ compatible with the three assumptions are those of the form $V = I \times S$, with $(S, \hat{g})$ a 3-dimensional Riemannian manifold with constant curvature $\kappa = 0$, $\kappa = 1$ or $\kappa = -1$, I an open interval of $\mathbb{R}$ (that can be the whole $\mathbb{R}$) and the metric $\tilde{g}$ of the form (II.15), where t is the coordinate of I. Such a cosmological model is called a *Robertson-Walker* model. It is supposed in these models that, moreover, the matter (different galaxies) constitutes a perfect fluid having mass-energy density ρ and pressure p that depend on t only.

If the additional assumption is made that the manifold $(S, \hat{g})$ is complete and simply connected, then it is well known that $(S, \hat{g})$ should necessarily be $\mathbb{R}^3$ with the Euclidean metric, the sphere $\mathbb{S}^3$ with the metric of curvature 1, or the hyperbolic space $\mathbb{H}^3$ with metric of curvature -1.

II.6 Appendix: a theorem in affine geometry

The objective of this Appendix is to give the theorem we have referred to in Remark II.2.

Theorem II.4. *Consider $\mathbb{R}^4$ endowed with the Minkowski metric $\tilde{\eta}$. Let f be a bijective map $\mathbb{R}^4 \to \mathbb{R}^4$ mapping straight lines having time-like vector (with respect to $\tilde{\eta}$) into straight lines having time-like vector. Then f is an affinity (that is, has the form $f(x) = F(x) + p$ where F is linear and $p \in \mathbb{R}^4$).*

Proof. Let us say that a plane or a hyperplane of $\mathbb{R}^4$ is *admissible* if it has the form $a + H$, where H is a vector subspace (of dimension 2 or 3) admitting a basis formed by time-like vectors and $a \in \mathbb{R}^4$.

Observe first that f maps admissible planes into admissible planes. Indeed, let $V = a + H$ be an admissible plane. We would like to see that $f(V)$ is also an admissible plane. Let $\{v_1, v_2\}$ be a basis of H formed by time-like vectors. The straight lines $a + \langle v_1 \rangle$ and $a + \langle v_2 \rangle$ will be mapped by f into two straight lines r_1 and r_2 having time-like vector. Each of these straight lines should contain $f(a)$. Therefore r_1 and r_2 are coplanary (because they intersect each other). Let W be the plane that contains them. Assume that x is an arbitrary point of V not situated on the straight lines $a + \langle v_1 \rangle$ and $a + \langle v_2 \rangle$. Let us prove that $f(x) \in W$. Let α be any straight line contained in V passing through x and having time-like vector u such that $u \notin \langle v_1 \rangle$, $u \notin \langle v_2 \rangle$. Then $f(\alpha)$ will be a straight line having time-like vector and containing $f(x)$. Now, since α intersects $a + \langle v_1 \rangle$ and $a + \langle v_2 \rangle$, the straight line $f(\alpha)$ will intersect r_1 and r_2. Therefore $f(\alpha) \subset W$ and hence $f(x) \in W$. This proves that $f(V) \subset W$. The same argument applied to f^{-1} finally proves that $f(V) = W$.

A similar argument would show that f maps admissible hyperplanes into admissible hyperplanes.

In order to prove that f is an affinity, it suffices to check that it maps any straight line into a straight line (then a well-known theorem can be applied claiming that any bijection of $\mathbb{R}^n$ to $\mathbb{R}^n$ that maps straight lines into straight lines is an affinity). Let r be a straight line. If we prove that r can be obtained as the intersection of three admissible hyperplanes V_1, V_2 and V_3, $r = V_1 \cap V_2 \cap V_3$, then $f(V_1)$, $f(V_2)$ and $f(V_3)$ will be hyperplanes whose intersection is necessarily a straight line (since f is bijective). This straight line will be $f(r)$, and we will be done.

Let C be the set of time-like vectors of $\mathbb{R}^4$. Let C^+ be the connected component of C containing the vector $(1, 0, 0, 0)$ (we use the notation where the first component is time). The straight line r will have the form $a + \langle v \rangle$. Choose three vectors u, u' and u'' of C^+ such that $\{v, u, u', u''\}$ is a basis of $\mathbb{R}^4$. Consider the three hyperplanes

$$V_1 = a + \langle v, u, u' \rangle$$
$$V_2 = a + \langle v, u, u'' \rangle$$
$$V_3 = a + \langle v, u', u'' \rangle .$$

Since the vectors v, u, u' and u'' are linearly independent, the intersection $V_1 \cap V_2 \cap V_3$ is the straight line r. The proof will be completed if we check that each of these hyperplanes is admissible. Let us see it in the case of V_1. Write

$$w = \lambda v + u + u' ,$$

where λ is a constant to be determined later on. If $\tilde{\eta}$ denotes the Minkowski inner product, we have

$$\begin{aligned}\tilde{\eta}(w,w) &= \lambda^2 \tilde{\eta}(v,v) + 2\lambda\tilde{\eta}(v,u) + 2\lambda\tilde{\eta}(v,u') \\ &\quad + \tilde{\eta}(u,u) + \tilde{\eta}(u',u') + 2\tilde{\eta}(u,u') .\end{aligned} \tag{II.16}$$

In the Minkowski space, if X and Y are vectors belonging to C^+, the inner product $\tilde{\eta}(X,Y)$ is negative. Therefore the sum $\tilde{\eta}(u,u)+\tilde{\eta}(u',u')+2\tilde{\eta}(u,u')$ in the equality (II.16) is a negative number $-A$. Therefore we can choose λ small enough to satisfy $\lambda^2\tilde{\eta}(v,v) + 2\lambda\tilde{\eta}(v,u) + 2\lambda\tilde{\eta}(v,u') < A$. Then $\tilde{\eta}(w,w) < 0$. Finally we have

$$V_1 = a + \langle v,u,u' \rangle = a + \langle w,u,u' \rangle ,$$

with three time-like vectors w, u and u'. □

Chapter III

Approximation of Einstein's Equation by the Wave Equation

In this chapter, we will study small perturbations of the Minkowski metric and will reveal the relation between Einstein's equation and the classical wave equation. We will use the notation adopted in Section I.5.2.

When there is no matter creating gravitational field, we are within the framework of special relativity. Therefore the stress-energy tensor T equals zero and the metric we have to use is the Minkowski metric $\tilde{\eta}$ on $\mathbb{R}^4$. Observe that Einstein's equation (II.14) holds in this case since the Ricci tensor of $\tilde{\eta}$ equals zero. Suppose now that we switch to the case where there is a gravitational field created by a matter whose tensor T is small and the derivatives of T are also small (in other words, the tensor T does not have sudden variations). One should believe that the Lorentzian metric $\tilde{g}$ associated with the gravitational field will be a small perturbation of the Minkowski metric. Therefore $\tilde{g}$ will be a Lorentzian metric on $\mathbb{R}^4$ whose form is $\tilde{g} = \tilde{\eta} + \tilde{h}$, with $\tilde{h}$ small and derivatives of $\tilde{h}$ small. All these concepts of "small tensors" will be made more precise in what follows. Now, we would like to study Einstein's equation (II.14) for such a metric $\tilde{g}$ as a function of $\tilde{h}$.

For this purpose, and in order to make use of calculations in similar situations, we will work in a slightly more general framework. We take a pseudo-Riemannian manifold (M, g) of arbitrary dimension. We will consider pseudo-Riemannian metrics on M having form $g' = g + h$ with h small, and will study the difference between Ricci tensors of g' and g as a function of h.

III.1 Perturbations of the Ricci tensor

Let (M, g) be a pseudo-Riemannian manifold of dimension n. Assume that g' is another pseudo-Riemannian metric. Denote by ∇ the covariant derivative associated with g, and by ∇' that associated with g'. Write

$$Q(X,Y) = \nabla'_X Y - \nabla_X Y \ ,$$

where X and Y are vector fields. If f is a function, one has

$$\begin{aligned} Q(X,fY) &= \nabla'_X(fY) - \nabla_X(fY) = X(f)Y + f\nabla'_X Y - X(f)Y - f\nabla_X Y \\ &= fQ(X,Y) \ . \end{aligned}$$

The facts that $Q(fX,Y) = fQ(X,Y)$ and $Q(X,Y) = Q(Y,X)$ are proved along similar lines. Therefore Q is a two-times covariant and one-time contravariant tensor field, symmetric in covariant factors. Denote by $R'(X,Y)Z$ the curvature tensor of g', and by $R(X,Y)Z$ that of g. Now we would like to express the difference between the two curvature tensors (that of g' and that of g) as functions of the tensor Q. We have

$$\begin{aligned} R'(X,Y)Z &= \nabla'_X\nabla'_Y Z - \nabla'_Y\nabla'_X Z - \nabla'_{[X,Y]}Z \\ &= \nabla'_X(\nabla_Y Z + Q(Y,Z)) - \nabla'_Y(\nabla_X Z + Q(X,Z)) - \nabla_{[X,Y]}Z - Q([X,Y],Z) \\ &= \nabla_X\nabla_Y Z + Q(X,\nabla_Y Z) + \nabla_X Q(Y,Z) + Q(X,Q(Y,Z)) \\ &\quad - \text{the same, with } X \text{ and } Y \text{ permuted} - \nabla_{[X,Y]}Z - Q([X,Y],Z) \\ &= R(X,Y)Z + Q(X,\nabla_Y Z) + \nabla_X Q(Y,Z) + Q(X,Q(Y,Z)) \\ &\quad - \text{the same, with } X \text{ and } Y \text{ permuted} - Q([X,Y],Z) \ . \end{aligned}$$

Recall that the expression for $\nabla_X Q$ is

$$(\nabla_X Q)(Y,Z) = \nabla_X(Q(Y,Z)) - Q(\nabla_X Y,Z) - Q(Y,\nabla_X Z) \ .$$

If we substitute the term $\nabla_X Q(Y,Z)$ in the latter expression for $R'(X,Y)Z$ by $(\nabla_X Q)(Y,\ Z) + Q(\nabla_X Y,Z) + Q(Y,\nabla_X Z)$, and the term $\nabla_Y Q(X,Z)$ by a similar expression, we obtain

$$\begin{aligned} R'(X,Y)Z = R(X,Y)Z &+ (\nabla_X Q)(Y,Z) - (\nabla_Y Q)(X,Z) \\ &+ Q(X,Q(Y,Z)) - Q(Y,Q(X,Z)) \ . \end{aligned}$$

Let us now write this formula once again in the component-wise form:

$$R'^r_{ijk} = R^r_{ijk} + \nabla_j Q^r_{ki} - \nabla_k Q^r_{ji} + Q^r_{jl}Q^l_{ki} - Q^r_{kl}Q^l_{ji} \ .$$

By making a contraction of r and j in this formula, we obtain the following formula relating Ricci tensors of g and g' as a function of the tensor Q (the difference between two covariant derivatives):

$$R'_{ik} = R_{ik} + \nabla_r Q^r_{ki} - \nabla_k Q^r_{ri} + Q^r_{rl}Q^l_{ki} - Q^r_{kl}Q^l_{ri} \ . \tag{III.1}$$

Let h be the difference $h = g' - g$. We would like to express the tensor Q in the latter formula as a function of the initial metric g and of h. By formula (I.4) giving the connection ∇' as a function of g', we obtain

$$\begin{aligned}2g'(Q(X,Y),Z) &= 2g'(\nabla'_X Y - \nabla_X Y, Z)\\ &= X(g'(Y,Z)) + Y(g'(Z,X)) - Z(g'(X,Y)) - g'(X,[Y,Z])\\ &\quad - g'(Y,[X,Z]) - g'(Z,[Y,X]) - 2g'(\nabla_X Y, Z)\ .\end{aligned}$$

Replace here g' by $g+h$ and use once again the Riemann formula (I.4) to obtain

$$\begin{aligned}2g'(Q(X,Y),Z) &= X(h(Y,Z)) + Y(h(Z,X)) - Z(h(X,Y))\\ &\quad - h(X,[Y,Z]) - h(Y,[X,Z]) - h(Z,[Y,X]) - 2h(\nabla_X Y, Z)\ .\end{aligned}$$

Now substitute the term $X(h(Y,Z))$ in the right side of this expression by $(\nabla_X h)(Y,Z) + h(\nabla_X Y, Z) + h(Y, \nabla_X Z)$ and the terms $Yh(Z,X)$ and $Zh(X,Y))$ by similar expressions. We finally have

$$2g'(Q(X,Y),Z) = (\nabla_X h)(Y,Z) + (\nabla_Y h)(Z,X) - (\nabla_Z h)(X,Y)\ .$$

The component-wise version of this formula is

$$Q^i_{jk} = \frac{1}{2} g'^{il}(\nabla_j h_{kl} + \nabla_k h_{lj} - \nabla_l h_{jk})\ , \tag{III.2}$$

where (g'^{il}) is, as usual, the inverse matrix of (g'_{il}). In order to have Q as a function of g and of h as we have wished, we should express the inverse matrix (g'^{il}) of (g'_{il}) in the latter formula as a function of the initial metric g and of h. We do not aim at an exact formula; an approximate expression will be sufficient for us since we suppose that the difference h between g' and g is small and has small derivatives (we will clarify the sense of these words in the sequel). The formula $g' = g + h = g(1 + g^{-1}h)$, where 1 stands for the identity matrix, implies $g'^{-1} = (1 + g^{-1}h)^{-1}g^{-1}$. Denote by H the matrix $g^{-1}h$. Since

$$(1+H)^{-1} = 1 - H + H^2 - H^3 + \dots\ ,$$

we obtain, by neglecting the terms whose order in h is > 1, that $g'^{-1} = (1+H)^{-1} \cdot g^{-1} \approx g^{-1} - Hg^{-1} = g^{-1} - g^{-1}hg^{-1}$. Therefore

$$g'^{jk} \approx g^{jk} - g^{jl}h_{li}g^{ik}\ . \tag{III.3}$$

From now on, we will use the initial metric g in order to identify each tangent space with its dual space. This will enable us to identify, when necessary, covariant and contravariant tensors (we will move indices up and down using the initial metric g). Then the latter formula takes the form

$$g'^{jk} \approx g^{jk} - h^{jk}\ . \tag{III.4}$$

Substitute this expression into (III.2) and delete the terms of order > 1 in h to obtain

$$Q^i_{jk} \approx \frac{1}{2} g^{il}(\nabla_j h_{kl} + \nabla_k h_{lj} - \nabla_l h_{jk})\ . \tag{III.5}$$

When we substitute this expression into (III.1), the terms containing products of two Q's have order > 1 in h. Therefore we should calculate as a function of h those terms that contain one covariant derivative of Q. After some calculations, we finally obtain

$$R'_{ik} \approx R_{ik} + \frac{1}{2} g^{rl}(\nabla_r \nabla_k h_{il} + \nabla_r \nabla_i h_{lk} - \nabla_r \nabla_l h_{ki}) - \frac{1}{2} \nabla_k \nabla_i \mathrm{tr}\, h \,, \qquad \text{(III.6)}$$

where tr h denotes the trace of h, that is, $g^{rl}h_{rl}$.

The following theorem states in a rigorous and elementary manner what we have done.

Theorem III.1. *Let (M, g) be a pseudo-Riemannian manifold. Assume that h is a 2-covariant symmetric tensor on M, C^∞, such that for any t belonging to an interval $(-\varepsilon, +\varepsilon)$ of $\mathbb{R}$ the 2-tensor $g + th$ is not singular (that is, a pseudo-Riemannian metric). Let $D(h)$ be the 2-covariant tensor given in any local chart by*

$$D(h)_{ik} = \frac{1}{2} g^{rl}(\nabla_r \nabla_k h_{il} + \nabla_r \nabla_i h_{lk} - \nabla_r \nabla_l h_{ki}) - \frac{1}{2} \nabla_k \nabla_i \mathrm{tr}\, h \,.$$

For any $t \in (-\varepsilon, \varepsilon)$ we have

$$Ric(g + th) = Ric(g) + tD(h) + T(t, h)$$

where $T(t, h)$ is a 2-tensor on M dependent on h (we will consider it to be fixed) and on t and such that at each point $x \in M$ one has

$$\lim_{t \to 0} \frac{T(t, h)_x}{t} = 0 \,.$$

Comment. In the case of Riemannian manifolds, we could state this theorem in a much more interesting way in the framework of convergence of 2-tensors in Sobolev spaces. However, in the pseudo-Riemannian case, the inner product of tensors $\langle\ ,\ \rangle$ is not positive definite and Sobolev norms of (I.15) are not real norms. Therefore we have chosen the above way of stating the theorem which is easy to prove.

Proof. We could say that this theorem requires no proof since it is a result of the above calculations. However, we will revisit these calculations in detail, especially at those points where we have neglected higher-order terms. First, observe that formula (III.2) is exact. The first approximate formula we find in the above calculations is (III.3). Let us focus on it. First of all, let us say that it is convenient to take the metric g' to be $g' = g + th$ (instead of $g' = g + h$ we have taken before). Consider the matrix-valued identity

$$(g + th)^{-1} = g^{-1} - g^{-1}(th)g^{-1} + (g + th)^{-1}(thg^{-1})^2 \,. \qquad \text{(III.7)}$$

In order to see that this is an identity, it suffices to multiply it on the left by $g+th$. We would like to see now that, in spite of the fact that formula (III.7) relates matrices whose entries depends on the components of g and h in a local chart, what we actually have is a formula relating contravariant 2-tensors that are globally defined. Any pseudo-Riemannian metric g' on M gives rise to a natural inner product in the dual $T_x(M)^*$ of each tangent space $T_x(M)$. That is, a contravariant 2-tensor on M. In each local chart, the components g'^{ij} of this contravariant tensor are the entries of the the inverse matrix of (g'_{ij}). Therefore the term $(g+th)^{-1}$ in (III.7) can be interpreted as the contravariant tensor $A(th)$ that corresponds to the metric $g' = g + th$ in the way we have already explained. The remaining terms in (III.7) are also global tensors. For example, the term $(g+th)^{-1}(thg^{-1})^2$, regarded as a matrix, is the matrix $B(th)$ given by

$$B(th)^{ik} = A(th)^{ij}(th)_{jr}(th)^{rk} = t^2 A(th)^{ij} h_{jr} h^{rk} , \tag{III.8}$$

where h^{rk} are the components of the contravariant tensor h^* obtained from h by means of the initial metric g. This expression shows that $B(th)$ is a global contravariant tensor obtained by appropriate contractions of the tensor product $A(th) \otimes h \otimes h^*$. Formula (III.7) has the following component-wise form:

$$A(th)^{ik} = g^{ik} - th^{ik} + B(th)^{ik} .$$

This is an exact formula that corresponds to the approximate formula (III.4), or to (III.3) as well. By expression (III.8) for $B(th)$ and by the fact that $A(0) = (g+0)^{-1} = g^{-1}$ for $t = 0$, we have

$$\lim_{t \to 0} \frac{B(th)_x}{t} = 0$$

for all $x \in M$.

What we have done up to now is nothing else than to give an exact sense to approximate formula (III.4). In a similar (but easier) way, an exact sense can be given to approximate equalities (III.5) and (III.6), which would complete the proof. □

III.2 Einstein's equation for small perturbations of the Minkowski metric

As we have already said at the beginning of this chapter, we will now study the case of a gravitational field generated by a matter whose stress-energy tensor T is small as well as its derivatives. We assume that the Lorentzian metric $\tilde{g}'$ on $\mathbb{R}^4$ associated with the gravitational field is a small perturbation of the Minkowski metric, $\tilde{g}' = \tilde{\eta} + \tilde{h}$, with $\tilde{h}$ small and the derivatives of $\tilde{h}$ small. Einstein's equation will have the form

$$\widetilde{Ric}(\tilde{g}') = \chi(T - \frac{1}{2}(\mathrm{tr}_{\tilde{g}'} T)\tilde{g}') .$$

Since $\tilde{g}' = \tilde{\eta} + \tilde{h}$ and by formula (III.4) giving us the matrix $(\tilde{g}'^{\alpha\beta})$, we have

$$
\begin{aligned}
(\mathrm{tr}_{\tilde{g}'} T)\tilde{g}' &= (\tilde{g}'^{\alpha\beta} T_{\alpha\beta})\tilde{g}' \\
&\approx ((\tilde{\eta}^{\alpha\beta} - \tilde{h}^{\alpha\beta})T_{\alpha\beta})(\tilde{\eta} + \tilde{h}) \approx (\tilde{\eta}^{\alpha\beta} T_{\alpha\beta})\tilde{\eta} = (\mathrm{tr}_{\tilde{\eta}} T)\tilde{\eta} \, .
\end{aligned}
$$

Here, we have neglected the terms containing products of $\tilde{h}$ and T or products of two $\tilde{h}$ since they are small.

Therefore Einstein's equation can be written approximately as

$$
\widetilde{Ric}(\tilde{\eta} + \tilde{h}) = \chi(T - \frac{1}{2}(\mathrm{tr}_{\tilde{\eta}} T)\tilde{\eta}) \, . \tag{III.9}
$$

Let $D(\tilde{h})$ be the covariant 2-tensor on $\mathbb{R}^4$ given by the formula:

$$
D(\tilde{h})_{\alpha\beta} = \frac{1}{2}\tilde{\eta}^{\lambda\mu}(\tilde{\nabla}_\lambda \tilde{\nabla}_\beta \tilde{h}_{\alpha\mu} + \tilde{\nabla}_\lambda \tilde{\nabla}_\alpha \tilde{h}_{\mu\beta} - \tilde{\nabla}_\lambda \tilde{\nabla}_\mu \tilde{h}_{\beta\alpha}) - \frac{1}{2}\tilde{\nabla}_\beta \tilde{\nabla}_\alpha \mathrm{tr}\, \tilde{h} \, ,
$$

where $\tilde{\nabla}$ denotes the covariant derivative associated with the initial metric (which in our case is the Minkowski metric).

By formula (III.6) (or by Theorem III.1), we have

$$
\widetilde{Ric}(\tilde{\eta} + \tilde{h}) \approx \widetilde{Ric}(\tilde{\eta}) + D(\tilde{h}) \, .
$$

Since the Ricci tensor of $\tilde{\eta}$ equals zero, equation (III.9) is approximately written as

$$
D(\tilde{h}) = \chi(T - \frac{1}{2}(\mathrm{tr}_{\tilde{\eta}} T)\tilde{\eta}) \, . \tag{III.10}
$$

Though this equation is approximate, it is still too complex. In order to simplify it, instead of considering all Lorentzian metrics having the form $\tilde{g}' = \tilde{\eta} + \tilde{h}$ with $\tilde{h}$ small we look at only those metrics $\tilde{g}' = \tilde{\eta} + \tilde{h}$ such that $\tilde{h}$ (in addition to being small) satisfies the extra condition

$$
\mathrm{div}\, \tilde{h} = \frac{1}{2} d\, \mathrm{tr}\, \tilde{h} \, . \tag{III.11}
$$

The component-wise form of this condition is

$$
\tilde{\nabla}^\mu \tilde{h}_{\mu\alpha} = \frac{1}{2}\tilde{\nabla}_\alpha \mathrm{tr}\, \tilde{h} \, .
$$

Since the covariant derivatives associated with the Minkowski metric are ordinary derivatives and since $\tilde{\nabla}_\alpha$ commutes with $\tilde{\nabla}_\beta$, one can readily see that for $\tilde{h}$ satisfying condition (III.11) we have

$$
D(\tilde{h})_{\alpha\beta} = -\frac{1}{2}\tilde{\nabla}^\mu \tilde{\nabla}_\mu \tilde{h}_{\beta\alpha} = -\frac{1}{2}\Box \tilde{h}_{\beta\alpha} \, ,
$$

where $\Box$ is the d'Alembert operator acting on functions $\tilde{f}$ in $\mathbb{R}^4$ in the following way:

$$\Box\tilde{f} = \frac{\partial^2 \tilde{f}}{\partial(x^1)^2} + \frac{\partial^2 \tilde{f}}{\partial(x^2)^2} + \frac{\partial^2 \tilde{f}}{\partial(x^3)^2} - \frac{1}{c^2}\frac{\partial^2 \tilde{f}}{\partial t^2} .$$

Therefore equation (III.10) for $\tilde{h}$ satisfying (III.11) takes the form

$$-\frac{1}{2}\Box\tilde{h}_{\alpha\beta} = \chi S_{\alpha\beta} , \qquad \text{(III.12)}$$

where S denotes the tensor $T - (1/2)(\mathrm{tr}_{\tilde{\eta}}T)\tilde{\eta}$ appearing on the right-hand side of (III.10). Observe that S vanishes in the regions of $\mathbb{R}^4$ where the tensor T vanishes (where there is no matter) and Einstein's equation is approximately written in these regions as

$$\Box\tilde{h}_{\alpha\beta} = 0 .$$

The latter is the wave equation for each of 10 components $\tilde{h}_{\alpha\beta}$ of $\tilde{h}$.

Summing up: an approximate Einstein equation corresponding to a tensor metric $\tilde{g}' = \tilde{\eta} + \tilde{h}$ and to a stress-energy tensor T (with small T and $\tilde{h}$) is (III.10). For the metrics $\tilde{g}'$ coming from an $\tilde{h}$ satisfying (III.11), an approximate Einstein's equation can be written in the most simple form (III.12). Therefore we have to be curious about the physical interpretation of condition (III.11). To this end, we have to study the action on equation (III.10) of diffeomorphisms that are close to identity.

III.3 Action on metrics of diffeomorphisms close to the identity

Here we consider a slightly more general situation than that of the previous section. Suppose that we are in a Lorentzian manifold $(V, \tilde{g})$ for which a Riemannian metric $\tilde{\gamma}$ exists such that the Riemannian manifold $(V, \tilde{\gamma})$ is complete. In this case, diffeomorphisms of V close to the identity can be parametrized by vector fields ξ close to zero by means of the geodesic exponential $\tilde{\gamma}$ in the way we are now going to explain.

Given a vector field ξ, denote by f_ξ the diffeomorphism

$$\begin{array}{rccc} f_\xi : & V & \longrightarrow & V \\ & x & \rightarrow & z(x,1,\xi) \end{array} ,$$

where $t \rightarrow z(x,t,\xi)$ stands for the unique geodesic of the Riemannian manifold $(V, \tilde{\gamma})$ such that $z(x,0,\xi) = x$ and $\dot{z}(x,0,\xi) = \xi$. Assume that $\tilde{g}' = \tilde{g} + \tilde{h}$ is a Lorentzian metric close to $\tilde{g}$. Write $\tilde{g}' \circ f_\xi = (f_\xi^{-1})^*\tilde{g}'$. Denote by $\tilde{\nabla}'$ the covariant derivative associated with the metric $\tilde{g}'$ and, by $\tilde{\nabla}$, the covariant derivative associated with $\tilde{g}$. If the metric $\tilde{g}'$ has the form $\tilde{g}' = \tilde{g} + \tilde{h}$ with small $\tilde{h}$, the metric $\tilde{g}' \circ f_\xi$ will also be close to $\tilde{g}$ when ξ is small. Therefore it will have the form $\tilde{g}' \circ f_\xi = \tilde{g} + \tilde{h}_\xi$ with $\tilde{h}_\xi$ small. The relation between $\tilde{h}_\xi$ and $\tilde{h}$ is given in the following statement.

Proposition III.2. *One has*

$$\tilde{h}_\xi \approx \tilde{h} - L_\xi \tilde{g} \ .$$

Hence, by Corollary I.8, *we have in any local chart*

$$(\tilde{h}_\xi)_{\alpha\beta} \approx \tilde{h}_{\alpha\beta} - \tilde{\nabla}_\alpha \xi_\beta - \tilde{\nabla}_\beta \xi_\alpha \ .$$

Proof. Apply Taylor's formula to the curve $t \to z(x,t,\xi)$ to obtain

$$z^\alpha(x,1,\xi) = z^\alpha(x,0,\xi) + \dot{z}^\alpha(x,0,\xi)(1-0) + \ldots .$$

This gives

$$f_\xi(x)^\alpha = x^\alpha + \xi^\alpha(x) + T^\alpha(\xi) \ , \tag{III.13}$$

where $T(\xi)$ denotes a term satisfying (for a fixed x) the condition

$$\lim_{\tau\to 0} \frac{T(\tau\xi(x))}{\tau} = 0 \ .$$

In an analogous way, applying Taylor's formula to the curve $t \to z(x,t,-\xi)$ we obtain a similar expression for f_ξ^{-1}:

$$f_\xi^{-1}(x)^\alpha = x^\alpha - \xi^\alpha(x) + T'^\alpha(\xi) \ . \tag{III.14}$$

Expression (III.13) enables us to calculate the Jacobian

$$\frac{\partial f_\xi^\alpha}{\partial x^\beta} = \delta_{\alpha\beta} + \frac{\partial \xi^\alpha}{\partial x^\beta} + \ldots .$$

The Jacobian of the inverse diffeomorphism will be

$$\frac{\partial (f_\xi^{-1})^\alpha}{\partial x^\beta} = \delta_{\alpha\beta} - \frac{\partial \xi^\alpha}{\partial x^\beta} + \ldots . \tag{III.15}$$

Since

$$((f_\xi^{-1})^*(\tilde{g}'))_x \left(\frac{\partial}{\partial x^\alpha}, \frac{\partial}{\partial x^\beta}\right) = \tilde{g}'_{f_\xi^{-1}(x)} \left((f_\xi^{-1})_* \frac{\partial}{\partial x^\alpha}, (f_\xi^{-1})_* \frac{\partial}{\partial x^\beta} \right) ,$$

we have, taking into account expression (III.15) for the Jacobian of the inverse diffeomorphism,

$$\begin{aligned}(\tilde{g}' \circ f_\xi)_{\alpha\beta}(x) &= \tilde{g}'_{\alpha\beta}(f_\xi^{-1}(x)) \\ &\quad - (\partial_\alpha \xi^\lambda)\tilde{g}'_{\lambda\beta}(f_\xi^{-1}(x)) - (\partial_\beta \xi^\lambda)\tilde{g}'_{\alpha\lambda}(f_\xi^{-1}(x)) + \ldots \end{aligned} \tag{III.16}$$

where ∂_α denotes, as usual, $\partial/\partial x^\alpha$. Taking into account (III.14) we have

$$\begin{aligned}\tilde{g}'_{\alpha\beta}(f_\xi^{-1}(x)) &= \tilde{g}'_{\alpha\beta}(x) + \sum_\lambda (f_\xi^{-1}(x)^\lambda - x^\lambda)\partial_\lambda \tilde{g}'_{\alpha\beta} + \dots \\ &= \tilde{g}'_{\alpha\beta}(x) + \sum_\lambda ((x^\lambda - \xi^\lambda) - x^\lambda)\partial_\lambda \tilde{g}'_{\alpha\beta} + \dots .\end{aligned}$$

This can be written as

$$\tilde{g}'_{\alpha\beta}(f_\xi^{-1}(x)) = \tilde{g}'_{\alpha\beta}(x) - \xi(\tilde{g}'_{\alpha\beta}) + \dots$$

where $\xi(\tilde{g}'_{\alpha\beta})$ denotes the field ξ acting (as derivative) on the function $\tilde{g}'_{\alpha\beta}$.

By substituting this expression into (III.16), we obtain

$$\begin{aligned}(\tilde{g}' \circ f_\xi)_{\alpha\beta}(x) = \tilde{g}'_{\alpha\beta}(x) - \xi(\tilde{g}'_{\alpha\beta}) \\ - (\partial_\alpha \xi^\lambda)\tilde{g}'_{\lambda\beta}(x) - (\partial_\beta \xi^\lambda)\tilde{g}'_{\alpha\lambda}(x) + \dots .\end{aligned} \tag{III.17}$$

Now, the terms $-(\partial_\alpha \xi^\lambda)\tilde{g}'_{\lambda\beta}(x) - (\partial_\beta \xi^\lambda)\tilde{g}'_{\alpha\lambda}(x)$ in the above expression can be written as

$$\tilde{g}'\left(\left[\xi, \frac{\partial}{\partial x^\alpha}\right], \frac{\partial}{\partial x^\beta}\right) + \tilde{g}'\left(\frac{\partial}{\partial x^\alpha}, \left[\xi, \frac{\partial}{\partial x^\beta}\right]\right) .$$

By substituting this into (III.17) and by taking into account expression (I.10) for the Lie derivative, we have

$$(\tilde{g}' \circ f_\xi)_{\alpha\beta}(x) = \tilde{g}'_{\alpha\beta}(x) - (L_\xi \tilde{g}')_{\alpha\beta}(x) + \dots . \tag{III.18}$$

Using the notation adopted in Section III.1, we obtain

$$\tilde{\nabla}'_{\partial/\partial x^\alpha}\xi = \tilde{\nabla}_{\partial/\partial x^\alpha}\xi + Q(\partial/\partial x^\alpha, \xi) = \tilde{\nabla}_{\partial/\partial x^\alpha}\xi + Q_{\alpha\beta}\xi^\beta .$$

Now, the terms $Q_{\alpha\beta}\xi^\beta$ contain products of $\tilde{h}$ and ξ (both small). Therefore we can neglect these terms. This is why (III.18) can be written as

$$\begin{aligned}(\tilde{g}' \circ f_\xi)_{\alpha\beta} &\approx \tilde{g}'_{\alpha\beta} - (\tilde{\nabla}_\alpha \xi^\lambda)\tilde{g}'_{\lambda\beta} - (\tilde{\nabla}_\beta \xi^\lambda)\tilde{g}'_{\alpha\lambda} \\ &= \tilde{g}_{\alpha\beta} + \tilde{h}_{\alpha\beta} - (\tilde{\nabla}_\alpha \xi^\lambda)(\tilde{g}_{\lambda\beta} + \tilde{h}_{\lambda\beta}) - (\tilde{\nabla}_\beta \xi^\lambda)(\tilde{g}_{\alpha\lambda} + \tilde{h}_{\alpha\lambda}) \\ &\approx \tilde{g}_{\alpha\beta} + \tilde{h}_{\alpha\beta} - (\tilde{\nabla}_\alpha \xi^\lambda)\tilde{g}_{\lambda\beta} - (\tilde{\nabla}_\beta \xi^\lambda)\tilde{g}_{\alpha\lambda} \\ &= \tilde{g}_{\alpha\beta} + \tilde{h}_{\alpha\beta} - \tilde{\nabla}_\alpha \xi_\beta - \tilde{\nabla}_\beta \xi_\alpha .\end{aligned}$$

This argument proves the statement. □

III.4 Continuing the calculation of Section III.2

Let us come back to the framework of Section III.2. Equation (III.10) was given as an approximate Einstein equation corresponding to a metric $\tilde{g}' = \tilde{\eta} + \tilde{h}$ and to a stress-energy tensor T (with T and $\tilde{h}$ being small). For the metrics $\tilde{g}'$ coming from an $\tilde{h}$ satisfying (III.11), the approximate Einstein equation has been written in a simpler form (III.12). We would like to see now that there we need not worry about solving equation (III.10), since it suffices to solve equation (III.12) which is simpler.

Let us first see that equation (III.10) is invariant with respect to the action of infinitesimal diffeomorphisms. That is, if an infinitesimal diffeomorphism is applied to this equation, **exactly** the same equation is obtained. Begin with the left-hand side of (III.10). If ξ is a vector field, we claim (using the notation of the previous section) that $D(\tilde{h})$=$D(\tilde{h}_\xi)$. Indeed, the definition of $D(\tilde{h})$ implies

$$D(\tilde{h}_\xi)_{\alpha\beta} = \frac{1}{2}\tilde{\eta}^{\lambda\mu}(\tilde{\nabla}_\lambda\tilde{\nabla}_\beta(\tilde{h}_\xi)_{\alpha\mu} + \tilde{\nabla}_\lambda\tilde{\nabla}_\alpha(\tilde{h}_\xi)_{\mu\beta} - \tilde{\nabla}_\lambda\tilde{\nabla}_\mu(\tilde{h}_\xi)_{\beta\alpha}) - \frac{1}{2}\tilde{\nabla}_\beta\tilde{\nabla}_\alpha \mathrm{tr}\ \tilde{h}_\xi\ .$$

Replace here $\tilde{h}_\xi$ by the expression $(\tilde{h}_\xi)_{\alpha\beta} = \tilde{h}_{\alpha\beta} - \tilde{\nabla}_\alpha\xi_\beta - \tilde{\nabla}_\beta\xi_\alpha$ given by Proposition III.2 to obtain, after some simple calculations, $D(\tilde{h}_\xi)$=$D(\tilde{h})$.

Naturally, if the left-hand side of (III.10) is invariant with respect to the action of an infinitesimal diffeomorphism, the right-hand side has the same property. In any case, this can be seen directly in the way we are going to explain now.

Recall that we have adopted the notation $S = T - (1/2)(\mathrm{tr}_{\tilde{\eta}}T)\tilde{\eta}$. The same calculations we have made in order to obtain formula (III.18) would lead us to the expression

$$(S \circ f_\xi) = S - L_\xi S + \ldots .$$

By (I.10), we have

$$(L_\xi S)_{\alpha\beta} = \xi(S_{\alpha\beta}) - S\left(\left[\xi, \frac{\partial}{\partial x^\alpha}\right], \frac{\partial}{\partial x^\beta}\right) - S\left(\frac{\partial}{\partial x^\alpha}, \left[\xi, \frac{\partial}{\partial x^\beta}\right]\right) .$$

Each of these terms contains products of ξ and of S, both of them small. Therefore $L_\xi S \approx 0$. Hence $(S \circ f_\xi) \approx S$.

The second step we are going to make is to see that, given a metric $\tilde{g}' = \tilde{\eta} + \tilde{h}$ which is close to $\tilde{\eta}$, we will be able to find a vector field ξ for which the metric $\tilde{g}' \circ f_\xi = \tilde{\eta} + \tilde{h}_\xi$ would satisfy condition (III.11). This is written as

$$\tilde{\nabla}^\rho(\tilde{h}_\xi)_{\rho\alpha} = \frac{1}{2}\tilde{\nabla}_\alpha \mathrm{tr}\ \tilde{h}_\xi\ . \qquad \text{(III.19)}$$

Since $(\tilde{h}_\xi)_{\alpha\rho} = \tilde{h}_{\alpha\rho} - \tilde{\nabla}_\alpha\xi_\rho - \tilde{\nabla}_\rho\xi_\alpha$, the left-hand side of (III.19) is

$$\tilde{\nabla}^\rho(\tilde{h}_\xi)_{\rho\alpha} = \tilde{\nabla}^\rho\tilde{h}_{\rho\alpha} - \tilde{\nabla}_\alpha \mathrm{div}\xi - \Box\xi_\alpha\ .$$

On the other hand, tr $\tilde{h}_\xi = $ tr $\tilde{h} - 2\,\mathrm{div}\,\xi$. Therefore (III.19) is finally written as

$$\Box\xi_\alpha = \tilde{\nabla}^\rho \tilde{h}_{\alpha\rho} - \frac{1}{2}\tilde{\nabla}_\alpha \mathrm{tr}\,\tilde{h}\ .$$

For a given $\tilde{h}$, this is an equation with respect to ξ having infinite solutions dependent on the initial condition that would be imposed on ξ. Any of these solutions gives us the required infinitesimal diffeomorphism.

This implies that the extra condition (III.11) that we have imposed in order to simplify the approximate Einstein equation (III.9) is physically irrelevant, since any metric transformed by an appropriate diffeomorphism would satisfy this condition. The equation transformed by means of this diffeomorphism (which would be the equation we need to solve) is exactly the same as the initial one. Summing up, in order to solve the approximate Einstein equation we need only solve the system

$$\begin{cases} -\dfrac{1}{2}\Box\tilde{h}_{\alpha\beta} = \chi S_{\alpha\beta} \\ \tilde{\nabla}^\rho \tilde{h}_{\rho\alpha} = \dfrac{1}{2}\tilde{\nabla}_\alpha \mathrm{tr}\,\tilde{h}\ . \end{cases} \tag{III.20}$$

Einstein gave explicit solutions to these equations in papers [30] and [31]. In order to see the way these equations can be solved, let us say that it is well known in classical mechanics that the gravitational potential (in $\mathbb{R}^3$) generated by a matter whose density is $\rho(x)$, is given by

$$V(x) = -K\int_{\mathbb{R}^3} \frac{\rho(y)}{|x-y|}\,dy\ ,$$

and that this corresponds to the solution of the Poisson equation

$$\Delta V = 4\pi K\rho$$

which vanishes at infinity.

One can prove along similar lines that the solution in $\mathbb{R}^4$ of

$$\Box V = 4\pi K\rho$$

that vanishes at infinity of $\{t\}\times\mathbb{R}^3$ is given by the delayed potential

$$V(x,t) = -K\int_{\mathbb{R}^3} \frac{\rho(y, t \pm \frac{1}{c}|x-y|)}{|x-y|}\,dy\ .$$

Therefore the solution $\tilde{h}$ of the first equation of (III.20) that vanishes at infinity of $\{t\}\times\mathbb{R}^3$ is given by

$$\tilde{h}_{\alpha\beta}(x,t) = 2\chi\int_{\mathbb{R}^3} \frac{S_{\alpha\beta}(y, t - \frac{1}{c}|x-y|)}{|x-y|}\,dy\ . \tag{III.21}$$

Now, we have to ask whether these $\tilde{h}_{\alpha\beta}$ also satisfy the second equation in (III.20). Let us see that the answer is in the affirmative. Indeed, we know that the stress-energy tensor T has divergence zero. This implies that the tensor $S = T - (1/2)(\mathrm{tr}_{\tilde{\eta}}T)\tilde{\eta}$ satisfies the condition

$$\tilde{\nabla}^{\rho} S_{\rho\alpha} = \frac{1}{2}\tilde{\nabla}_{\alpha}\mathrm{tr}\ S\ .$$

Using this condition we can see that the solutions $\tilde{h}_{\alpha\beta}$ given by (III.21) automatically satisfy the second equation in (III.20).

III.5 Comparison with classical gravitation

As we have already explained in Section II.4.1, Einstein's equation (II.14)

$$\widetilde{Ric}(\tilde{g}) = \chi(T - \frac{1}{2}(\mathrm{tr}_{\tilde{g}}T)\tilde{g})$$

is the equation that replaces in relativity the classical Poisson equation

$$\Delta V = 4\pi K\rho\ . \tag{III.22}$$

Therefore we should be able to approximately obtain Poisson's equation from that of Einstein. We have already said in Chapter II that the components $\tilde{g}_{\alpha\beta}$ of the metric that is associated with the gravitational field in relativity, play the same role as that played by the gravitational potential V in classical physics. However, there are ten components $\tilde{g}_{\alpha\beta}$, while the classical gravitational potential V is only one. Therefore we should clarify with more precision to what object the classical gravitational potential V should be related.

Let us consider the case of a weak gravitational field and suppose, as we have already done in previous sections, that the Lorentzian metric $\tilde{g}$ associated with the field is a small perturbation of that of Minkowski, $\tilde{g} = \tilde{\eta} + \tilde{h}$. In order to be able to compare the relativistic objects and the classical potential V, let us assume that the latter is independent of time and the components $\tilde{g}_{\alpha\beta}$ of the relativistic metric (on $\mathbb{R}^4$) do not depend on the coordinate t.

If $x(\tau) = (x^0(\tau), x^1(\tau), x^2(\tau), x^3(\tau))$ is the curve of $\mathbb{R}^4$ representing the motion of a particle having unit mass released in liberty under the action of the field, where τ is the proper time, the curve $x(\tau)$ will be a geodesic of $(\mathbb{R}^4, \tilde{g})$ and we will have

$$\frac{d^2x^{\alpha}}{d\tau^2} = -\tilde{\Gamma}^{\alpha}_{\beta\gamma}\frac{dx^{\beta}}{d\tau}\frac{dx^{\gamma}}{d\tau}\ .$$

If we now use the convention that the Latin indices $(i, j, k\ldots)$ range from 1 to 3 only, we obtain

$$\frac{d^2x^{i}}{d\tau^2} = -\tilde{\Gamma}^{i}_{jk}\frac{dx^{j}}{d\tau}\frac{dx^{k}}{d\tau} - 2\tilde{\Gamma}^{i}_{j0}\frac{dx^{j}}{d\tau}\frac{dx^{0}}{d\tau} - \tilde{\Gamma}^{i}_{00}\left(\frac{dx^0}{d\tau}\right)^2\ .$$

If we assume that the velocity of the particle is small as compared to the speed of light, the proper time τ will almost coincide with that of Newton and we will have

$$\frac{dx^0}{d\tau} = c\,\frac{dt}{d\tau} \approx c\;.$$

Therefore

$$\frac{d^2x^i}{d\tau^2} \approx -\tilde{\Gamma}^i_{jk}\,\frac{dx^j}{d\tau}\,\frac{dx^k}{d\tau} - 2c\,\tilde{\Gamma}^i_{j0}\,\frac{dx^j}{d\tau} - c^2\tilde{\Gamma}^i_{00}\;.$$

The acceleration of a particle slightly moving as acted upon by a gravitational field is independent of its speed in classical mechanics. This makes us assume that, at small velocities with respect to c, the first and second terms on the right-hand side of the latter equation are small with respect to the last term. Hence

$$\frac{d^2x^i}{d\tau^2} \approx -c^2\tilde{\Gamma}^i_{00}\;.$$

Now

$$\tilde{\Gamma}^i_{00} = \frac{1}{2}\,\tilde{g}^{i\alpha}(2\partial_0\tilde{g}_{0\alpha} - \partial_\alpha\tilde{g}_{00}) = -\frac{1}{2}\,\tilde{g}^{i\alpha}\partial_\alpha\tilde{g}_{00}\;,$$

since we have supposed that $\tilde{g}_{\alpha\beta}$ are independent of time. Since $\tilde{g} = \tilde{\eta} + \tilde{h}$, we have

$$\tilde{\Gamma}^i_{00} \approx -\frac{1}{2}\tilde{\eta}^{i\alpha}\partial_\alpha\tilde{h}_{00} = -\frac{1}{2}\partial_i\tilde{h}_{00}\;.$$

Therefore

$$\frac{d^2x^i}{d\tau^2} \approx \frac{c^2}{2}\,\partial_i\tilde{h}_{00}\;.$$

Since in classical mechanics

$$\frac{d^2x^i}{dt^2} = -\partial_i V\;,$$

we see that $\tilde{h}_{00}$ has to be approximately $-2V/c^2 + C$, where C is a constant to be determined. Since the classical potential vanishes at infinity, by imposing that $\tilde{h}_{00}$ be vanishing for each t at infinity in $\{t\} \times \mathbb{R}^3$, we obtain

$$\tilde{h}_{00} \approx -\frac{2V}{c^2}\;. \qquad \text{(III.23)}$$

In order to proceed with comparing the classical and relativistic cases, assume that the gravitational field is originated by a perfect fluid. Recall that the stress-energy tensor of such a fluid is

$$T = \frac{1}{c^2}\left(\rho + p/c^2\right)\vec{u}\otimes\vec{u} + \frac{p}{c^2}\tilde{g}\;,$$

where $\tilde{g}$ is the metric corresponding to the gravitational field and $\vec{u}$ is the 4-dimensional velocity field of the fluid. Naturally, in order to compare, we suppose that the 3-dimensional velocity of the fluid is small implying that the 4-dimensional vector $\vec{u}$ will be close to the vector field $\partial/\partial t$ on $\mathbb{R}^4$. Assume also that the pressure p is small with respect to c^2. That is,

$$T \approx \frac{1}{c^2}\,\rho\,\frac{\partial}{\partial t}\otimes\frac{\partial}{\partial t}\,.$$

If we interpret this tensor as a covariant tensor (by means of the Minkowski metric), we obtain

$$T \approx c^2 \rho\, dt \otimes dt\ .$$

In order to calculate the tensor S appearing on the right-hand side of Einstein's equation, we first observe that tr ${}_{\tilde\eta}T = \tilde\eta_{\alpha\beta}T^{\alpha\beta} = \tilde\eta_{00}T^{00} = -\rho$. Therefore

$$S = T - \frac{1}{2}(\mathrm{tr}_{\tilde g}T)\tilde g \approx T - \frac{1}{2}(\mathrm{tr}_{\tilde\eta}T)\tilde\eta = c^2\rho\, dt\otimes dt + \frac{1}{2}\,\rho\,\tilde\eta\ .$$

Hence the first equation in (III.20) corresponding to $\alpha = \beta = 0$ will be

$$-\frac{1}{2}\,\Box\tilde h_{00} = \chi S_{00} = \frac{1}{2}\,\chi c^2 \rho\ . \qquad \text{(III.24)}$$

If we substitute here the expression for $\tilde h_{00}$ given by (III.23), we obtain

$$\Box V = \frac{c^4\chi}{2}\,\rho\ .$$

Since we have supposed V to be independent of time, $\Box V = \Delta V$. Therefore

$$\Delta V = \frac{c^4\chi}{2}\,\rho\ . \qquad \text{(III.25)}$$

By comparing this equation with that of Poisson (III.22), we arrive at the conclusion that the constant χ should approximately be

$$\chi = \frac{8\pi K}{c^4}\ ,$$

where K is the Newton gravitational constant.

We should say that, in all our exposition about the approximation of the Poisson equation by that of Einstein, we have considered a potential V and components $\tilde g_{\alpha\beta}$ of the metric that are independent of time. This has enabled us to switch from a hyperbolic equation (III.24) to an elliptic one (III.25). If we do not adopt the simplified viewpoint of considering $\tilde h_{\alpha\beta}$ to be independent of time, then the equation for which $\tilde h_{\alpha\beta}$ are solutions is (III.20). This equation is of a very different nature as compared to the (elliptic) Poisson's equation governing the classical gravitational potential. The fact that, in regions where there is no matter, the $\tilde h_{\alpha\beta}$'s satisfy the wave equation (gravitational waves) is a characteristic feature of general relativity having nothing to do with classical gravitation.

Chapter IV

Cauchy Problem for Einstein's Equation with Matter

As we have already mentioned in Chapter II, general relativity is a classic gravitation theory where the physical space is interpreted as a continuum, mathematically modeled by means of a time-oriented Lorentzian manifold. Einstein's equation relating the gravitational field with the matter that creates this field is

$$G(\tilde{g}) = \chi T \ ,$$

where $G(\tilde{g})$ is Einstein's tensor of the Lorentzian metric $\tilde{g}$, T is the stress-energy tensor describing the matter, and χ is a universal constant. Moreover, we have seen in Chapter II that the tensor T should have divergence zero. In many cases, the expression of T contains the metric $\tilde{g}$ (as it happens in the case of perfect fluids in which the expression of T is given by $T = (\rho + p)\,\vec{u} \otimes \vec{u} + p\tilde{g}$, employing units where the speed of light equals 1). Since T often contains $\tilde{g}$, Einstein's equation $G(\tilde{g}) = \chi T$ cannot be interpreted as an equation where one would like to find $\tilde{g}$ given T. Both T and $\tilde{g}$ should be constructed simultaneously in order to satisfy the equation and in such a way that their restrictions to a given hypersurface M of V satisfy certain initial conditions (Cauchy data) given in advance. Since the tensor T that should be constructed simultaneously with $\tilde{g}$ should have divergence zero, what we really should solve is the system

$$\begin{cases} G(\tilde{g}) = \chi T \\ \mathrm{div}_{\tilde{g}} T = 0 \ , \end{cases}$$

with the initial conditions given over M. This is called solving the Cauchy problem for Einstein's equation.

This chapter aims at giving a rather self-contained presentation of the problem. To reach this goal we have chosen, from an immense sea of literature concerning the topic, a point of view that bring us in a simple manner to a result which is sufficient for our purposes in the remaining part of the book. This result is Theorem IV.7. Therefore this chapter is oriented towards proving this theorem, including all preliminary material required for the proof. The statement of the theorem involves an assumption on the stress-energy tensor T which is very technical and difficult to be checked (to be precise, this is the assumption that for a fixed Lorentzian metric $\tilde{g}$, the equation $\mathrm{div}_{\tilde{g}} T = 0$ can be written at some local chart of M as a first-order symmetric, hyperbolic system for which the hypersurface M is nowhere characteristic). We have devoted Section IV.9 to proving that the stress-energy tensor of perfect fluids, under certain constraints, just satisfies this assumption in Theorem IV.7. We also prove in this section that, more generally, the stress-energy tensor of holonomic media (media introduced by Lichenerowicz in [44]) satisfies, under certain constraints, the above mentioned assumption of the theorem.

The line we follow in order to prove the fundamental theorem is basically inspired by the article by Y. Choquet-Bruhat [21] published in 1952, even though this article only considered the Cauchy problem for Einstein's equation in the vacuum. An extensive and up-to-date bibliography on the topic can be found in the article by H. Friedrich and A. Rendall [37].

A reader who is not much interested in the proof of the fundamental theorem of this chapter, but rather in the statement of this theorem (which will be used in the remaining part of the book) should first read Section IV.4, in order to understand what class of stress-energy tensors is used, then look for the statement of Theorem IV.7, and finally read the statements of the proposition and the corollary in Section IV.9.

IV.1 Differential operators in an open set of $\mathbb{R}^{n+1}$

Denote by $(x^0, \dots, x^n)$ the canonical coordinates of $\mathbb{R}^{n+1}$. Let U be an open set of $\mathbb{R}^{n+1}$ and U' an open interval of $\mathbb{R}$. Denote by $\mathcal{F}(U, U')$ the set of C^∞ functions from U to U'. A map $D : \mathcal{F}(U, U') \to \mathcal{F}(U, \mathbb{R})$ is said to be a quasilinear differential operator of order k if it has the following form:

$$D(f)(x) = \sum_{\alpha_1=0}^{n} \dots \sum_{\alpha_k=0}^{n} a^{\alpha_1 \dots \alpha_k}(x, [f]_{k-1}) \partial_{\alpha_1} \dots \partial_{\alpha_k} f + F(x, [f]_{k-1}) \ ,$$

where ∂_α stands as usual for $\partial / \partial x^\alpha$, and where the notation $a^{\alpha_1 \dots \alpha_k}(x, [f]_{k-1})$ and $F(x, [f]_{k-1})$ means that the coefficients $a^{\alpha_1 \dots \alpha_k}$ and the term F are C^∞ functions of x and f whose values at x depend only on the values at x of f and its derivatives up to order $k - 1$ in the sense that we are going to explain.

For each integer $r \geq 1$ denote by n_r the number of r-tuples $(\alpha_1, \dots, \alpha_r)$ with α_i integer and such that $0 \leq \alpha_i \leq n$ and $\alpha_1 \leq \alpha_2 \dots \leq \alpha_r$. For each such r-tuple consider the differential operator $\partial_{\alpha_1} \dots \partial_{\alpha_r}$. Arrange all these operators of order

r in some fixed order and denote them by $P_1^{(r)} \ldots P_{n_r}^{(r)}$. Here the super-index (r) stands for the order of these operators. For a fixed positive integer k let F be a C^∞ function

$$\begin{array}{rccc} F: & U \times U' \times \mathbb{R}^{n_1} \times \ldots \times \mathbb{R}^{n_k} & \longrightarrow & \mathbb{R} \\ & (x, y, y^{(1)}, \ldots, y^{(k)}) & \to & F(x, y, y^{(1)}, \ldots, y^{(k)}) \end{array} .$$

As each $y^{(i)} \in \mathbb{R}^{n_i}$, we have $y^{(i)} = (y_1^{(i)}, \ldots, y_{n_i}^{(i)})$ so that

$$F(x, y, y^{(1)}, \ldots, y^{(k)}) = F(x, y, y_1^{(1)}, \ldots, y_{n_1}^{(1)}, \ldots, y_1^{(k)}, \ldots, y_{n_k}^{(k)}) .$$

The function F induces a function (still denoted by F)

$$\begin{array}{rccc} F: & U \times \mathcal{F}(U, U') & \longrightarrow & \mathcal{F}(U, \mathbb{R}) \\ & (x, f) & \to & F(x, f) \end{array}$$

defined by

$$F(x, f) = F(x, f(x), (P_1^{(1)} f)(x), \ldots, (P_{n_1}^{(1)} f)(x), \ldots, (P_1^{(k)} f)(x), \ldots, (P_{n_k}^{(k)} f)(x)) .$$

A function $F: U \times \mathcal{F}(U, U') \to \mathcal{F}(U, \mathbb{R})$ constructed in this way is said to be a C^∞ function of x and f whose values at x depend only on the values at x of f and its derivatives up to order k. We will write $F(x, [f]_k)$ in order to explicitly recall this fact, but when $k = 0$ we shall write $F(x, f)$ instead of $F(x, [f]_0)$ in order to simplify the notation.

Let us return to quasilinear differential operators. In the sequel we shall deal with operators acting on $\mathbb{R}^m$-valued functions rather than operators acting on ordinary functions. Now let U be an open set of $\mathbb{R}^{n+1}$ and U' an open set of $\mathbb{R}^m$. Denote by $\mathcal{F}(U, U')$ the set of C^∞ maps $f : U \to U' \subset \mathbb{R}^m$. An element $f \in \mathcal{F}(U, U')$ will have the form $f = (f^1, \ldots, f^m)$, where the components f^A, with A ranging from 1 to m, are ordinary functions on U. A map $D : \mathcal{F}(U, U') \to \mathcal{F}(U, \mathbb{R}^m)$ is said to be a vector-valued quasilinear differential operator of order k if it has the following form:

$$D(f)^A(x) = \sum_{\alpha_1=0}^{n} \cdots \sum_{\alpha_k=0}^{n} \sum_{B=1}^{m} (a(x, [f]_{k-1})_B^A)^{\alpha_1 \ldots \alpha_k} \partial_{\alpha_1} \ldots \partial_{\alpha_k} f^B + F^A(x, [f]_{k-1}) ,$$

where A ranges from 1 to m and where the coefficients $(a_B^A)^{\alpha_1 \ldots \alpha_k}$ and the terms F^A are C^∞ functions of x and f whose values at x depend only on the values at x of f and its derivatives up to order $k - 1$. From now on, we will drop summation signs in an expression similar to the previous one (by application of Einstein's convention on repeated indices).

Since we will only work with differential operators of order 1 and 2, it is convenient to mention that, in the case of quasilinear operators of order 1 taking values on $\mathbb{R}^m$, the latter expression for $D(f)$ reduces to (dropping the summation signs)

$$D(f)^A(x) = (a(x,f)^A_B)^\alpha\, \partial_\alpha f^B + F^A(x,f)\ .$$

If we denote by a^α the matrix $((a^A_B)^\alpha)$, the latter equality can be written in vector form as

$$D(f)(x) = a^\alpha(x,f)\partial_\alpha f + F(x,f)\ , \tag{IV.1}$$

where now $f = (f^A)$ and $F = (F^A)$ are column matrices containing m elements and a^α are matrices $m \times m$. In the case where the matrices a^α in the latter expression depend on x, but are independent of f, the differential operator is called linear.

In a similar way, in the particular case of quasilinear operators D of order 2 taking values on $\mathbb{R}^m$, the expression for $D(f)$ is

$$D(f)(x) = a^{\alpha\beta}(x,f,\partial_\gamma f)\partial_\alpha\partial_\beta f + F(x,f,\partial_\gamma f)\ , \tag{IV.2}$$

where now each coefficient $a^{\alpha\beta}$ is an $m \times m$-matrix, f and F are vectors belonging to $\mathbb{R}^m$. On the other hand, since $\partial_\alpha\partial_\beta = \partial_\beta\partial_\alpha$, the coefficients $a^{\alpha\beta}$ in (IV.2) can be considered to be symmetric with respect to α, β.

The point that we consider operators D acting on functions f which take values on a fixed open set U' of $\mathbb{R}^m$ is important. In fact, in the case of Einstein's equation we will deal with operators that, over a certain local chart, make sense for functions taking values on a certain open set of $\mathbb{R}^m$ (and make no sense for functions taking values on the whole of $\mathbb{R}^m$).

The above considered differential operators are related to systems of m equations in partial derivatives on the open set U. For example, a quasilinear system of m equations in partial derivatives of order 1 or 2 on an open set U of $\mathbb{R}^{n+1}$ is an expression having the form $D(f) = 0$, where D is as in (IV.1) or (IV.2). The unknowns of the system are the m functions $f^1 \dots f^m$ that are components of f.

Let us define the *leading symbol* of such a quasilinear operator. Denote by $\mathcal{M}(\mathbb{R})_m$ the set of real-valued $m \times m$-matrices. Let f be a solution of the (vector-valued) equation $D(f) = 0$, where D is a quasilinear operator of order 1 taking values on $\mathbb{R}^m$ and having form (IV.1). The leading symbol of D associated to the solution f is defined as the map

$$\begin{array}{rccc} \sigma_L(D,f): & \mathbb{R}^{n+1} \times \mathbb{R}^{n+1} & \longrightarrow & \mathcal{M}(\mathbb{R})_m \\ & (x,\xi) & \to & ((a(x,f)^A_B)^\alpha \xi_\alpha)\ . \end{array}$$

If D is a quasilinear operator of order 2 taking values on $\mathbb{R}^m$ having form (IV.2), the leading symbol is defined in a similar way by

$$\begin{array}{rcl} \sigma_L(D,f): \ \mathbb{R}^{n+1} \times \mathbb{R}^{n+1} & \longrightarrow & \mathcal{M}(\mathbb{R})_m \\ (x,\xi) & \to & ((a(x,f,\partial_\gamma f)^A_B)^{\alpha\beta}\xi_\alpha\xi_\beta) \ . \end{array}$$

In the case where the operators are linear the coefficients a^α or $a^{\alpha\beta}$ do not depend on f and therefore the leading symbol will not depend on any concrete solution f of the equation $D(f) = 0$.

Let us now relate the leading symbol we have just defined with the Cauchy problem for the equation $D(f) = 0$. Let U be an open set of $\mathbb{R}^{n+1}$, where the operator D is defined. Suppose that a hypersurface S of $\mathbb{R}^{n+1}$ contained in U is given. Suppose that a differentiable vector field N of $\mathbb{R}^{n+1}$, with support in S, is also given in such a way that N is transversal to S. That is, we have $N_x \in \mathbb{R}^{n+1}$ for each $x \in S$ such that N_x differentiably depends on x and such that $\langle N_x\rangle \oplus T_x(S) = \mathbb{R}^{n+1}$. If the operator D has order k (with $k = 1,2$), then we suppose that k functions, $u_0 \dots u_{k-1}$, are given on S, taking values in $\mathbb{R}^m$. These functions are called the *Cauchy data*. The Cauchy problem consists then in asking whether a solution f of the equation $D(f) = 0$ exists such that for each point $x \in S$, the r-th order derivative of f in direction of N_x would coincide with $u_r(x)$, for $r = 0 \dots k-1$.

Let us focus for a while on the simplest case where the operator D is (quasi-linear) of order 1. Then it can be expressed by (IV.1). Assume that ξ is the 1-form on $\mathbb{R}^{n+1}$ with support in S defined for each $x \in S$ by the conditions $\xi_x \subset (\mathbb{R}^{n+1})^*$ such that $\xi_x(N_x) = 1$ and $\xi_x(v) = 0$ if $v \in T_x(S)$. Fix the point $x \in S$ and choose a basis $v_1 \dots v_n$ of $T_x(S)$. Then $\{N_x, v_1 \dots v_n\}$ is a basis of $\mathbb{R}^{n+1}$. Consider these vectors as operators (acting by directional derivation on functions defined in a neighborhood of x). For notational homogeneity, write $v_0 = N_x$. The operators ∂_α at a point x are related to the operators v_α through the formulas of the change of basis

$$\partial_\alpha = A^\beta_\alpha v_\beta \ .$$

The dual basis of the basis $\{\partial_\alpha\}$ is $\{dx^\alpha\}$. Denote by $\{\theta^\alpha\}$ the dual basis of $\{v_\alpha\}$. It is well known that the change of basis between these two dual bases is given by

$$\theta^\beta = A^\beta_\alpha dx^\alpha \ . \tag{IV.3}$$

Let f be a solution of $D(f) = 0$. At the point $x \in S$ we have already fixed, this is written as

$$D(f)(x) = a^\alpha(x,f)\partial_\alpha f + F(x,f) = 0 \ .$$

Replace here ∂_α by their expressions with respect to v_α. We obtain

$$a^\alpha(x,f)A^\beta_\alpha v_\beta(f) + F(x,f) = 0 \ .$$

All coefficients in the above equation ($a^\alpha(x,f)A^\beta_\alpha$ and $F(x,f)$) are defined on S by the Cauchy data (reduced in this case to the vector value of f on S only). Since $v_1 \dots v_n$ are tangent to the hypersurface S, the values of $v_i(f)$ (for $i = 1 \dots n$) are also defined by the Cauchy data. The unique term that may not be defined by the Cauchy data is the transversal derivative $v_0(f) = N_x(f)$. If the matrix $a^\alpha(x,f)A^0_\alpha$ were non-singular, then the equation

$$a^\alpha(x,f)A^0_\alpha v_0(f) + a^\alpha(x,f)A^i_\alpha v_i(f) + F(x,f) = 0 \tag{IV.4}$$

could be solved with respect to $v_0(f)$, enabling us to express it as a function of the Cauchy data. To summarize, if the matrix $a^\alpha(x,f)A^0_\alpha$ is non-singular, the transversal derivative of any solution f at $x \in S$ is defined by the Cauchy data, and is not defined by this data otherwise.

Let us calculate the components of ξ with respect to the canonical basis $\{dx^\alpha\}$. By (IV.3), we obtain $\xi = \theta^0 = A^0_\alpha dx^\alpha$. Therefore the component ξ_α of ξ is A^0_α. Hence $a^\alpha A^0_\alpha = a^\alpha \xi_\alpha$ which is exactly $\sigma_L(f,D)(x,\xi)$. Therefore if the matrix $\sigma_L(f,D)(x,\xi)$ is non-singular then the transversal derivative at $x \in S$ of any solution is determined by the Cauchy data, and is not defined by this data otherwise.

Consider the first case where the matrix $\sigma_L(f,D)(x,\xi)$ is non-singular. The transversal derivative $N_x(f) = v_0(f)$ of any solution f is defined by the Cauchy data. By applying the derivation v_0 to equation (IV.4), we would see by the same argument that the second-order derivative $v_0^2(f) = N^2(f)$ is also defined by the Cauchy data. By repeating the process, all transversal derivatives $N^r(f)$ are defined by the Cauchy data. Therefore all the terms of the formal Taylor series of f at the point x corresponding to the normal direction are defined by the Cauchy data and, at least formally, one can calculate the solution f at a neighborhood of x by means of the Cauchy data. A hypersurface S of $\mathbb{R}^{n+1}$ together with a transversal vector N defined on S is called *nowhere characteristic* with respect to the solution f if the matrix $\sigma_L(f,D)(x,\xi)$ is non-singular at any point $x \in S$, where ξ denotes the 1-form of $\mathbb{R}^{n+1}$ having support on S defined at each $x \in S$ by the conditions $\xi_x \in (\mathbb{R}^{n+1})^*$ such that $\xi_x(N_x) = 1$ and $\xi_x(v) = 0$ if $v \in T_x(S)$. In a similar manner, S together with the transversal field N is called a *characteristic* with respect to the solution f if the matrix $\sigma_L(f,D)(x,\xi)$ is singular at every point $x \in S$.

Let us enlarge on this latter case and assume that the kernel of the matrix of $\sigma_L(f,D)(x,\xi)$ has dimension k at any point $x \in S$. We would like to see that the Cauchy data on S cannot be arbitrary and have to satisfy k *constraint equations*. Recall that as we deal with square matrices, the dimension of the kernel of $\sigma_L(f,D)(x,\xi)$ coincides with that of the kernel of the transposed matrix $\sigma_L(f,D)^*(x,\xi)$. We have seen that the matrix $\sigma_L(f,D)(x,\xi)$ is just the matrix $a^\alpha(x,f)A^0_\alpha$ in the first term of equation (IV.4). For brevity, denote this matrix by B. Suppose that u is a vector belonging to the kernel of the transposed matrix ${}^tB$. Consider u as a column matrix. Multiply matrix equality (IV.4) on the left by the transposed matrix ${}^tu$ and transpose the result. The first term of this equality equals zero since ${}^t({}^tuBv_0(f)) = {}^tv_0(f){}^tBu = 0$, taking into account that u belongs to the kernel of ${}^tB$. Therefore the first term vanishes and the remaining terms depend on the Cauchy data. Hence the Cauchy data cannot be arbitrary and have to satisfy the obtained relation. The same argument remains valid for any vector belonging to the kernel of ${}^tB$; this will give us k constraint equations that the Cauchy data should satisfy.

What we have done for quasilinear operators of order 1 can be repeated without any problem for any order. The definitions of the *characteristic hypersurface* and *nowhere characteristic hypersurface* are identical.

Assume that D is a quasilinear differential operator of order k on an open set U of $\mathbb{R}^{n+1}$ taking values in $\mathbb{R}^m$. Let f be a solution of $D(f) = 0$. Write $Q_{(x,f)}(\xi) = \det \sigma_L(f, D)(x, \xi)$ which is a homogeneous polynomial in ξ of degree km. The equation $D(f) = 0$ is called *elliptic* at the point x with respect to a solution f if the polynomial equation $Q_{(x,f)}(\xi) = 0$ has the only (trivial) solution $\xi = 0$. It is called *hyperbolic* at the point x with respect to the solution f if there exists some $u \in (\mathbb{R}^{n+1})^*$ such that, for any vector v that is not a scalar multiple of u, the equation $Q_{(x,f)}(\lambda u + v) = 0$ has km solutions with respect to λ, counting each of them according to its multiplicity. Therefore, for example, on $\mathbb{R}^4$ the wave equation

$$-\frac{\partial^2 f}{\partial x_0^2} + \frac{\partial^2 f}{\partial x_1^2} + \frac{\partial^2 f}{\partial x_2^2} + \frac{\partial^2 f}{\partial x_3^2} = 0$$

is hyperbolic (with respect to any solution). Indeed, $Q_{(x,f)}(\xi) = -\xi_0^2 + \xi_1^2 + \xi_2^2 + \xi_3^2$ (not depending here on x nor on f). If we take $u = (1, 0, 0, 0)$ and arbitrary v, we have $Q(\lambda u + v) = -(\lambda + v_0)^2 + \sum_{i=0}^3 v_i^2$. Then the equation $Q(\lambda u + v) = 0$ is equivalent to $(\lambda + v_0)^2 = P$, with the notation of $P = \sum_i v_i^2$, having two solutions $\lambda = \pm\sqrt{P} - v_0$.

We are going to deal here with hyperbolic systems and we will neeed, in the particular case of first-order systems, a stronger notion of hyperbolicity than that given above. First, a system of m quasilinear differential equations of order 1 having form $D(f) = 0$, with D given by (IV.1), will be called *hyperbolic, symmetric* with respect to a solution f if each of the matrices $a^\alpha(x, f)$ is symmetric and there exists a linear combination $\sum \xi_\alpha a^\alpha$ of the matrices a^α that is positive definite. By making a change of coordinates on $\mathbb{R}^{n+1}$, we can always assume that the coefficients ξ of one of these combinations are $\xi = (1, 0, \ldots, 0)$. Therefore we can suppose that it is the matrix a^0 that is positive definite. Let us show that in the particular case of first-order systems this notion of hyperbolicity is stronger than the one we have defined before. Assume that $a^0(x, f)$ is positive definite and that the remaining $a^i(x, f)$ are symmetric. Take $u = (1, 0, \ldots, 0)$ and any v. Then

$Q_{(x,f)}(\lambda u + v) = \det((\lambda + v_0)a^0 + \sum_i v_i a^i)$. Write $\mu = \lambda + v_0$. The equation $\det(\mu a^0 + \sum_i v_i a^i) = 0$ is the one determining the eigenvalues μ of the symmetric matrix $\sum_i v_i a^i$ with respect to the positive symmetric matrix a^0. This problem has m real-valued solutions.

The following theorem, which can be found in the papers [24],[33],[35],[41], will play a fundamental role in our study of the Cauchy problem for Einstein's equation.

Theorem IV.1. *Assume that $\tilde{U}$ is an open set of $\mathbb{R}^{n+1}$ having form $\tilde{U} = (-\varepsilon, \varepsilon) \times U$, where U is an open set of $\mathbb{R}^n$. Denote the coordinates of U by x^i, $i = 1 \dots n$, and distinguish the coordinate x^0 of $\mathbb{R}^{n+1}$ from the remaining coordinates by putting $x^0 = t$. Consider a quasilinear system of order 1:*

$$\sum_{\alpha=0}^{n} a^\alpha(x^\beta, f)\frac{\partial f}{\partial x^\alpha} + b(x^\beta, f) = 0 \,, \tag{IV.5}$$

where f is a function defined on $\tilde{U}$ and taking values in an open set U' of $\mathbb{R}^m$, a^α are $m \times m$ matrices and b is an order m column. Let $\phi(x^i)$ be a function taking values in $U' \subset \mathbb{R}^m$ and defined on the hypersurface $S = \{0\} \times U$ of $\tilde{U}$ only (that is, a function on an open set U of $\mathbb{R}^n$) whose Sobolev regularity order is s. That is, $\phi \in \mathcal{W}^s(U, \mathbb{R}^m)$. Suppose that

- *For each $x \in \tilde{U}$ and every solution f of* (IV.5) *the matrices $a^\alpha(x, f)$ are symmetric and a^0 is positive definite.*
- $s > \frac{n}{2} + 1$.

Then the problem is well posed, that is, there exists $\delta < \varepsilon$ satisfying the following conditions:

- *Equation* (IV.5) *has a unique solution f defined on $(-\delta, \delta) \times U$, with $f = \phi$ on S.*
- *The solutions and the Cauchy data belong to the same Sobolev spaces, that is for $t \in (-\delta, \delta)$, with f_t denoting the function on U given by $f_t(x) = f(t, x)$, we have $f_t \in \mathcal{W}^s(U, \mathbb{R}^m)$.*
- *The solutions depend continuously on the Cauchy data: the mapping $\phi \to f_t$, $t \in (-\delta, \delta)$, is continuous with respect to the norm of $\mathcal{W}^s(U, \mathbb{R}^m)$.*

IV.2 Differential operators in vector bundles

In order to deal with the Cauchy problem for Einstein's equation, it will not be sufficient to work with the concept of a quasilinear differential operator on an open set of $\mathbb{R}^{n+1}$ taking values on $\mathbb{R}^m$, which we discussed in the previous section. We will need to use operators defined in vector bundles.

Assume that V is a differentiable manifold of dimension $n+1$. Let $\pi : E \to V$ and $\pi' : F \to V$ be vector bundles on V whose fibers are $\mathbb{R}$-vector spaces of dimension m (the same m for the two bundles). Denote by $\Gamma(E)$ and $\Gamma(F)$ the sets of C^∞ sections of E and F, respectively. Assume that W is an open set of E satisfying $\pi(W) = V$. Denote by $\Gamma(W)$ the set of elements $\gamma \in \Gamma(E)$ whose images belong to W, $\gamma : V \to W \subset E$.

We will say that a mapping $D : \Gamma(W) \to \Gamma(F)$ is a quasilinear differential operator of order 1 if for any domain of a local chart U of V such that $\pi^{-1}(U)$ and $\pi'^{-1}(U)$ become trivial (that is $E|_U = \pi^{-1}(U) \cong U \times \mathbb{R}^m$, $F|_U = \pi'^{-1}(U) \cong U \times \mathbb{R}^m$) and for any section $f \in \Gamma(W)$ expressed in the trivialisation $E|_U \cong U \times \mathbb{R}^m$ by $f = (f^1 \dots f^m)$, $D(f)$ is expressed on U in the corresponding trivialisation by $D(f) = (D(f)^1 \dots D(f)^m)$, whose components $D(f)^A$ are as follows:

$$D(f)^A(x) = (a(x,f)^A_B)^\alpha \, \partial_\alpha f^B + F^A(x,f) \,, \tag{IV.6}$$

where the functions $(a(x,f)^A_B)^\alpha$ and $F^A(x,f)$ are C^∞ and depend on f, but do not depend on the derivatives of f. As we have already seen in the previous section, the latter equation can be rewritten in matrix form as

$$D(f)(x) = a^\alpha(x,f)\partial_\alpha f + F(x,f) \,, \tag{IV.7}$$

where now both $f = (f^A)$ and $F = (F^A)$ are column matrices having m components and a^α are $m \times m$ matrices.

In a similar manner, the operator D is called quasilinear of order 2 if the component-wise expression of $D(f)$ in the above local charts and trivialisations has the form

$$D(f)(x) = a^{\alpha\beta}(x,f,\partial_\gamma f)\partial_\alpha\partial_\beta f + F(x,f,\partial_\gamma f) \,, \tag{IV.8}$$

where now every coefficient $a^{\alpha\beta}$ is an $m \times m$ matrix, f and F are vectors belonging to $\mathbb{R}^m$, and the entries of the matrix $a^{\alpha\beta}(x,f,\partial_\gamma f)$ and the components of the vector $F(x,f,\partial_\gamma)$ are C^∞ functions of x and f whose values at x depend only on the values at x of f and its first derivatives.

Assume now, for example, that D is a quasilinear operator of order 1. For any $p \in V$ and $\xi \in T_p(V)^*$ we define the leading symbol of D associated to a section $f \in \Gamma(W)$, at the point p and corresponding to the covector $\xi \in T_p(V)^*$, as the linear map

$$\sigma_L(D,f)(p,\xi) : E_p \longrightarrow F_p$$

associating to every $u \in E_p$ the element $v \in F_p$ defined by the following procedure: We take a local chart $(U, x^0 \dots x^n)$ of V such that $p \in U$ and $E|_U$ and $F|_U$ become trivial. We take trivialisations of $E|_U$ and $F|_U$ given respectively by m elements $\gamma^1 \dots \gamma^m$ of $\Gamma(E|_U)$ and m elements $\gamma'^1 \dots \gamma'^m$ of $\Gamma(F|_U)$ that are linearly independent at each point of U. If the expressions of $D(f)$ in this local chart (using these trivialisations) is given by (IV.6), then the coordinates v^A of $v = \sigma_L(D,f)(p,\xi)u$ are

$$v^A = (a(p,f)^A_B)^\alpha \xi_\alpha u^B ,$$

where the ξ_α are the components of ξ.

One can check that this definition of $\sigma_L(D,f)(p,\xi)u$ does not depend either on the local chart $(U, x^0 \dots x^n)$ or on the trivialisations of $E|_U$ and $F|_U$ that are used. Let us show this for the local chart. If $(U', y^0 \dots y^n)$ is another local chart containing p, we will have in the intersection $U \cap U'$:

$$\frac{\partial}{\partial x^\alpha} = \sum_\beta \frac{\partial y^\beta}{\partial x^\alpha} \frac{\partial}{\partial y^\beta} .$$

One should substitute this coordinate change into expression (IV.6) in order to obtain the expression of $D(f)$ with respect to the coordinates (y^β). Therefore

$$D(f)^A = (a(y,f)^A_B)^\alpha \frac{\partial y^\beta}{\partial x^\alpha} \frac{\partial f^B}{\partial y^\beta} + F^A .$$

We deduce from this expression that if the components of $\xi \in T_p(V)^*$ with respect to the coordinates (y^β) are $\tilde{\xi}_\beta$, then the expression of $v = \sigma_L(D,f)(p,\xi)u$ in the coordinates (y^β) will be

$$v^A = (a^A_B)^\alpha \frac{\partial y^\beta}{\partial x^\alpha} \tilde{\xi}_\beta u^B .$$

But, if (ξ_α) are the components of ξ with respect to the coordinates $(x^0 \dots x^n)$, then it is well known that

$$\xi_\alpha = \frac{\partial y^\beta}{\partial x^\alpha} \tilde{\xi}_\beta ,$$

and the latter expression of v^A in terms of the coordinates (y^β) coincides with that we had before in terms of the coordinates (x^α).

In the case where D is a quasilinear operator of order 2 and the expression in components of $D(f)$ in the local charts and the respective trivialisations have form (IV.8), the leading symbol $\sigma_L(D,f)(p,\xi)$ of D associated to a section f at a point p and corresponding to a covector $\xi \in T_p(V)^*$, is defined by the linear map $E_p \longrightarrow F_p$ assigning to each $u \in E_p$ the $v \in F_p$ given in coordinates and trivialisations as follows:

$$v^A = (a(x,f)^A_B)^{\alpha\beta} \xi_\alpha \xi_\beta u^B ,$$

not depending either on the coordinates or on the trivialisations that we choose.

The above introduced concepts of ellipticity and hyperbolicity for quasilinear systems $D(f) = 0$ on an open set of $\mathbb{R}^{n+1}$ can be easily extended to the current situation where we deal with quasilinear operators on fiber bundles over a differentiable manifold. For example, if now $D : \Gamma(W) \to \Gamma(F)$ is a quasilinear operator, we say that the equation $D(f) = 0$ is hyperbolic with respect to a solution f of this equation if in any local chart $(U, x^0 \dots x^n)$ such that $E|_U$ and $F|_U$ become trivial, the quasilinear system of m equations which is obtained on the open set U is hyperbolic in the sense of the previous section. In a similar manner, for quasilinear operators of first order in fiber bundles there is the notion of a hyperbolic, symmetric operator with respect to a solution f of $D(f) = 0$.

IV.3 Harmonic maps

Here, we will recall the concept of a harmonic map between two pseudo-Riemannian manifolds introduced by J. Eells and J. M. Sampson [29]; we will need this concept in what follows. Assume that (V, g) and $(\hat{V}, \hat{g})$ are two pseudo-Riemannian manifolds and Φ is a differentiable map from V to $\hat{V}$. Though the concept of a harmonic map can be defined when dimensions of V and $\hat{V}$ are different, we are only interested in the case where V and $\hat{V}$ have the same dimension. Therefore, suppose that both manifolds have the same dimension n. Denote by E the fiber bundle $\Phi^*(T(\hat{V}))$ on V, inverse image of the tangent bundle $T(\hat{V})$ by Φ on $\hat{V}$:

$$\begin{array}{ccc} E = \Phi^*(T(\hat{V})) & \longrightarrow & T(\hat{V}) \\ \downarrow & & \downarrow \\ V & \xrightarrow{\Phi} & \hat{V} \end{array} .$$

At any $x \in V$ the fiber E_x of E is the tangent space $T_{\Phi(x)}(\hat{V})$. A differentiable section s of E is nothing else than a map assigning to each $x \in V$ an element $s(x) \in E_x = T_{\Phi(x)}(\hat{V})$ such that, if f is a differentiable function on $\hat{V}$, the function $x \to s(x)(f)$ is also a differentiable function on V. The linear tangent map $D\Phi_x$ of Φ at a point x,

$$T_x(V) \xrightarrow{D\Phi_x} T_{\Phi(x)}(\hat{V}) ,$$

maps X_x to the element $(D\Phi_x)(X_x)$ belonging to $T_{\Phi(x)}(\hat{V})$ such that for any differentiable function f on $\hat{V}$, $(D\Phi_x)(X_x)(f) = X_x(f \circ \Phi)$. This makes clear that, for each differentiable field X on V, the function $x \to (D\Phi_x)(X_x)(f)$ is differentiable and generates a differentiable section of E which will be denoted by $(D\Phi)(X)$,

$$\begin{array}{ccc} \mathcal{X}(V) & \xrightarrow{D\Phi} & \Gamma(E) \\ X & \to & (x \to (D\Phi_x)(X_x)) \end{array} ,$$

where $\mathcal{X}(V)$ denotes the set of differentiable fields on V.

In this way, $D\Phi$ can be interpreted as a differentiable 1-form on V with coefficients in the fiber bundle E, that is, as an element of $\Gamma(T(V)^* \otimes E)$.

If $(x^1 \dots x^n)$ is a coordinate system in a neighbourhood of a point $x_0 \in V$ and $(y^1 \dots y^n)$ is a coordinate system in a neighborhood of the point $\Phi(x_0)$, the sections

$s_j(x) = (\partial/\partial y^j)_{\Phi(x)}$ are differentiable sections of E in a neighborhood of x_0 that at each x form a basis of E_x. Using the coordinates (x^i) and this trivialisation of E in a neighborhood of x_0, the 1-form $D\Phi$ with coefficients in E is written as

$$D\Phi = \frac{\partial \Phi^j}{\partial x^i}\, dx^i \otimes \left(\frac{\partial}{\partial y^j}\right)_{\Phi(x)} .$$

The natural connection $\hat{\nabla}$ of $\hat{V}$ induces a connection in the fiber bundle E which, with an abuse of notation, we will denote also by $\hat{\nabla}$, as we will define in what follows. If Y is a differentiable vector field on $\hat{V}$ defined on a neighborhood of $\Phi(x_0)$, let s_Y be the section of E defined by $s_Y(x) = Y_{\Phi(x)}$. If X is a vector field on V in a neighborhood of x_0 and Y is a vector field on $\hat{V}$ in a neighborhood of $\Phi(x_0)$, we define

$$(\hat{\nabla}_X s_Y)_x = \left(\hat{\nabla}_{D\Phi_x(X_x)} Y\right)_{\Phi(x)} . \tag{IV.9}$$

The natural connection ∇ that exists in the fiber bundle $T(V)^*$ together with the connection $\hat{\nabla}$ we have just defined in $E = \Phi^*(T(\hat{V}))$ induce a connection ∇' in the fiber bundle $T(V)^* \otimes E$ defined by the condition

$$\nabla'_X(\omega \otimes s) = \nabla_X \omega \otimes s + \omega \otimes \hat{\nabla}_X s .$$

This connection ∇' enables us to define the operator δ associating to each 1-form α with coefficients in E locally expressed by

$$\alpha = \alpha_i^j dx^i \otimes \left(\frac{\partial}{\partial y^j}\right)_{\Phi(x)} ,$$

the 0-form with coefficients in E (that is, the section of E) given by

$$\delta\alpha = -g^{ij}(\nabla'_i \alpha_j^k) \left(\frac{\partial}{\partial y^k}\right)_{\Phi(x)} ,$$

and this expression is independent of the coordinate system.

The map Φ from V to $\hat{V}$ is called *harmonic* if $\delta(D\Phi) = 0$. We will have

$$\nabla'_{\frac{\partial}{\partial x^i}} D\Phi = \nabla'_{\frac{\partial}{\partial x^i}} \left(\frac{\partial \Phi^k}{\partial x^j} dx^j \otimes \frac{\partial}{\partial y^k}\right) .$$

If we calculate in this expression $\left(\hat{\nabla}_{\frac{\partial}{\partial x^i}} \frac{\partial}{\partial y^k}\right)_x$ according to definition (IV.9), we obtain

$$\left(\hat{\nabla}_{\frac{\partial}{\partial x^i}} \frac{\partial}{\partial y^k}\right)_x = \left(\hat{\nabla}_{\frac{\partial \Phi^r}{\partial x^i}\frac{\partial}{\partial y^r}} \frac{\partial}{\partial y^k}\right)_{\Phi(x)} = \frac{\partial \Phi^r}{\partial x^i}\, \hat{\Gamma}^s_{rk}(\Phi(x)) \left(\frac{\partial}{\partial y^s}\right)_{\Phi(x)} .$$

Therefore the expression of $\left(\nabla'_{\frac{\partial}{\partial x^i}} D\Phi\right)_x$ becomes

$$\left(\frac{\partial^2\Phi^s}{\partial x^i \partial x^j} - \frac{\partial \Phi^s}{\partial x^l}\Gamma^l_{ij}(x) + \frac{\partial \Phi^k}{\partial x^j}\frac{\partial \Phi^r}{\partial x^i}\hat{\Gamma}^s_{rk}(\Phi(x))\right) dx^j \otimes \left(\frac{\partial}{\partial y^s}\right)_{\Phi(x)} .$$

Hence the expression of $\delta D\Phi$ is

$$-g^{ij}\left(\frac{\partial^2\Phi^s}{\partial x^i \partial x^j} - \frac{\partial \Phi^s}{\partial x^l}\Gamma^l_{ij} + \frac{\partial \Phi^k}{\partial x^j}\frac{\partial \Phi^r}{\partial x^i}\hat{\Gamma}^s_{rk}(\Phi(x))\right)\left(\frac{\partial}{\partial y^s}\right)_{\Phi(x)} .$$

We are interested in the case where the manifolds V and $\hat{V}$ are the same and the map Φ is the identity,

$$I : (V, g) \longrightarrow (V, \hat{g}) .$$

Then δDI, in a local coordinate system $(x^1 \dots x^n)$, is

$$\delta DI = g^{ij}(\Gamma^s_{ij} - \hat{\Gamma}^s_{ij})\frac{\partial}{\partial x^s} . \tag{IV.10}$$

This is a globally defined vector field we will denote by H. The identity is harmonic (with respect to the initial metric g and the final metric $\hat{g}$) when the vector field H vanishes.

Denote by Γ^s the expression $g^{ij}\Gamma^s_{ij}$ appearing in (IV.10). Observe that the fields $\Gamma^s\frac{\partial}{\partial x^s}$ and $g^{ij}\hat{\Gamma}^s_{ij}\frac{\partial}{\partial x^s}$ appearing in (IV.10) have local sense only (they depend on the coordinates chosen), while the difference, H, is a globally defined vector field (since it is δDI).

We will need the following proposition.

Proposition IV.2. *Let $\Phi : V \longrightarrow \hat{V}$ be a diffeomorphism of two manifolds. Assume that there is a pseudo-Riemannian metric g on V and a pseudo-Riemannian metric $\hat{g}$ on $\hat{V}$. Let $\bar{g}$ be the metric on $\hat{V}$ given by $\bar{g} = (\Phi^{-1})^* g$. The diffeomorphism $\Phi : (V, g) \longrightarrow (\hat{V}, \hat{g})$ is harmonic if and only if the identity $I : (\hat{V}, \bar{g}) \longrightarrow (\hat{V}, \hat{g})$ is harmonic.*

Proof. The proof of the proposition follows from definitions. In the same way that the diffeomorphism Φ generates a morphism Φ^* of differential forms, associating to any p-form α on $\hat{V}$ a p-form $\Phi^*(\alpha)$ on V, the diffeomorphism Φ also generates a morphism Φ^* for differential forms with coefficients in a fiber bundle. In the case we are dealing with, where the fiber bundle we consider on $\hat{V}$ is the tangent bundle, if α is a p-form on $\hat{V}$ with coefficients in $T(\hat{V})$, $\Phi^*(\alpha)$ is in a natural way a p-form on V with coefficients in the fiber bundle $\Phi^*(T(\hat{V}))$. In the case of 1-forms, Φ^* is a map

$$\Phi^* : \Gamma(T(\hat{V})^* \otimes T(\hat{V})) \longrightarrow \Gamma(T(V)^* \otimes \Phi^*(T(\hat{V}))) .$$

The canonical 1-form on $\hat{V}$ with coefficients in the tangent bundle $T(\hat{V})$, associating to any vector field X on $\hat{V}$ the same field X (interpreted as a section of the bundle $T(\hat{V})$) is the one we have denoted before by DI (the 1-form associated to the linear tangent map to the identity $I : \hat{V} \longrightarrow \hat{V}$). This 1-form DI with coefficients in $T(\hat{V})$ is an element of $\Gamma(T(\hat{V})^* \otimes T(\hat{V}))$. Going back to definitions, it is clear that $\Phi^*(DI) = D\Phi$. On the other hand, consider the connection ∇'' in the fiber bundle $T(\hat{V})^* \otimes T(\hat{V})$ induced by the connection $\bar{\nabla}$ in the first factor $T(\hat{V})^*$ associated with the metric $\bar{g}$ and by the connection $\hat{\nabla}$ in the second factor $T(\hat{V})$ associated to the metric $\hat{g}$. In the fiber bundle $T(V)^* \otimes \Phi^*(T(\hat{V}))$, consider the usual connection ∇' induced by the connection ∇ of $T(V)^*$ associated to the metric g and by the connection $\hat{\nabla}$ in $\Phi^*(T(\hat{V}))$ associated to the metric $\hat{g}$ by means of Φ. It follows from the definitions that Φ^* commutes with these connections of $T(\hat{V})^* \otimes T(\hat{V})$ and $T(V)^* \otimes \Phi^*(T(\hat{V}))$, respectively. If we denote by δ'' the operator δ for 1-forms on $\hat{V}$ with coefficients in $T(\hat{V})$ associated to the connection ∇'' and to the metric $\bar{g}$, and by δ' the operator for 1-forms on V with coefficients in $\Phi^*(T(\hat{V}))$ associated to the connection ∇' and to the metric g, the map Φ^* will also commute with δ'' and δ'. Then we will have $\Phi^*(\delta'' DI) = \delta'\Phi^*(DI) = \delta' D\Phi$. Therefore $\delta' D\Phi$ vanishes if and only if $\delta'' DI$ vanishes. □

IV.4 Admissible classes of stress-energy tensors

We have seen in Chapter II that studying physical phenomena related to a gravitational field requires us to consider geometry of a Lorentzian manifold $(V, \tilde{g})$ where the metric $\tilde{g}$ is related to the matter generating the gravitational field by means of Einstein's equation (II.12) written as $G(\tilde{g}) = \chi T$, where $G(\tilde{g})$ is Einstein's tensor of the metric $\tilde{g}$, χ is a constant and T is the stress-energy tensor of the matter that creates the gravitational field.

Any study of Einstein's equation that could be made from the point of view of partial differential equations requires first that we specify the form of the stress-energy tensor T. If we are interested in studying Einstein's equation in the vacuum, we have to take $T = 0$. If our interest is in the gravitational fields generated by perfect fluids, we would have to say that T is of the form

$$T = (\rho + p)u \otimes u + p\tilde{g} \ , \tag{IV.11}$$

where u is a vector field satisfying $\tilde{g}(u, u) = -1$, and where the pressure p and the density ρ are related by means of a known function (state function). It goes without saying that in the above expression of T we have taken units such that the speed of light c equals 1. We should also specify that if we want to interpret T as a covariant tensor, we would replace u in the last formula by the 1-form u^* associated to the vector u through the metric $\tilde{g}$.

Let us fix our attention on the fact that saying that the tensor T corresponds to a perfect fluid does not define the tensor T completely; this only specifies the form of the tensor. In reality, the class of stress-energy tensors of perfect fluids is that of tensors T parametrised by u, ρ and $\tilde{g}$, having form (IV.11), where u is a vector field on V, $\tilde{g}$ is a Lorentzian metric, ρ is a function, and where $\tilde{g}(u,u) = -1$ holds and p is a given function of ρ. In order to adopt a sufficiently general point of view, we are going to define the concept of an *admissible class of stress-energy tensors* which would encompass all types of stress-energy tensors that we will consider in the remaining part of this book (we will see that the class of tensors of perfect fluids is admissible).

Let V be a 4-dimensional differentiable manifold. Assume that $T(V)$ is the tangent bundle. Denote by $S_2(V)$ the subbundle of $\otimes^2 T(V)^*$ that at each point x has the space of covariant *symmetric* 2-tensors at the point x as a fiber. Suppose that $F \to V$ is an $\mathbb{R}$-vector bundle on V of rank 4 (the fibres F_x are 4-dimensional $\mathbb{R}$-vector spaces). Consider the bundle $E = S_2(V) \oplus F$. Assume that W is an open set of the bundle E having the following properties: 1) The restriction of the canonical projection $\pi : E \to V$ to W (mapping W to V) is surjective. 2) If $w_x \in W$ is in the fiber E_x and $w_x = \tilde{g}_x + f_x$, with $\tilde{g}_x \in S_2(V)_x$ and $f_x \in F_x$, then $\tilde{g}_x$ is a Lorentzian metric on the vector space $T_x(V)$. Denote, as we usually do, by $\Gamma(W)$ the set of sections γ of E belonging to the class C^∞ such that $\gamma(x) \in W$ for all $x \in V$.

Definition IV.3. Let F be a rank 4 vector bundle over V and set $E = S_2(V) \oplus F$. Assume that

$$\begin{array}{rccc} T: & \Gamma(W) & \longrightarrow & \Gamma(S_2(V)) \\ & w & \to & T(w) \end{array}$$

is a map, where W is an open set of E satisfying the above properties 1) and 2). Then the map T will be called an *admissible class of stress-energy tensors* if it satisfies the property that the value $T(w)_x$ of $T(w)$ at every $x \in V$ depends only on the value of w on x (and does not depend on those of the derivatives).

Let us note that the stress-energy tensors of perfect fluids can be described using this procedure and therefore constitute an *admissible* class of stress-energy tensors. Suppose that the 4-dimensional manifold V is such that one can find a rank 3 subbundle F' of the tangent bundle and a vector field N on V such that at each $x \in V$ we have $T_x(V) = F'_x \oplus \langle N_x \rangle$. For any Lorentzian metric $\tilde{g}$ on V such that N_x is time-like at each point and F'_x is space-like, and for each $v \in \Gamma(F')$ consider the vector field u given at any point x by

$$u_x = v_x + rN_x \ ,$$

where r is given by the expression

$$r = \frac{-\tilde{g}(v_x, N_x) - \sqrt{\tilde{g}(v_x, N_x)^2 - \tilde{g}(N_x, N_x)(\tilde{g}(v_x, v_x) + 1)}}{\tilde{g}(N_x, N_x)} \ .$$

First, observe that since $\tilde{g}(N_x, N_x) < 0$ and $\tilde{g}(v_x, v_x) > 0$ by assumption, the expression under the square root sign is positive. On the other hand, from

$$\tilde{g}(v_x, N_x)^2 - \tilde{g}(N_x, N_x)(\tilde{g}(v_x, v_x) + 1) > \tilde{g}(v_x, N_x)^2 ,$$

we deduce

$$\sqrt{\tilde{g}(v_x, N_x)^2 - \tilde{g}(N_x, N_x)(\tilde{g}(v_x, v_x) + 1)} > |\tilde{g}(v_x, N_x)| \geq -\tilde{g}(v_x, N_x) .$$

This implies that the numerator in the expression for r is negative. Since the denominator is also negative, we obtain $r > 0$. By this argument, for every $v \in \Gamma(F')$ and for any Lorentzian metric $\tilde{g}$ we have constructed a vector field u on V equal to $u_x = v_x + rN_x$ for every x and satisfying $\tilde{g}(u, u) = -1$ by construction. Denote by $\mathcal{R}$ the trivial rank 1 vector bundle on V given by $\mathcal{R} = V \times \mathbb{R}$. Let F be the direct sum $F = F' \oplus \mathcal{R}$ (which is a rank 4 vector bundle). By means of this bundle F, we construct the vector bundle E (as we have already mentioned) by the formula $E = S_2(V) \oplus F = S_2(V) \oplus F' \oplus \mathcal{R}$. Assume that W is an open set of E such that if $w \in W \subset E$ belongs to the fiber E_x, then w has the form $w = (\tilde{g}_x, v_x, p_x)$, where $\tilde{g}_x$ is a Lorentzian metric in the vector space $T_x(V)$ for which N_x is time-like and F'_x is space-like. Using the first two components $\tilde{g}_x$ and v_x of this w we construct the vector $u_x = v_x + rN_x$ using the above procedure. Then we consider the 2-covariant tensor field T given at each point x as follows:

$$T_x = (\rho(p_x) + p_x)u_x^* \otimes u_x^* + p_x\tilde{g}_x ,$$

where a function $\rho(p)$ is supposed to be known (and given) representing the state function, and where u_x^* denotes the 1-form at the point x associated to the vector u_x through the metric $\tilde{g}_x$. In this way we can think of the class of stress-energy tensors that correspond to perfect fluids as a certain parametrization of $\Gamma(S_2(V))$ by $\Gamma(W)$, where W is an open set of a bundle E having the form $E = S_2(V) \oplus F$, F being a vector bundle of rank 4.

The concept of *admissible* class of stress-energy tensors is very general and encompasses many other cases different from perfect fluids we have considered above.

Remark IV.4. There may be various fields describing the properties of the matter of each particular problem taken into consideration (for example, the electromagnetic field, the fields corresponding to a perfect fluid or a holonomic medium, etc.) and it is not clear at all how the class of stress-energy tensors associated to these fields must be constructed. A way to proceed is the so-called Lagrangian formulation, which derives this class from a suitable function or density associated to the particular problem (the *Lagrangian*). See for example [40]. But the way to obtain the stress-energy tensor from a particular Lagrangian is nowadays a vast area of research. See for example [39], [19], [20].

IV.5 Differential operator associated to Einstein's equation

As we have already mentioned, from now on we will suppose that we have an admissible class of stress-energy tensors given by a mapping

$$\begin{array}{ccc} \Gamma(W) & \longrightarrow & \Gamma(S_2(V)) \\ w & \rightarrow & T(w) \end{array} ,$$

where W is an open set of a bundle E having the form $E = S_2(V) \oplus F$, F being a vector bundle of rank 4.

Consider the system of equations

$$\begin{cases} \widetilde{Ric} - (1/2)\tilde{R}\tilde{g} - \chi T = 0 \\ \mathrm{div}_{\tilde{g}} T = 0 , \end{cases}$$

where T is a stress-energy tensor parametrised in the way we have just described. As we have seen in Chapter II, Einstein's equation can also be written in form (II.14). Therefore we can rewrite the above system in the equivalent form

$$\begin{cases} \widetilde{Ric}(\tilde{g}) - \chi(T - \frac{1}{2}(\mathrm{tr}_{\tilde{g}} T)g) = 0 \\ \mathrm{div}_{\tilde{g}} T = 0 . \end{cases}$$

Let us specify now what differential operator $\mathcal{E}$ is the one that is associated to this system. Using the above introduced notation, this operator is given by the mapping

$$\begin{array}{ccc} \Gamma(S_2(V)) \oplus \Gamma(F) & & \\ \| & & \\ \Gamma(S_2(V) \oplus F) & & \\ \cup & & \\ \Gamma(W) & \xrightarrow{\mathcal{E}} & \Gamma(S_2(V) \oplus T(V)^*) \\ w = (\tilde{g}, f) & \rightarrow & (\mathcal{E}_1, \mathcal{E}_2) \end{array} ,$$

with

$$\mathcal{E}_1 = \widetilde{Ric}(\tilde{g}) - \chi \left\{ T(w) - \tfrac{1}{2}(\mathrm{tr}_{\tilde{g}} T(w))\tilde{g} \right\} ,$$
$$\mathcal{E}_2 = \mathrm{div}_{\tilde{g}} T(w) .$$

Observe that the original bundle $S_2(V) \oplus F$ and the final bundle $S_2(V) \oplus T(V)^*$ have the same rank $(10 + 4 = 14)$ according to what we have requested in the definitions of section IV.2.

We would like to study now the first component $\mathcal{E}_1$ of the operator $\mathcal{E}$. If we consider $\mathcal{E}_1$ as a mapping of an open set W of $\Gamma(S_2(V)) \oplus F)$ to $\Gamma(S_2(V))$, then the original bundle and the final one would not have the same rank. In order to avoid this situation, we would need to fix $f \in \Gamma(F)$ and to consider the operator

$$\begin{array}{ccc} \Gamma(W') & \stackrel{\mathcal{E}_1}{\longrightarrow} & \Gamma(S_2(V)) \\ \tilde{g} & \to & \mathcal{E}_1(\tilde{g}, f) \end{array} ,$$

where W' is an open set of $S_2(V)$ having the property that if $\tilde{g}_x \in W'_x$, then $(\tilde{g}_x, f_x) \in W_x$.

Let us write now the principal part of this operator coordinate-wise. The terms of the expression of $\mathcal{E}_1$ that contain second derivatives come from $\widetilde{Ric}(\tilde{g})$ since, by definition of the *admissible class of stress-energy tensors*, $T(\tilde{g}, f)$ contains no derivatives of either $\tilde{g}$ or f. In Section I.2, there is a coordinate-wise expression of the Ricci tensor. We will employ the formulas given in that section and use the symbol $\cong$ to indicate equality of the terms whose order of derivation in both expressions is the highest one. If $\tilde{R}_{\alpha\beta}$ stands for the components of the Ricci tensor of a Lorentzian metric $\tilde{g}$, we have

$$\begin{aligned} \tilde{R}_{\alpha\beta} = \tilde{R}^{\lambda}_{\alpha\lambda\beta} \cong \partial_\lambda \tilde{\Gamma}^{\lambda}_{\beta\alpha} - \partial_\beta \tilde{\Gamma}^{\lambda}_{\lambda\alpha} &\cong \frac{1}{2}\tilde{g}^{\lambda\mu}(\partial_\lambda\partial_\beta\tilde{g}_{\alpha\mu} + \partial_\lambda\partial_\alpha\tilde{g}_{\beta\mu} - \partial_\lambda\partial_\mu\tilde{g}_{\beta\alpha}) \\ &\quad - \frac{1}{2}\tilde{g}^{\lambda\mu}(\partial_\beta\partial_\lambda\tilde{g}_{\alpha\mu} + \partial_\beta\partial_\alpha\tilde{g}_{\lambda\mu} - \partial_\beta\partial_\mu\tilde{g}_{\lambda\alpha}) \,. \end{aligned}$$

The first and the fourth terms on the right-hand of this expression cancel out and we finally obtain

$$\tilde{R}_{\alpha\beta} \cong \frac{1}{2}\tilde{g}^{\lambda\mu}(\partial_\lambda\partial_\alpha\tilde{g}_{\beta\mu} + \partial_\beta\partial_\mu\tilde{g}_{\lambda\alpha} - \partial_\lambda\partial_\mu\tilde{g}_{\beta\alpha} - \partial_\beta\partial_\alpha\tilde{g}_{\lambda\mu}) \,. \tag{IV.12}$$

We can write this expression (by isolating the common factor $\partial_\nu\partial_\rho\tilde{g}_{\lambda\mu}$) in the following way:

$$\tilde{R}_{\alpha\beta} \cong \frac{1}{2}(\tilde{g}^{\nu\mu}\delta^{\rho}_{\alpha}\delta^{\lambda}_{\beta} + \tilde{g}^{\nu\lambda}\delta^{\rho}_{\beta}\delta^{\mu}_{\alpha} - \tilde{g}^{\nu\rho}\delta^{\lambda}_{\beta}\delta^{\mu}_{\alpha} - \tilde{g}^{\lambda\mu}\delta^{\nu}_{\beta}\delta^{\rho}_{\alpha})\partial_\nu\partial_\rho\tilde{g}_{\lambda\mu} \,.$$

For each pair of indices (ν, ρ), we denote by $a^{\nu\rho}$ the endomorphism of $S_2(V)_x$ given by the matrix

$$(a^{\nu\rho})^{\lambda\mu}_{\alpha\beta} = \frac{1}{2}(\tilde{g}^{\nu\mu}\delta^{\rho}_{\alpha}\delta^{\lambda}_{\beta} + \tilde{g}^{\nu\lambda}\delta^{\rho}_{\beta}\delta^{\mu}_{\alpha} - \tilde{g}^{\nu\rho}\delta^{\lambda}_{\beta}\delta^{\mu}_{\alpha} - \tilde{g}^{\lambda\mu}\delta^{\nu}_{\beta}\delta^{\rho}_{\alpha}) \,.$$

With this notation,

$$\tilde{R}_{\alpha\beta} \cong (a^{\nu\rho})^{\lambda\mu}_{\alpha\beta}\partial_\nu\partial_\rho\tilde{g}_{\lambda\mu} \,.$$

Therefore the operator $\mathcal{E}_1$ can be locally written as

$$\mathcal{E}_1 = a^{\nu\rho}\partial_\nu\partial_\rho + A \,,$$

where A denotes a term containing no second-order derivatives. Since the matrices $a^{\nu\rho}$ depend only on the metric $\tilde{g}$ (and not on the second derivatives of $\tilde{g}$), the operator $\mathcal{E}_1$ is quasilinear. Let us calculate the leading symbol of this operator.

For fixed f, let $\tilde{g}$ be a solution of the equation $\mathcal{E}_1(\tilde{g}, f) = 0$. If we fix some $p \in V$ and some $\xi \in T_p(V)^*$, the leading symbol of $\mathcal{E}_1$ associated to the solution $\tilde{g}$, corresponding to the point p and the covector ξ, is the mapping

$$\begin{array}{rccc} \sigma_L(\mathcal{E}_1, \tilde{g}, f)(p, \xi): & S_2(V)_p & \longrightarrow & S_2(V)_p \\ & s & \rightarrow & v \end{array}$$

with

$$v_{\alpha\beta} = \frac{1}{2}\tilde{g}^{\lambda\mu}(\xi_\lambda \xi_\alpha s_{\beta\mu} + \xi_\beta \xi_\mu s_{\lambda\alpha} - \xi_\lambda \xi_\mu s_{\beta\alpha} - \xi_\beta \xi_\alpha s_{\lambda\mu})\,. \tag{IV.13}$$

Now we are able to prove the following theorem.

Theorem IV.5. *Let $\tilde{g}$ be a solution of $\mathcal{E}_1(\tilde{g}, f) = 0$. Assume a hypersurface M of V is given. Suppose that M is space-like with respect to $\tilde{g}$ and that there exists a unitary vector field N with support on M, being normal to M. That is, a differentiable vector field N of V having support on M, such that N_x at each $x \in M$ is orthogonal to $T_x(M)$ with respect to $\tilde{g}_x$ and such that $\tilde{g}_x(N_x, N_x) = -1$. Let ξ be the 1-form on V with support on M associated to the normal field N. Then the dimension of the kernel of $\sigma_L(\mathcal{E}_1, \tilde{g}, f)(p, \xi)$ at each point $p \in M$ is equal to 4.*

Proof. We can take a basis $\{N_p, e_1, e_2, e_3\}$ in the vector space $T_p(V)$ whose first vector is N_p and such that the metric $\tilde{g}_p$ diagonalises in this basis and that the vectors e_i are unitary with respect to $\tilde{g}$. We know that $s \in S_2(V)_p$ is mapped by $\sigma_L(\mathcal{E}_1, \tilde{g}, f)(p, \xi)$ into the element v given by (IV.13). But we have $\tilde{g}^{0i} = 0$, $\tilde{g}^{ij} = \delta_{ij}$, $\tilde{g}^{00} = -1$, $\xi_0 = \tilde{g}_{0\alpha} n^\alpha = -1$, $\xi_i = 0$ in this basis. Therefore we obtain

$$v_{ij} = s_{ij}\ , \ v_{0i} = 0\ , \ v_{00} = 0\,.$$

Hence s belongs to the kernel if $s_{ij} = 0$. Since the four components $s_{0,\alpha}$ can be arbitrary (while the components s_{ij} are zeros), the kernel has dimension 4. □

This theorem implies that if a hypersurface M of V is given and if we would like to solve the equation $\mathcal{E}_1(\tilde{g}, f) = 0$, with $f \in \Gamma(F)$ fixed, and we would like the hypersurface to be space-like with respect to the solutions $\tilde{g}$, the Cauchy data on M should satisfy four constraint equations (and the hypersurface M will always be characteristic).

IV.6 Constraint equations

In this section, we will assume that V is a 4-dimensional differentiable manifold and M is a hypersurface of V admitting a differentiable vector field N of V with support on M, transversal to M. With the notation adopted in Section IV.5, for fixed $f \in \Gamma(F)$, let $\tilde{g}$ be a solution of $\mathcal{E}_1(\tilde{g}, f) = 0$ at a small neighborhood of the hypersurface M and assume that the following conditions are satisfied: $\tilde{g}$ is a Lorentzian metric satisfying $\tilde{g}_x(N_x, N_x) = -1$ at any point $x \in M$, and such that $T_x(M)$ is perpendicular to N_x with respect to $\tilde{g}$. For each $x \in M$, let $\gamma_x(t)$ be the geodesic, with respect to the metric $\tilde{g}$, passing through x and having N_x as a tangent vector at x. Suppose that U' is the neighborhood of $M \times \{0\}$ in $M \times \mathbb{R}$ formed by those (x, t) for which $\gamma_x(t)$ is defined. The mapping

$$\begin{array}{cccc} \gamma: & U' & \longrightarrow & V \\ & (x,t) & \rightarrow & \gamma_x(t) \end{array}$$

gives a diffeomorphism of U' (an open set of $M \times \mathbb{R}$) and some neighborhood U of M in V. For every $t \in \mathbb{R}$, we denote by M_t the image under γ of $(M \times \{t\}) \cap U'$ which will be a hypersurface of V (if $(M \times \{t\}) \cap U' \neq \emptyset$). Let us observe that the family $\{M_t\}_t$ of hypersurfaces of V is given by the diffeomorphism γ (and so, it depends on γ).

Every vector field X of M naturally gives rise to a vector field in the product manifold $M \times \mathbb{R}$, still denoted by X, taking at every $(x, t) \in M \times \mathbb{R}$ the value $X_x \in T_x(M) \subset T_{(x,t)}(M \times \mathbb{R})$. By means of the diffeomorphism γ, we can think of the vector field X of M as generating a vector field (still denoted by X) of V defined at the neighborhood U of M in V (a vector field which at each point p is tangent to the hypersurface M_t passing through p). In a similar way, the vector field $\frac{\partial}{\partial t}$ of $M \times \mathbb{R}$ gives rise by means of the diffeomorphism γ to a field of V defined in the neighborhood U (we will still denote it by $\frac{\partial}{\partial t}$) coinciding for $t = 0$ with the transversal field N given on M. Since the field $\frac{\partial}{\partial t}$ is geodesic (that is $\tilde{\nabla}_{\partial/\partial t}\partial t = 0$) and satisfies for $t = 0$ the condition $\tilde{g}(\frac{\partial}{\partial t}, \frac{\partial}{\partial t}) = -1$, it will satisfy this condition for all t. On the other hand, the well-known Gauss lemma assures that the field $\frac{\partial}{\partial t}$ is perpendicular to all hypersurfaces M_t. Indeed, let X be a field of $U \subset V$ generated by a field X of M by means of the above described procedure (at each point, X will be tangent to the hypersurface M_t passing through that point). We will have

$$\frac{\partial}{\partial t}\tilde{g}(\frac{\partial}{\partial t}, X) = \tilde{g}(\frac{\partial}{\partial t}, \tilde{\nabla}_X \frac{\partial}{\partial t}) = \frac{1}{2}X(\tilde{g}(\frac{\partial}{\partial t}, \frac{\partial}{\partial t})) = 0 \ .$$

Therefore $\tilde{g}(\frac{\partial}{\partial t}, X)$ is constant. Since it vanishes for $t = 0$, then it will vanish for all t.

The second fundamental form S of the hypersurface M is defined by $S(X, Y) = -\tilde{g}(\tilde{\nabla}_X Y, N)$, where X and Y are vector fields tangent to M. If we think of the fields X and Y as being extended to the neighborhood U of M in V in the way we have described above, then $S(X, Y) = -\tilde{g}(\tilde{\nabla}_X Y, \frac{\partial}{\partial t})$ is defined in the whole U and represents at each point the second fundamental form S_t of the hypersurface M_t passing through that point. Let g_t be the Riemannian metric on M_t obtained by restriction of $\tilde{g}$ to tangent vectors to M_t. Suppose that ∇_t is the connection on the manifold M_t associated to the Riemannian metric g_t. For notational simplicity, we will drop the sub-index t and denote by g and ∇ the Riemannian metric and the corresponding connection on any M_t. Let X and Y be vector fields of M. We will interpret them to be extended to the neighborhood U of M in the way we have already mentioned. At each point, these fields are tangent to the corresponding M_t, and we have the splitting $\tilde{\nabla}_X Y = \nabla_X Y + S(X,Y)\frac{\partial}{\partial t}$. Denote by P the endomorphism associated to the second fundamental form S. For the fields X and Y of M extended to the neighborhood U of M in the way we have already mentioned, we have $g(P(X),Y)) = S(X,Y)$. Since $\tilde{g}(\frac{\partial}{\partial t},Y) = 0$, we obtain $\tilde{g}(\tilde{\nabla}_X\frac{\partial}{\partial t},Y) = -\tilde{g}(\frac{\partial}{\partial t},\tilde{\nabla}_X Y) = S(X,Y)$. Therefore $P(X) = \tilde{\nabla}_X\frac{\partial}{\partial t}$ (remark that $\tilde{\nabla}_X\frac{\partial}{\partial t}$ is tangent to M_t as can be deduced from the identity $\tilde{g}\left(\tilde{\nabla}_X\frac{\partial}{\partial t},\frac{\partial}{\partial t}\right) = 0$). In the calculations we are going to carry out now, the following facts (that are trivial corollaries of our construction) should be taken into account: $[X,\frac{\partial}{\partial t}] = 0$ if X is a field of M extended to the neighborhood U of M in the above described way, and $\tilde{\nabla}_{\partial/\partial t}\frac{\partial}{\partial t} = 0$.

Denote by $R_t(X,Y)Z$, $Ric_t(X,Y)$ and R_t, respectively, the curvature tensor, Ricci tensor and the scalar curvature of he Riemannian metrics g_t (on M_t). We would like to relate the Ricci tensor, $\widetilde{Ric}$, of the Lorentzian metric $\tilde{g}$ with the Ricci tensor Ric_t and the second fundamental form S_t of these hypersurfaces. As above we will drop the sub-index t. In order to obtain this relation, we will see that it is sufficient to express $\tilde{R}(X,Y)Z$ and $\tilde{R}(X,\frac{\partial}{\partial t})\frac{\partial}{\partial t}$ as a function of the curvature and the second fundamental form of M_t when X and Y are fields in a neighborhood of M coming from fields on M. Then we begin by writing $\tilde{R}(X,Y)Z$. We know that $\tilde{R}(X,Y)Z = \tilde{\nabla}_X\tilde{\nabla}_Y Z - \tilde{\nabla}_Y\tilde{\nabla}_X Z - \tilde{\nabla}_{[X,Y]}Z$. Let us calculate now the term $\tilde{\nabla}_X\tilde{\nabla}_Y Z$ in this expression:

$$\begin{aligned}\tilde{\nabla}_X\tilde{\nabla}_Y Z &= \tilde{\nabla}_X\left(\nabla_Y Z + S(Y,Z)\frac{\partial}{\partial t}\right)\\ &= \nabla_X\nabla_Y Z + S(X,\nabla_Y Z)\frac{\partial}{\partial t} + X(S(Y,Z))\frac{\partial}{\partial t} + S(Y,Z)P(X)\,.\end{aligned}$$

By a similar calculation carried out for other terms of $\tilde{R}(X,Y)Z$ and by adding up the results, we obtain:

$$\begin{aligned}\tilde{R}(X,Y)Z &= R(X,Y)Z + S(Y,Z)P(X)\\ &\quad - S(X,Z)P(Y) + \{(\nabla_X S)(Y,Z) - (\nabla_Y S)(X,Z)\}\frac{\partial}{\partial t}\,, \qquad \text{(IV.14)}\end{aligned}$$

where X, Y and Z are fields of M extended to the neighborhood U of M in the way we have described above. Let us make now a similar calculation for $\tilde{R}(X,\frac{\partial}{\partial t})\frac{\partial}{\partial t}$. We have

$$\tilde{R}(X,\frac{\partial}{\partial t})\frac{\partial}{\partial t} = \tilde{\nabla}_X\tilde{\nabla}_{\partial/\partial t}\frac{\partial}{\partial t} - \tilde{\nabla}_{\partial/\partial t}\tilde{\nabla}_X\frac{\partial}{\partial t}\,.$$

The first term is zero since the field $\frac{\partial}{\partial t}$ is geodesic. Then we obtain

$$\tilde{R}(X, \frac{\partial}{\partial t})\frac{\partial}{\partial t} = -\tilde{\nabla}_{\partial/\partial t} P(X) \ . \tag{IV.15}$$

At this point, one could think that $-\tilde{\nabla}_{\partial/\partial t} P(X) = -\tilde{\nabla}_{P(X)}\frac{\partial}{\partial t} = -P(P(X))$. But this is incorrect since even though X is a field in a neighborhood of the hypersurface M originated by a field in M, the same does not happen to $P(X)$. With respect to a coordinate system (x^1, x^2, x^3, t) in a certain neighborhood of M in V generated by a coordinate system of M by means of the diffeomorphism γ, the field X would be expressed in the form $X = \sum X^i(x^1, x^2, x^3)\partial/\partial x^i$, while the field $P(X)$, having no component $\partial/\partial t$, may however depend on t (its components in $\partial/\partial x^i$ may depend on t). Therefore $P(X)$ in general does not commute with $\partial/\partial t$ (while X does commute with $\partial/\partial t$).

Let us use expressions (IV.14) and (IV.15) in order to calculate $\widetilde{Ric}(X,Y)$, $\widetilde{Ric}(\frac{\partial}{\partial t}, X)$ and $\widetilde{Ric}(\frac{\partial}{\partial t}, \frac{\partial}{\partial t})$. With respect to any coordinate system, expression (I.6) implies that

$$\widetilde{Ric}(A,B) = \tilde{g}^{\lambda\mu}\tilde{g}\left(\tilde{R}(\frac{\partial}{\partial x^\lambda}, B)A, \frac{\partial}{\partial x^\mu}\right) \ .$$

By using (as before) an adapted coordinate system (t, x^1, x^2, x^3), if X and Y are vector fields generated by fields of M, we have

$$\begin{aligned} \widetilde{Ric}(X,Y) = {} & g^{ij}\tilde{g}(\tilde{R}(\frac{\partial}{\partial x^i}, Y)X, \frac{\partial}{\partial x^j}) \\ & - \tilde{g}(\tilde{R}(\frac{\partial}{\partial t}, Y)X, \frac{\partial}{\partial t}) \ . \end{aligned} \tag{IV.16}$$

Let us calculate each term in this expression on its own. Before proceeding to do this calculation, we introduce the tensor $S \times S$ defined in the following way: $S \times S$ is the bilinear mapping assigning the function $S(X, P(Y))$ to each pair (X, Y) of vector fields tangent to all M_t. That is, $(S \times S)(X,Y) = S(X, P(Y)) = g(P(X), P(Y))$. It follows from the definition that $S \times S$ is symmetric. In the component-wise form, $(S \times S)_{ij} = g^{rs}S_{ri}S_{sj}$. By formula (IV.14), one can readily see that the first term in expression (IV.16) can be written as

$$\begin{aligned} g^{ij}\tilde{g}(\tilde{R}(\frac{\partial}{\partial x^i}, Y)X, \frac{\partial}{\partial x^j}) = {} & Ric(X,Y) + S(X,Y) \\ & + \mathrm{tr}_g(S)\, S(X,Y) - (S \times S)(X,Y) \ . \end{aligned}$$

Let us calculate the last term in (IV.16). By properties of the curvature tensor, we have $-\tilde{g}(\tilde{R}(\frac{\partial}{\partial t}, Y)X, \frac{\partial}{\partial t}) = \tilde{g}(\tilde{R}(\frac{\partial}{\partial t}, Y)\frac{\partial}{\partial t}, X)$. By (IV.15), this term can be written as $\tilde{g}(\tilde{\nabla}_{\partial/\partial t} P(Y), X)$. Then

$$\tilde{g}(\tilde{\nabla}_{\partial/\partial t} P(Y), X) = \frac{\partial}{\partial t}\tilde{g}(P(Y), X) - \tilde{g}(P(Y), \tilde{\nabla}_{\partial/\partial t} X) \ .$$

The last term is $-\tilde{g}(P(X), P(Y))$ since $\tilde{\nabla}_{\partial/\partial t}X = \tilde{\nabla}_X \frac{\partial}{\partial t} = P(X)$. By a simple calculation based on the definition of $S \times S$, we can finally see that

$$\tilde{g}(\tilde{\nabla}_{\partial/\partial t}P(Y), X) = \frac{\partial}{\partial t} S(X, Y) - (S \times S)(X, Y) .$$

Adding up the expressions for two terms of (IV.16) gives

$$\begin{aligned} \widetilde{Ric}(X, Y) = & \frac{\partial S(X, Y)}{\partial t} - 2(S \times S)(X, Y) \\ & + Ric(X, Y) + \operatorname{tr}(S) S(X, Y) . \end{aligned} \tag{IV.17}$$

Let us now calculate $\widetilde{Ric}(\frac{\partial}{\partial t}, X)$ if X is a vector field at a neighborhood of M in V generated by a vector field on M. We have

$$\widetilde{Ric}(\frac{\partial}{\partial t}, X) = g^{ij}\tilde{g}(\tilde{R}(\frac{\partial}{\partial x^i}, X)\frac{\partial}{\partial t}, \frac{\partial}{\partial x^j}) - \tilde{g}(\tilde{R}(\frac{\partial}{\partial t}, X)\frac{\partial}{\partial t}, \frac{\partial}{\partial t}) .$$

The last term equals zero since the curvature tensor is antisymmetric with respect to its two last factors. Therefore

$$\widetilde{Ric}(\frac{\partial}{\partial t}, X) = -g^{ij}\tilde{g}(\tilde{R}(\frac{\partial}{\partial x^i}, X)\frac{\partial}{\partial x^j}, \frac{\partial}{\partial t}) .$$

We deduce from expression (IV.14) that

$$\widetilde{Ric}(\frac{\partial}{\partial t}, X) = g^{ij}(\nabla_{\partial/\partial x^i} S)(X, \frac{\partial}{\partial x^j}) - g^{ij}(\nabla_X S)(\frac{\partial}{\partial x^i}, \frac{\partial}{\partial x^j}) .$$

Next, we can write the last term as

$$-g^{ij}(\nabla_X S)_{ij} = -X(g^{ij} S_{ij}) + S_{ij}(\nabla_X g^{ij}) ,$$

but the last term equals zero since the covariant derivative of the metric tensor with respect to any vector field equals zero. Thus $-g^{ij}(\nabla_X S)(\frac{\partial}{\partial x^i}, \frac{\partial}{\partial x^j}) = -X(\operatorname{tr}_g S)$. Hence we finally obtain

$$\widetilde{Ric}(\frac{\partial}{\partial t}, X) = g^{ij}(\nabla_{\partial/\partial x^i} S)(X, \frac{\partial}{\partial x^j}) - X(\operatorname{tr}_g S) . \tag{IV.18}$$

Let us now calculate $\widetilde{Ric}(\frac{\partial}{\partial t}, \frac{\partial}{\partial t})$. We have

$$\widetilde{Ric}(\frac{\partial}{\partial t}, \frac{\partial}{\partial t}) = g^{ij}\tilde{g}(\tilde{R}(\frac{\partial}{\partial x^i}, \frac{\partial}{\partial t})\frac{\partial}{\partial t}, \frac{\partial}{\partial x^j}) .$$

By formula (IV.15), we obtain

$$\begin{aligned} \widetilde{Ric}(\frac{\partial}{\partial t}, \frac{\partial}{\partial t}) &= -g^{ij}\tilde{g}(\tilde{\nabla}_{\frac{\partial}{\partial t}} P(\frac{\partial}{\partial x^i}), \frac{\partial}{\partial x^j}) \\ &= -g^{ij}\frac{\partial}{\partial t}\tilde{g}(P(\frac{\partial}{\partial x^i}), \frac{\partial}{\partial x^j}) + g^{ij}\tilde{g}(P(\frac{\partial}{\partial x^i}), \tilde{\nabla}_{\frac{\partial}{\partial t}}\frac{\partial}{\partial x^j}) . \end{aligned}$$

Further, we can write the last term as

$$g^{ij} P_i^k P_j^s g_{ks} = g^{ij} S_{is} P_j^s = g^{ij} g^{sr} S_{is} S_{jr} .$$

At any point of a hypersurface M_t this is exactly the inner product of the 2-tensor S with itself. Denote this inner product by $S.S$. With this notation, we obtain

$$\widetilde{Ric}(\frac{\partial}{\partial t}, \frac{\partial}{\partial t}) = -g^{ij} \frac{\partial S_{ij}}{\partial t} + S.S . \tag{IV.19}$$

Before discussing the Cauchy problem, we still need another identity coming from the Lorentzian geometry related to the family of hypersurfaces $\{M_t\}_t$ of V originated by the diffeomorphism γ. Assume that X and Y are two vector fields of M and consider them as extended to vector fields at a neighborhood of M in V. We have

$$\frac{\partial}{\partial t} g(X,Y) = \tilde{g}(\tilde{\nabla}_{\partial/\partial t} X, Y) + \tilde{g}(X, \tilde{\nabla}_{\partial/\partial t} Y) .$$

Since X and $\frac{\partial}{\partial t}$ commute, $\tilde{g}(\tilde{\nabla}_{\partial/\partial t} X, Y) = g(P(X), Y) = S(X,Y)$. Therefore the latter identity can be rewritten to the form

$$\frac{\partial}{\partial t} g(X,Y) = 2S(X,Y) .$$

Using the sub-index t of the metrics g_t and of the second fundamental forms S_t of the hypersurfaces M_t, this identity takes the form

$$\frac{\partial g_t}{\partial t} = 2S_t . \tag{IV.20}$$

Let us now discuss the Cauchy problem for the equation $\mathcal{E}_1 = 0$. Recall that we supposed at the beginning of this section that we deal with a 4-dimensional differentiable manifold V and that M is a hypersurface of V admitting a differentiable vector field N of V with support on M, transversal to M. For fixed $f \in \Gamma(F)$, we are interested in solutions $\tilde{g}$ of the equation $\mathcal{E}_1(\tilde{g}, f) = 0$ such that $\tilde{g}$ is a Lorentzian metric for which we have $\tilde{g}_x(N_x, N_x) = -1$ at any point $x \in M$, $T_x(M)$ is orthogonal to N_x, and that moreover some initial conditions hold on M (we have already discussed some of these conditions) that we are now going to talk about. Every solution $\tilde{g}$ of the equation that satisfies the above condition (to be a Lorentzian metric making N normal to M with $\tilde{g}(N,N) = -1$) will determine a diffeomorphism γ between an open set of the product $M \times \mathbb{R}$ and a neighborhood of the hypersurface M in V as we have already described, determining the family of hypersurfaces $\{M_t\}_t$ and that of Riemannian metrics $\{g_t\}_t$ obtained by restriction of $\tilde{g}$ to M_t. Then the derivative $\frac{\partial g_t}{\partial t}$ for $t = 0$ will be the derivative of g in the normal direction N. We should observe, however, that the concept of derivative of g in the direction of N depends on the solution $\tilde{g}$ (since both the family $\{M_t\}$ and the family $\{g_t\}$ depend on this solution). By identity (IV.20), this normal derivative is just 2 times the second fundamental form S of M with respect to the metric $\tilde{g}$.

In all bibliography related to the topic, the quadratic form $2S$ is denoted by k. Since the equation we have to solve is of second order, giving in advance on M the metric g and the derivative of g in the direction of N is equivalent to giving the pair of quadratic forms (g, k) as Cauchy data on M. We know (by virtue of the previous section) that the pair of Cauchy data (g, k) on M cannot be arbitrary, but has to satisfy certain constraint equations that we are now going to obtain.

The equation $\mathcal{E}_1(\tilde{g}, f) = 0$ is written as follows:

$$\widetilde{Ric}(\tilde{g}) = \chi \left\{ T(\tilde{g}, f) - \frac{1}{2}(\mathrm{tr}_{\tilde{g}} T(\tilde{g}, f))\tilde{g} \right\} . \tag{IV.21}$$

Let X be a vector field on M and let us extend it to a neighborhood of M in V in the way we have already described. Apply both sides of (IV.21) to the pair $(X, \frac{\partial}{\partial t})$. Since $\tilde{g}(X, \frac{\partial}{\partial t}) = 0$, we have

$$\widetilde{Ric}(X, \frac{\partial}{\partial t}) = \chi T(X, \frac{\partial}{\partial t}) .$$

By formula (IV.18), we now obtain

$$\frac{1}{2} \left\{ g^{ij} (\nabla_{\frac{\partial}{\partial x^i}} k)(X, \frac{\partial}{\partial x^j}) - X(\mathrm{tr}_g k) \right\} = \chi T(X, \frac{\partial}{\partial t}) . \tag{IV.22}$$

The stress-energy tensor T appearing in this equation should be *admissible* in the sense of the definition given in Section IV.4. Since the metric $\tilde{g}$ is determined at points $x \in M$ by the knowledge of the restriction g of $\tilde{g}$ to M (since $\tilde{g}_x(N_x, N_x) = -1$), this implies that though T may in general depend on $\tilde{g}$ (and on the given f), the tensor T at points belonging to M depend on g (and on the given f) only. That is, the restriction of T to M depends on the *Cauchy data* g only. Therefore equation (IV.22) on M gives a constraint condition between the Cauchy data g and k. Denote by $\gamma(g, k)$ the 1-form on M assigning to every vector field X,

$$\gamma(g, k)(X) = \frac{1}{2} \left\{ g^{ij} (\nabla_{\frac{\partial}{\partial x^i}} k)(X, \frac{\partial}{\partial x^j}) - X(\mathrm{tr}_g k) \right\} . \tag{IV.23}$$

Since the right-hand side of (IV.22) can be written at point $x \in M$ as $\chi(i(N)T)(X)$, where $i(N)$ denotes the interior contraction of T by the transversal vector N, the constraint equation (IV.22) will be briefly written as

$$\gamma(g, k) = \chi \, i(N) T(g, f) .$$

This constraint equation is an equation relating 1-forms on a 3-dimensional manifold. Therefore it is equivalent to three scalar equations. Since we have seen at the end of Section IV.5 that there should be four equations, we miss one. By (IV.21), we have (at points $x \in M$):

$$\widetilde{Ric}(N, N) = \chi (T(N, N) + \frac{1}{2}(\mathrm{tr}_{\tilde{g}} T)) .$$

But, we have seen in Chapter II that $\chi \mathrm{tr}_{\tilde{g}} T = -\tilde{R}$ (recall that $\tilde{R}$ is the scalar curvature of the metric $\tilde{g}$). Therefore we have at points belonging to M,

$$\widetilde{Ric}(N,N) = \chi T(N,N) - \frac{\tilde{R}}{2} \,. \tag{IV.24}$$

On the other hand, by definition of the scalar curvature, we have in an adapted coordinate system

$$\tilde{R} = g^{ij}\tilde{R}_{ij} - \tilde{R}_{00} \,.$$

By (IV.17) and the last equality,

$$\tilde{R} + \tilde{R}_{00} = g^{ij}\tilde{R}_{ij} = g^{ij}\frac{\partial S_{ij}}{\partial t} - 2S.S + R + (\mathrm{tr}_g S)^2 \,.$$

Therefore

$$g^{ij}\frac{\partial S_{ij}}{\partial t} = \tilde{R} + \tilde{R}_{00} + 2S.S - R - (\mathrm{tr}_g S)^2 \,.$$

By (IV.19) and by the last equality, we now obtain

$$\tilde{R}_{00} = \frac{1}{2}(-\tilde{R} - S.S + R + (\mathrm{tr}_g S)^2) \,.$$

By equating the right-hand sides of the last equality and of (IV.24) at points $x \in M$, we obtain

$$\frac{1}{2}(-S.S + R + (\mathrm{tr}_g S)^2) = \chi T(N,N)$$

which is the required constraint equation we were missing. From now on, we set

$$\mathcal{H}(g,k) = \frac{1}{8}(k.k - 4R - (\mathrm{tr}_g k)^2) \,, \tag{IV.25}$$

and write the last constraint equation as

$$\mathcal{H}(g,k) = -\chi\, i(N)T(g,f)(N) \,.$$

IV.7 Hyperbolic reduction

We continue using the notation introduced in Section IV.5. For fixed $f \in \Gamma(F)$, consider the operator $\tilde{g} \to \mathcal{E}_1(\tilde{g}, f)$. Since we assume f to be fixed, we will sometimes write $\mathcal{E}_1(\tilde{g})$ instead of $\mathcal{E}_1(\tilde{g}, f)$.

In a local chart, the terms containing second-order derivatives of $\mathcal{E}_1$ come from the components of the Ricci tensor and are given by expression (IV.12). In principle, this expression does not imply that the operator $\mathcal{E}_1$ is hyperbolic. We are going now to introduce appropriate changes into the coordinate-wise expression of $\mathcal{E}_1$ in order to be able to apply the theory of hyperbolic operators on $\mathbb{R}^4$ to this expression.

Let us use the concept of a harmonic map introduced in Section IV.3.

Definition IV.6. In our manifold V, we fix a reference Lorentzian metric $\hat{g}$. Any other Lorentzian metric $\tilde{g}$ will be called harmonic with respect to $\hat{g}$ if the identity

$$I: \quad (V, \tilde{g}) \quad \longrightarrow \quad (V, \hat{g})$$

is harmonic, which is equivalent to saying, with the notation of Section IV.3, that the vector field H whose components are $H^\mu = \tilde{g}^{\lambda\nu}\tilde{\Gamma}^\mu_{\lambda\nu} - \tilde{g}^{\lambda\nu}\hat{\Gamma}^\mu_{\lambda\nu}$ vanishes.

For notational simplicity denote by $\tilde{\Gamma}^\mu$ the term $\tilde{g}^{\lambda\nu}\tilde{\Gamma}^\mu_{\lambda\nu}$ appearing in the expression of H^μ. Let us now make the derivatives of $\tilde{\Gamma}^\mu$ to appear in expression (IV.12) of terms containing second-order derivatives of the Ricci tensor. We will use the sign $\cong$ to indicate equality of terms containing second-order derivatives of $\tilde{g}$ in both expressions. We have

$$\begin{aligned}
\tilde{g}_{\alpha\mu}\partial_\beta\tilde{\Gamma}^\mu &= \tilde{g}_{\alpha\mu}\partial_\beta\{\tilde{g}^{\lambda\nu}\tilde{\Gamma}^\mu_{\lambda\nu}\} \cong \tilde{g}_{\alpha\mu}\tilde{g}^{\lambda\nu}\partial_\beta\tilde{\Gamma}^\mu_{\lambda\nu} \\
&\cong \frac{1}{2}\tilde{g}_{\alpha\mu}\tilde{g}^{\lambda\nu}\tilde{g}^{\mu\gamma}(\partial_\beta\partial_\lambda\tilde{g}_{\nu\gamma} + \partial_\beta\partial_\nu\tilde{g}_{\lambda\gamma} - \partial_\beta\partial_\gamma\tilde{g}_{\lambda\nu}) \\
&= \frac{1}{2}\tilde{g}^{\lambda\nu}(\partial_\beta\partial_\lambda\tilde{g}_{\nu\alpha} + \partial_\beta\partial_\nu\tilde{g}_{\lambda\alpha} - \partial_\beta\partial_\alpha\tilde{g}_{\lambda\nu}) \,.
\end{aligned}$$

Now, the first and the second term on the right-hand side of this equation are equal. Therefore

$$\tilde{g}_{\alpha\mu}\partial_\beta\tilde{\Gamma}^\mu \cong \frac{1}{2}\tilde{g}^{\lambda\nu}(2\partial_\beta\partial_\lambda\tilde{g}_{\nu\alpha} - \partial_\beta\partial_\alpha\tilde{g}_{\lambda\nu}) \,.$$

Then, in second order, the combination $\tilde{g}_{\alpha\mu}\partial_\beta\tilde{\Gamma}^\mu + \tilde{g}_{\beta\mu}\partial_\alpha\tilde{\Gamma}^\mu$ is written as

$$\tilde{g}_{\beta\mu}\partial_\alpha\tilde{\Gamma}^\mu + \tilde{g}_{\alpha\mu}\partial_\beta\tilde{\Gamma}^\mu \cong \tilde{g}^{\mu\nu}(\partial_\alpha\partial_\mu\tilde{g}_{\beta\nu} + \partial_\beta\partial_\mu\tilde{g}_{\alpha\nu} - \partial_\alpha\partial_\beta\tilde{g}_{\mu\nu}) \,.$$

If we compare this expression to (IV.12), we see that the terms of the Ricci tensor containing second-order derivatives can be written as

$$\tilde{R}_{\alpha\beta} \cong \frac{1}{2}\{\tilde{g}_{\alpha\mu}\partial_\beta\tilde{\Gamma}^\mu + \tilde{g}_{\beta\mu}\partial_\alpha\tilde{\Gamma}^\mu\} - \frac{1}{2}\tilde{g}^{\mu\nu}\partial_\mu\partial_\nu\tilde{g}_{\alpha\beta} \,.$$

Substitute into this equation $\tilde{\Gamma}^\mu$ as a function of the field H. If we set $Y^\mu = \tilde{g}^{\lambda\nu}\hat{\Gamma}^\mu_{\lambda\nu}$, then $\tilde{\Gamma}^\mu = H^\mu + Y^\mu$ and

$$\begin{aligned}
\tilde{R}_{\alpha\beta} \cong &-\frac{1}{2}\tilde{g}^{\mu\nu}\partial_\mu\partial_\nu\tilde{g}_{\alpha\beta} + \frac{1}{2}\{\tilde{g}_{\alpha\mu}\partial_\beta Y^\mu + \tilde{g}_{\beta\mu}\partial_\alpha Y^\mu\} \\
&+ \frac{1}{2}\{\tilde{g}_{\alpha\mu}\partial_\beta H^\mu + \tilde{g}_{\beta\mu}\partial_\alpha H^\mu\} \,.
\end{aligned}$$

In this expression, the terms in Y^μ contain no second derivatives of $\tilde{g}$ (since the Christoffel symbols $\hat{\Gamma}^\mu_{\lambda\nu}$ contain derivatives of the reference metric $\hat{g}$ only). Therefore we can write

$$\tilde{R}_{\alpha\beta} \cong -\frac{1}{2}\tilde{g}^{\mu\nu}\partial_\mu\partial_\nu\tilde{g}_{\alpha\beta} + \frac{1}{2}\{\tilde{g}_{\alpha\mu}\partial_\beta H^\mu + \tilde{g}_{\beta\mu}\partial_\alpha H^\mu\} \,. \tag{IV.26}$$

In the case where H is zero (this would occur if $\tilde{g}$ were harmonic with respect to the reference metric $\hat{g}$), the terms containing second-order derivatives of $\tilde{g}$ in the latter expression would be $-\frac{1}{2}\tilde{g}^{\mu\nu}\partial_\mu\partial_\nu\tilde{g}_{\alpha\beta}$, and the operator $\mathcal{E}_1$ would be *hyperbolic* (since the principal part of this operator would coincide with the classical wave operator).

For $f \in \Gamma(F)$ fixed, define an operator $\mathcal{E}'$ associating to each $\tilde{g}$ the following:

$$\mathcal{E}'(\tilde{g}) = \mathcal{E}_1(\tilde{g}) - \frac{1}{2}L_H\tilde{g} = \widetilde{Ric}(\tilde{g}) - \frac{1}{2}L_H\tilde{g} - \chi\left\{T(\tilde{g}, f) - \frac{1}{2}(\mathrm{tr}_{\tilde{g}}T(\tilde{g}, f))\tilde{g}\right\},$$

where $L_H\tilde{g}$ denotes, as usual, the Lie derivative of the metric $\tilde{g}$ with respect to the vector field H (H also depends on $\tilde{g}$). By Corollary I.8, the coordinate-wise expression of $L_H\tilde{g}$ is

$$(L_H\tilde{g})_{\alpha\beta} = \tilde{g}_{\alpha\mu}\tilde{\nabla}_\beta H^\mu + \tilde{g}_{\beta\mu}\tilde{\nabla}_\alpha H^\mu \cong \tilde{g}_{\alpha\mu}\partial_\beta H^\mu + \tilde{g}_{\beta\mu}\partial_\alpha H^\mu .$$

By formula (IV.26), the principal part of the operator $\mathcal{E}'$ is $-\frac{1}{2}\tilde{g}^{\mu\nu}\partial_\mu\partial_\nu\tilde{g}_{\alpha\beta}$. This shows that the operator $\mathcal{E}'$ is globally defined and *hyperbolic* in each local chart. On the other hand, if the metric $\tilde{g}$ is harmonic with respect to the reference metric, then the operator $\mathcal{E}'$ coincides with $\mathcal{E}_1$.

IV.8 Fundamental theorem

We are now ready to prove the fundamental theorem of this chapter.

Theorem IV.7. *Let V be a manifold of dimension 4 and $F \to V$ a vector bundle of rank 4 on V. Consider an open set W in the fiber bundle $E = S_2(V) \oplus F$ (where $S_2(V)$ is the bundle on V of covariant symmetric 2-tensors) satisfying the assumptions described in Section* IV.4. *Let T be an admissible class of stress-energy tensors (for the definition of an admissible class, see Section* IV.4*),*

$$\begin{array}{rccc} T: & \Gamma(W) & \longrightarrow & \Gamma(S_2(V)) \\ & (\tilde{g}, f) & \rightarrow & T(\tilde{g}, f) \end{array} .$$

Assume that a hypersurface M of V is given, admitting a vector field N of V with support on M, transversal to M. The transversal field N is supposed to be given. Let $s > 3$ be an integer. Assume that we have on M:

- *A pair of quadratic forms (g, k), with $g \in \mathcal{W}^{s+1}_{loc}(M)_2$ and $k \in \mathcal{W}^s_{loc}(M)_2$ such that g and k are symmetric and g is positive definite (the spaces $\mathcal{W}^s_{loc}(M)$ are the local Sobolev spaces introduced in Section* I.4*).*
- *A section ω of the fiber bundle $F|_M$ (restriction of F to M) belonging to the Sobolev space $\mathcal{W}^{s-1}_{loc}(\Gamma(F|_M))$ (Sobolev spaces for sections of fiber bundles are similar to those of p-tensors introduced in Section* I.4*).*

It is supposed that the admissible class of stress-energy tensors, T, fulfils the following technical assumption:

For any Lorentzian metric $\tilde{g}$ on V satisfying at any point $x \in M$ the condition $\tilde{g}_x(N_x, N_x) = -1$ and the condition that N_x is perpendicular to $T_x(M)$, equation $div_{\tilde{g}} T(\tilde{g}, f) = 0$ can be rewritten in a certain neighborhood of each point of M, being the domain of a local chart, as a symmetric, hyperbolic system of order 1 with respect to f, for which the hypersurface M is nowhere characteristic.

It is assumed that the data $\{g, k, \omega\}$ on M satisfy the constraint equations

$$\begin{cases} \gamma(g,k) = \chi\, i(N)T(g,\omega) \\ \mathcal{H}(g,k) = -\chi\, i(N)T(g,\omega)(N) \ , \end{cases}$$

where $\gamma(g,k)$ is defined in (IV.23) *and $\mathcal{H}(g,k)$ is defined in* (IV.25).

Then at a neighborhood $\mathcal{P}$ of M in V, there exist a Lorentzian metric $\tilde{g} \in \mathcal{W}^{s+1}_{loc}(\mathcal{P})_2$ and a section f of the vector bundle F on $\mathcal{P}$, with $f \in \mathcal{W}^{s-1}_{loc}(\Gamma(\mathcal{P}, F))$, that are solutions of Einstein's equation $\mathcal{E}(\tilde{g}, f) = 0$, that is

$$\begin{cases} \widetilde{Ric}(\tilde{g}) - \chi \left\{ T(\tilde{g}, f) - \frac{1}{2}(tr_{\tilde{g}} T(\tilde{g}, f))\tilde{g} \right\} = 0 \\ div_{\tilde{g}} T(\tilde{g}, f) = 0 \ , \end{cases}$$

and such that the following conditions are satisfied:

- *At each point $x \in M$, we have $\tilde{g}(N_x, N_x) = -1$ and N_x is perpendicular to $T_x(M)$ with respect to $\tilde{g}$.*
- *The restriction of $\tilde{g}$ to M is g. Two times the second fundamental form of the hypersurface M of V with respect to $\tilde{g}$ is k. At points $x \in M$ we have $f(x) = \omega(x)$.*

The solution $(\tilde{g}, f)$ is unique up to diffeomorphisms defined on a certain neighborhood of M in V, equal to the identity on M and preserving the vector field N.

Remark IV.8. Since the significance of $T(g,\omega)$ on the right-hand sides of the constraint equations relating the Cauchy data is not evident, we now explain it: Given a Riemannian metric g on M, let $\tilde{g}_x$ be, for each $x \in M$, the Lorentzian metric on the vector space $T_x(V)$ defined by saying that it equals g_x on $T_x(M)$, that $\tilde{g}_x(N_x, N_x) = -1$, and N_x is perpendicular to $T_x(M)$. Since, by the definition of an admissible class of stress-energy tensors, the value of $T(\tilde{g}, f)$ at each point x depends on $\tilde{g}_x$ and f_x only, for this $\tilde{g}$ defined at points of M the value of $T(\tilde{g}, f)$ will only depend on g and ω. This is what $T(g,\omega)$ means.

Proof. It follows from the well-known *Tubular Neighborhood Theorem* that there exists a diffeomorphism φ of a neighborhood $\tilde{U}'$ of $\{0\} \times M$ in $\mathbb{R} \times M$ on a neighborhood $\tilde{U}$ of M in V such that $\varphi(0,x) = x$ for all $x \in M$,

$$\begin{array}{cccc} & \mathbb{R}\times M & & V \\ & \cup & & \cup \\ \varphi: & \tilde{U}' & \longrightarrow & \tilde{U} \end{array} .$$

Let us now give a short and independent proof of this fact. First, observe that N (defined at points belonging to the hypersurface M only) admits an extension $\tilde{N}$ to a neighborhood $\tilde{W}$ of M in V. Indeed, for any $p \in M$ let $(\tilde{U}_p, x^0 \dots x^3)$ be a local chart in a neighborhood of p in V such that $M \cap \tilde{U}_p$ be given by $x^0 = 0$. Let $\tilde{W} = \cup_{p\in M}\tilde{U}_p$. Consider the covering $\{\tilde{U}_p\}$ of $\tilde{W}$. Let $\{\tilde{U}_i\}_{i\in I}$ be a locally finite refinement of $\{\tilde{U}_p\}_{p\in M}$ and $\{\nu_i\}$ a partition of unity subordinated to the above covering. On each $\tilde{U}_i$ (which is the domain of a local chart), choose an extension $\tilde{N}_i$ of N. Define on $\tilde{W}$ the vector field $\tilde{N} = \sum_{i\in I}\nu_i\tilde{N}_i$. This gives an extension $\tilde{N}$ of N. By the theorems on existence and uniqueness of ordinary differential equations, for each $x \in M$ there exists an interval $(-\varepsilon_x, \varepsilon_x)$ and a differentiable curve $t \to \gamma_x(t)$,

$$\gamma_x : (-\varepsilon_x, \varepsilon_x) \longrightarrow \tilde{W} ,$$

such that $t \to \gamma_x(t)$ is an integral curve of $\tilde{N}$ with the initial condition $\gamma_x(0) = x$. In the product manifold $\mathbb{R}\times M$, let $\tilde{U}'$ be the open set consisting of the pairs (t,x) with $x \in M$ and $t \in (-\varepsilon_x, \varepsilon_x)$. The map

$$\begin{array}{ccc} \tilde{U}' & \xrightarrow{\varphi} & \tilde{W} \\ (t,x) & \to & \gamma_x(t) \end{array}$$

gives a diffeomorphism of $\tilde{U}'$ on a neighborhood $\tilde{U}$ of M in V contained in $\tilde{W}$. This is the required diffeomorphism.

Fix now an arbitrary Riemannian metric a on the manifold M and consider the Lorentzian metric $\hat{g}$ on $\mathbb{R}\times M$ given by $\hat{g} = -dt^2 + a$. This metric generates, by means of the above diffeomorphism φ, a Lorentzian metric on a neighborhood $\tilde{U}$ of M in V which will still be denoted by $\hat{g}$ and called the *reference metric*. For each Lorentzian metric $\tilde{g}$ on the neighborhood $\tilde{U}$, denote by H the vector field defined in every local chart by $H^\mu = \tilde{\Gamma}^\mu - \tilde{g}^{\lambda\nu}\hat{\Gamma}^\mu_{\lambda\nu}$ (here, we use the notation introduced in Section IV.7). By the results in Section IV.7, we know that this field is globally defined on $\tilde{U}$. It is clear that this field depends on $\tilde{g}$ (and on the reference metric which is fixed once and forever).

Consider now the operator $\mathcal{E}'$ defined in Section IV.7:

$$\mathcal{E}'(\tilde{g}, f) = \widetilde{Ric}(\tilde{g}) - \frac{1}{2}L_H\tilde{g} - \chi\left\{T(\tilde{g}, f) - \frac{1}{2}(\mathrm{tr}_{\tilde{g}}T(\tilde{g}, f))\tilde{g}\right\} .$$

Instead of solving the system

$$\begin{cases} \mathcal{E}_1(\tilde{g}, f) = 0 \\ \mathrm{div}_{\tilde{g}}T(\tilde{g}, f) = 0 \end{cases} \tag{IV.27}$$

mentioned in the theorem, let us first solve the system

$$\begin{cases} \mathcal{E}'(\tilde{g}, f) = 0 \\ \mathrm{div}_{\tilde{g}} T(\tilde{g}, f) = 0 \end{cases} \tag{IV.28}$$

under appropriate initial conditions. The reason for doing so is that, as we have already seen in Section IV.7, in any local chart this second system will be written as a hyperbolic system while the first one cannot be written in this way. By virtue of what we have said in Section IV.7, the latter system is written in any local chart as follows:

$$\begin{cases} -\frac{1}{2}\tilde{g}^{\lambda\mu}\partial_\lambda\partial_\mu\tilde{g}_{\alpha\beta} + F_{\alpha\beta}(\tilde{g}, \partial\tilde{g}, f) = 0 \\ \mathrm{div}_{\tilde{g}} T(\tilde{g}, f) = 0 \ , \end{cases}$$

where the $F_{\alpha\beta}$ are functions depending on $\tilde{g}$, the first derivatives of $\tilde{g}$ and f. The unknowns in this system are $\tilde{g}_{\alpha\beta}$ (they are 10) and the four components of f. The system contains 14 equations. The first 10 equations are of order 2, while the last four equations are of order 1. We can transform it to a first-order system by introducing some auxiliary variables. In a coordinate system (t, x^1, x^2, x^3) adapted to the diffeomorphism φ, take $\partial_t\tilde{g}_{\alpha\beta} = \tilde{s}_{\alpha\beta}$ and $\partial_i\tilde{g}_{\alpha\beta} = \tilde{r}_{\alpha\beta i}$. The first-order system we obtain in this way (containing 54 equations with 54 unknowns) has the following form:

$$\begin{cases} \partial_t\tilde{g}_{\alpha\beta} - \tilde{s}_{\alpha\beta} = 0 \\ \partial_i\tilde{g}_{\alpha\beta} - \tilde{r}_{\alpha\beta i} = 0 \\ -\tilde{g}^{00}\partial_t\tilde{s}_{\alpha\beta} - \tilde{g}^{ij}\partial_i\tilde{r}_{\alpha\beta j} - 2\tilde{g}^{0i}\partial_i\tilde{s}_{\alpha\beta} + 2F_{\alpha\beta} = 0 \\ \mathrm{div}_{\tilde{g}} T(\tilde{g}, f) = 0 \ . \end{cases} \tag{IV.29}$$

Since, by definition of an *admissible class of stress-energy tensors*, the expression of $T(\tilde{g}, f)$ does not contain derivatives either of $\tilde{g}$ or f, we see that neither of the first 50 equations contains any derivative of f, while the last four equations contain no derivative of $\tilde{g}$ (because $\tilde{\nabla}\tilde{g} = 0$). Therefore this system can be split into two independent parts (an uncoupled system). Let us check that this system (on $\mathbb{R}^4$) satisfies all assumptions of Theorem IV.1. The first thing we should see is that it is quasilinear. The concept of a quasilinear system is related to a solution of the system. Consider any solution of the system $(\tilde{g}_{\alpha\beta}, \tilde{s}_{\alpha\beta}, \tilde{r}_{\alpha\beta i}, f)$ such that in our local chart, $\tilde{g}_{\alpha\beta}$ would define a Lorentzian metric $\tilde{g}$ such that $\tilde{g}(N_x, N_x) = -1$ would hold at points $x \in M$ and N_x would be perpendicular to $T_x(M)$ with respect to $\tilde{g}$. Let us see that the system is quasilinear for these solutions in a neighborhood of the hypersurface M (given by the equation $t = 0$). For this purpose, we should only look at the terms containing first derivatives. The part of the system that corresponds to the four equations $\mathrm{div}_{\tilde{g}} T(\tilde{g}, f) = 0$, which is the only part where derivatives of the variables f appear, is already quasilinear by assumption. Therefore we should focus on the first 50 equations (containing no derivatives of other variables). Since $\tilde{r}_{\alpha\beta i} = \partial_i\tilde{g}_{\alpha\beta}$, we have $\partial_t\tilde{r}_{\alpha\beta i} = \partial_t\partial_i\tilde{g}_{\alpha\beta} = \partial_i\partial_t\tilde{g}_{\alpha\beta} = \partial_i\tilde{s}_{\alpha\beta}$.

Hence $\tilde{g}^{ji}\partial_t \tilde{r}_{\alpha\beta i} = \tilde{g}^{ji}\partial_i \tilde{s}_{\alpha\beta}$. Since the matrix $(\tilde{g}^{ij})$ is non-singular, the first 50 equations in (IV.29) can be written in the following equivalent form:

$$\begin{cases} \partial_t \tilde{g}_{\alpha\beta} - \tilde{s}_{\alpha\beta} = 0 \\ \tilde{g}^{ji}\partial_t \tilde{r}_{\alpha\beta i} - \tilde{g}^{ji}\partial_i \tilde{s}_{\alpha\beta} = 0 \\ -\tilde{g}^{00}\partial_t \tilde{s}_{\alpha\beta} - \tilde{g}^{ij}\partial_i \tilde{r}_{\alpha\beta j} - 2\tilde{g}^{0i}\partial_i \tilde{s}_{\alpha\beta} + 2F_{\alpha\beta} = 0 \ . \end{cases} \tag{IV.30}$$

Consider the following matrices:

$$A^0 = \begin{pmatrix} I_{10} & 0_{10} & 0_{10} & 0_{10} & 0_{10} \\ 0_{10} & \tilde{g}^{11}I_{10} & \tilde{g}^{12}I_{10} & \tilde{g}^{13}I_{10} & 0_{10} \\ 0_{10} & \tilde{g}^{21}I_{10} & \tilde{g}^{22}I_{10} & \tilde{g}^{23}I_{10} & 0_{10} \\ 0_{10} & \tilde{g}^{31}I_{10} & \tilde{g}^{32}I_{10} & \tilde{g}^{33}I_{10} & 0_{10} \\ 0_{10} & 0_{10} & 0_{10} & 0_{10} & -\tilde{g}^{00}I_{10} \end{pmatrix},$$

$$A^i = \begin{pmatrix} 0_{10} & 0_{10} & 0_{10} & 0_{10} & 0_{10} \\ 0_{10} & 0_{10} & 0_{10} & 0_{10} & -\tilde{g}^{1i}I_{10} \\ 0_{10} & 0_{10} & 0_{10} & 0_{10} & -\tilde{g}^{2i}I_{10} \\ 0_{10} & 0_{10} & 0_{10} & 0_{10} & -\tilde{g}^{3i}I_{10} \\ 0_{10} & -\tilde{g}^{i1}I_{10} & -\tilde{g}^{i2}I_{10} & -\tilde{g}^{i3}I_{10} & -2\tilde{g}^{0i}I_{10} \end{pmatrix},$$

$$h = \begin{pmatrix} \tilde{g}_{\alpha\beta} \\ \tilde{r}_{\alpha\beta j} \\ \tilde{s}_{\alpha\beta} \end{pmatrix} \quad , \quad A = \begin{pmatrix} \tilde{s}_{\alpha\beta} \\ 0_{30} \\ -2F_{\alpha\beta} \end{pmatrix}$$

where I_n and 0_n stand for the identity matrix and the zero matrix of order n, respectively. With this notation, system (IV.30) can be written using matrices as follows:

$$A^0\partial_t h + A^i\partial_i h + P = 0 \ ,$$

where P is some matrix containing no derivatives. This shows that system (IV.30) is quasilinear for the solutions that contain the $\tilde{g}$ we have chosen. System (IV.28), we would like to solve, can be written, in any adapted local chart, in the following equivalent form:

$$\begin{cases} A^0\partial_t h + A^i\partial_i h + P = 0 \\ \operatorname{div}_{\tilde{g}} T(\tilde{g}, f) = 0 \ . \end{cases} \tag{IV.31}$$

The matrix A^0 and the matrices A^i are symmetric. Let us check that the matrix A^0 is positive definite. Indeed, observe that the elements of this matrix are given by

$$a^{10i+k,10j+l} = \begin{cases} \delta^{kl} & i = j = 0 \\ \tilde{g}^{ij}\delta^{kl} & i, j = 1, 2, 3 \\ -\tilde{g}^{00}\delta^{kl} & i = j = 4 \end{cases} \qquad k, l = 1, \ldots, 10.$$

Then the inner product $(v, A^0 v)$ is written as follows:

$$\begin{aligned}(v, A^0 v) &= v_\alpha a^{\alpha\beta} v_\beta \\ &= \sum_{k,l=1}^{10} v_k \delta^{kl} v_l + \sum_{i,j=1}^{3} \sum_{k,l=1}^{10} v_{10i+k} \tilde{g}^{ij} \delta^{kl} v_{10j+l} - \sum_{k,l=1}^{10} v_k \tilde{g}^{00} \delta^{kl} v_l \\ &= \sum_{k=1}^{10} v_k^2 + \sum_{i,j=1}^{3} \sum_{k=1}^{10} v_{10i+k} \tilde{g}^{ij} v_{10j+k} - \tilde{g}^{00} \sum_{k=1}^{10} v_k^2 .\end{aligned}$$

This expression is positive if $v \neq 0$. Since, on the other hand, by assumptions of the theorem, the equation $\mathrm{div}_{\tilde{g}} T(\tilde{g}, f) = 0$ is a first-order hyperbolic system with respect to f, quasilinear and symmetric, system (IV.31) satisfies the majority of assumptions in Theorem IV.1.

In order for us to apply Theorem IV.1 to this system (IV.31), we should still check that the hypersurface M is nowhere characteristic for any solution containing some Lorentzian metric $\tilde{g}$ such that $\tilde{g}(N_x, N_x) = -1$ holds at points $x \in M$ and that N_x is perpendicular to $T_x(M)$ with respect to $\tilde{g}$. The variables of the system (IV.31) are h (appearing in the first part and containing $\tilde{g}_{\alpha\beta}$, $\tilde{r}_{\alpha\beta j}$ and $\tilde{s}_{\alpha\beta}$), $\tilde{g}$ and f (appearing in the second part), although the second part contains derivatives of f only (an uncoupled system). Since M is nowhere characteristic for the second part of the system by assumption and, moreover, the system is uncoupled, it remains to check that M is nowhere characteristic for the first part. Let ξ be a 1-form of V with support on M defined at each $x \in M$ by $\xi_x(N_x) = 1$ and $\xi_x(v) = 0$ for all $v \in T_x(M)$. In any coordinate system (t, x^1, x^2, x^3) that is adapted to the diffeomorphism φ, we have $\xi_0 = 1$ and $\xi_i = 0$. The leading symbol of the second part of (IV.31) associated to ξ is the endomorphism whose matrix is $A^0\xi_0 + A^i\xi_i - A^0$. Since this matrix is positive definite, its determinant will in particular be $\neq 0$. Therefore M is nowhere characteristic.

Let us specify with what Cauchy data on M the system (IV.28) will be solved in an adapted coordinate system. It is clear that we should impose the condition that the restriction of f to M be ω, the restriction of $\tilde{g}$ to tangent vectors of M be g, that $\tilde{g}_{00}$ be -1 on M, and that $\tilde{g}_{0i}$ be 0 on M. We will also impose the condition that two times the second fundamental form of M with respect to $\tilde{g}$ be k. The latter, by (IV.20), is written as $\partial_t \tilde{g}_{ij} = k_{ij}$. We have not yet imposed what values should be taken by the four normal derivatives $\partial_t \tilde{g}_{0\alpha}$ in an adapted coordinate system. Denote these values by A_α and let us specify them later on. That is, $A_\alpha = \partial_t \tilde{g}_{0\alpha}$ on M.

All these initial conditions on M, or Cauchy data, have an appropriate translation in terms of variables of first-order system (IV.31) which are formalized in the following conditions on M: $f = \omega$, $\tilde{g}_{00} = -1$, $\tilde{g}_{0i} = 0$, $\tilde{g}_{ij} = g_{ij}$, $\tilde{s}_{ij} = k_{ij}$, $\tilde{s}_{0\alpha} = A_\alpha$ (to be determined), $\tilde{r}_{ijk} = \partial_k g_{ij}$, $\tilde{r}_{0\alpha k} = 0$. Once we say what A_α we take, we will apply Theorem IV.1 to system (IV.31) with the above mentioned Cauchy data. Now, in order for the theorem to be applied, the Cauchy data should belong to certain Sobolev spaces. We would like to see that (even with A_α still to be chosen) the data we have already fixed (they are a wide majority) satisfy those conditions. Let us do it, for example, for g_{ij}'s.

Take some $x \in M$ and choose neighborhoods U_1, U_2, U_3 of x in M such that U_3 has compact closure and is a domain of the local chart of M, $\bar{U}_1 \subset U_2$ and $\bar{U}_2 \subset U_3$. Let α be a C^∞ function on M equal to 1 on $\bar{U}_1$ and equal to zero outside U_2. Since, by assumption, the Cauchy data g on M belongs to $\mathcal{W}^{s+1}_{loc}(M)_2$, the product αg belongs to $\mathcal{W}^{s+1}(M)_2$. This implies that the functions g_{ij} in the local chart U_1 belong to $\mathcal{W}^{s+1}(U_1)$. The only thing we should do in order for the assumptions of Theorem IV.1 concerning the Cauchy data g_{ij} to hold, is to take domains of coordinates in M according to the above procedure. What we have said with respect to the data g_{ij} can be repeated concerning other data.

In order to apply Theorem IV.1 to system (IV.31) (which is equivalent to system (IV.28)), the only thing that remains to be done is to specify what A_α should be taken on M as values of the normal derivatives $\partial_t \tilde{g}_{0\alpha}$. We will choose these values in order for the vector field H associated to the solution $\tilde{g}$ to be zero on M, and we will see then that the fact that the data g and k satisfy the constraint equations (that remain unused till now) implies that this field H becomes zero everywhere. Therefore the solution of system (IV.28) (that we are solving now) will coincide with that of system (IV.27) we would like to solve.

Hence our immediate objective is to see in what way the derivatives $\partial_t \tilde{g}_{0\alpha}$ on M could be fixed in order for H be equal to zero on M. Recall that $H^\mu = \tilde{g}^{\alpha\beta}(\tilde{\Gamma}^\mu_{\alpha\beta} - \hat{\Gamma}^\mu_{\alpha\beta})$. Let us calculate some of the symbols $\hat{\Gamma}^\mu_{\alpha\beta}$. Recall that the reference metric $\hat{g}$ has been defined as $\hat{g} = -dt^2 + a$, where a was a fixed arbitrary Riemannian metric on M. Observe that, by definition of the reference metric $\hat{g}$, in an adapted coordinate system the field $\partial/\partial t$ is geodesic with respect to the metric $\hat{g}$. Indeed, $\hat{g}(\partial_t, \partial_t) = -1$ implies that $\hat{g}(\hat{\nabla}_{\partial_t}\partial_t, \partial_t) = 0$ and $\hat{g}(\hat{\nabla}_{\partial_i}\partial_t, \partial_t) = 0$. The equality $\hat{g}(\partial_t, \partial_i) = 0$ gives $\hat{g}(\hat{\nabla}_{\partial_t}\partial_t, \partial_i) = -\hat{g}(\partial_t, \hat{\nabla}_{\partial_t}\partial_i) = -\hat{g}(\partial_t, \hat{\nabla}_{\partial_i}\partial_t) = 0$. Therefore $\hat{g}(\hat{\nabla}_{\partial_t}\partial_t, \partial_\alpha) = 0$, which implies $\hat{\nabla}_{\partial_t}\partial_t = 0$ and consequently the four Christoffel symbols $\hat{\Gamma}^\mu_{00}$ vanish. The symbols $\hat{\Gamma}^0_{ij}$ are also zeros. Indeed,

$$\hat{\Gamma}^0_{ij} = \frac{1}{2}\hat{g}^{00}\{\partial_i \hat{g}_{j0} + \partial_j \hat{g}_{i0} - \partial_t a_{ij}\} = 0 \ ,$$

since $\hat{g}_{i0} = 0$ and a is a metric on M that does not depend on t.

Let us calculate some of symbols $\tilde{\Gamma}^\mu_{\alpha\beta}$ on M. By definition of the second fundamental form, we have $\tilde{\nabla}_{\partial_i}\partial_j = \nabla_{\partial_i}\partial_j + S_{ij}\partial_t$. In the component-wise form, we have

$$\tilde{\Gamma}^\mu_{ij}\partial_\mu = \Gamma^l_{ij}\partial_l + S_{ij}\partial_t$$

implying

$$\tilde{\Gamma}^l_{ij}|_M = \Gamma^l_{ij} \ , \ \tilde{\Gamma}^0_{ij}|_M = S_{ij} = \frac{1}{2}k_{ij} \ .$$

Since $\tilde{\nabla}_{\partial_t}\partial_j = \tilde{\nabla}_{\partial_j}\partial_t$ on M, we have: $\tilde{\nabla}_{\partial_t}\partial_j|_M = g^{il}S_{lj}\partial_i$. Therefore

$$\tilde{\Gamma}^i_{0j} = \frac{1}{2}g^{il}k_{lj}\ , \ \tilde{\Gamma}^0_{0j} = 0\ .$$

On the other hand,

$$\tilde{\Gamma}^r_{00}|_M = g^{rl}\partial_t\tilde{g}_{0l},$$
$$\tilde{\Gamma}^0_{00}|_M = \frac{1}{2}\tilde{g}^{00}\partial_t\tilde{g}_{00} = -\frac{1}{2}\partial_t\tilde{g}_{00}\ .$$

Now we are able to write down the conditions for H to be zero on M, $\tilde{g}^{\alpha\beta}(\tilde{\Gamma}^\mu_{\alpha\beta}-\hat{\Gamma}^\mu_{\alpha\beta}) = 0$. The first three conditions (corresponding to $\mu = r$) are written as follows: $g^{ij}(\tilde{\Gamma}^r_{ij} - \hat{\Gamma}^r_{ij}) + \tilde{g}^{00}(\tilde{\Gamma}^r_{00} - \hat{\Gamma}^r_{00}) = 0$. This gives $g^{rl}\partial_t\tilde{g}_{0l} = g^{ij}(\Gamma^r_{ij} - \hat{\Gamma}^r_{ij})$. Therefore

$$\begin{aligned}\partial_t\tilde{g}_{0l}|_M &= g_{rl}g^{ij}(\Gamma^r_{ij} - \hat{\Gamma}^r_{ij})\\ &= g^{ij}(\partial_i g_{jl} - \frac{1}{2}\partial_l g_{ij}) - g_{rl}g^{ij}\hat{\Gamma}^r_{ij}\ .\end{aligned} \tag{IV.32}$$

The other condition making H zero, which is $g^{ij}(\tilde{\Gamma}^0_{ij} - \hat{\Gamma}^0_{ij}) + \tilde{g}^{00}(\tilde{\Gamma}^0_{00} - \hat{\Gamma}^0_{00}) = 0$, can be immediately written as

$$\partial_t\tilde{g}_{00}|_M = -g^{ij}k_{ij}\ . \tag{IV.33}$$

Therefore taking as functions A_α on M (in the adapted local chart we are using here) the functions A_l given by the right-hand side of (IV.32) and the function A_0 given by the right-hand side of (IV.33), we make H equal to zero on M in the domain of this local chart.

Let us now prove that in the local chart we have used H become zero not only on M, but also on the whole domain of the local chart. For this purpose, the following lemma is useful.

Lemma IV.9. *For any solution $(\tilde{g}, f)$ of* (IV.28), *we have*

$$\mathrm{div}_{\tilde{g}}(L_H\tilde{g}) - \frac{1}{2}d(\mathrm{tr}_{\tilde{g}}L_H\tilde{g}) = 0\ , \tag{IV.34}$$

where d appearing in the last term on the left-hand side of the equality stands for the exterior derivative.

Proof of the lemma. We begin with the equality

$$\widetilde{Ric} - \frac{1}{2}L_H\tilde{g} - \chi\left(T - \frac{1}{2}(\mathrm{tr}_{\tilde{g}}T)\tilde{g}\right) = 0\ . \tag{IV.35}$$

By taking divergences with respect to $\tilde{g}$ in this equality and by the fact that the divergence of T is zero, as well as by the fact that (by Proposition I.6) the divergence of the Ricci tensor is $\frac{1}{2}d\tilde{R}$, and since the covariant differentiation of $\tilde{g}$ equals zero, we obtain

$$\frac{1}{2}d\tilde{R} - \frac{1}{2}\text{div } L_H\tilde{g} + \frac{1}{2}\chi d(\text{tr } T) = 0 \ . \tag{IV.36}$$

By making a contraction of the product of (IV.35) by $\tilde{g}^{\alpha\beta}$, we have

$$\tilde{R} - \frac{1}{2}\text{tr } L_H\tilde{g} - \chi\text{tr } T + 2\chi\text{tr } T = 0 \ .$$

Therefore $\chi\text{tr } T = -\tilde{R} + \frac{1}{2}\text{tr } L_H\tilde{g}$. By taking exterior derivatives, $\chi d(\text{tr } T) = -d\tilde{R} + \frac{1}{2}d(\text{tr } L_H\tilde{g})$. By substituting this into (IV.36), we obtain the required identity (IV.34). □

Let us see in what way this identity is written in a local chart. By Corollary I.8, the components of $L_H\tilde{g}$ are

$$(L_H\tilde{g})_{\alpha\beta} = \tilde{\nabla}_\alpha H_\beta + \tilde{\nabla}_\beta H_\alpha \ .$$

This readily implies that the equality (IV.34) is written (in components) in the form

$$\tilde{g}^{\lambda\mu}\tilde{\nabla}_\lambda\tilde{\nabla}_\mu H_\beta - \tilde{g}^{\lambda\mu}(\tilde{\nabla}_\beta\tilde{\nabla}_\lambda - \tilde{\nabla}_\lambda\tilde{\nabla}_\beta)H_\mu = 0 \ .$$

By definitions of the curvature and Ricci tensors, we write

$$\tilde{g}^{\lambda\mu}\tilde{\nabla}_\lambda\tilde{\nabla}_\mu H_\beta + \tilde{R}_{\nu\beta}H^\nu = 0 \ .$$

The terms containing second-order derivatives of H are therefore

$$\tilde{g}^{\lambda\mu}\partial_\lambda\partial_\mu H_\beta \ .$$

Hence equation (IV.34) generates in each local chart a second-order system with respect to the variables H_β, homogeneous and hyperbolic, that is known to have a unique solution, once the values of H_β and $\partial_t H_\beta$ are fixed on the hypersurface M whose equation is $t = 0$. Since our H becomes zero on M, if we manage to prove that the derivatives $\partial_t H_\beta$ of our H also become zero on M, we will have that our H (by uniqueness of the solution of (IV.34)) will be equal to zero in all local charts where the solution $\tilde{g}$ of system (IV.28) exists.

In an adapted local chart, consider the components $\{0, \alpha\}$ of identity (IV.34) on the hypersurface M:

$$\tilde{R}_{0\alpha} - \frac{1}{2}(L_H\tilde{g})_{0\alpha} - \chi\left(T_{0\alpha} - \frac{1}{2}(\text{tr}_{\tilde{g}}T)\tilde{g}_{0\alpha}\right) = 0 \ . \tag{IV.37}$$

Now, when the constraint equations have been obtained in Section IV.6, this has been done exactly by applying the tensor identity

$$\widetilde{Ric} - \chi\left(T - \frac{1}{2}(\text{tr}_{\tilde{g}}T)\tilde{g}\right) = 0$$

(in terminology used in that section) to pairs having form (X, N) and (N, N) on the manifold M. Since we now suppose that the constraint equations are satisfied, we obtain:

$$\tilde{R}_{0\alpha} - \chi \left(T_{0\alpha} - \frac{1}{2}(\mathrm{tr}_{\tilde{g}} T)\tilde{g}_{0\alpha} \right) = 0 \ .$$

By comparing this to equation (IV.37), we see that we will have on M,

$$(L_H \tilde{g})_{0\alpha} = 0 \ .$$

This can be expressed as

$$\tilde{\nabla}_\alpha H_0 + \tilde{\nabla}_0 H_\alpha = 0 \ ,$$

which can be written as

$$\partial_\alpha H_0 + \partial_0 H_\alpha - 2\tilde{\Gamma}^\lambda_{\alpha 0} H_\lambda = 0 \ .$$

Now, since the field H becomes zero on M, the term $\tilde{\Gamma}^\lambda_{\alpha 0} H_\lambda$ in the last equation equals zero. Therefore

$$\partial_\alpha H_0 + \partial_0 H_\alpha = 0$$

holds on M.

By putting $\alpha = 0$ we have $\partial_t H_0 = 0$. By making $\alpha = i$ we obtain $\partial_t H_i = -\partial_i H_0$. This equals zero since H is zero on M. Therefore all derivatives $\partial_t H_\alpha$ equal zero on M as is required.

Let us summarize what we have done up to now. We have set a Lorentzian reference metric $\hat{g}$ on a neighborhood of the hypersurface M. Then, for each point $x \in M$, we have managed to find a solution $(\tilde{g}, f)$ of system (IV.28) in a certain neighborhood of x in V whose closure is contained in a local chart having compact domain in such a way that the solution satisfies the initial conditions imposed on the initial system (IV.27). Moreover, the field H associated to $\tilde{g}$ and to the reference metric $\hat{g}$ vanishes. This last condition can be formulated in an equivalent form saying that $\tilde{g}$ is harmonic with respect to the reference metric $\hat{g}$. Since H vanishes, $(\tilde{g}, f)$ which in principle is a solution of (IV.28), will also be a solution of (IV.27) (since the difference between these two systems is only in the term $L_H \tilde{g}$ that equals zero in this case). Since all this has been done by applying Theorem IV.1 to the corresponding local chart, this theorem also tells us that the local solution $(\tilde{g}, f)$ of (IV.28) we have found is unique among the solutions $(\tilde{g}, f)$ for which $\tilde{g}$ is harmonic with respect to the reference metric (and which satisfy the initial conditions on M imposed in the statement of the theorem). All that we have done is local, in any neighborhood of $x \in M$ under the above assumptions. Now, in the intersection of two of these neighborhoods the local solutions $(\tilde{g}, f)$ obtained in each of them should coincide by the uniqueness argument, since both solutions are harmonic with respect to the reference metric. This implies that the local solutions constructed in each local chart give rise to a global solution $(\tilde{g}, f)$ defined in a certain neighborhood of M in V.

Since Theorem IV.1 used for constructing local solutions in each local chart also asserts that the local solutions we have obtained belong to Sobolev spaces of that local chart, it is clear that every element in the pair $(\tilde{g}, f)$ of the global solution will belong to a local Sobolev space, as is stated by the theorem.

The only thing we still have to do in order to complete the proof is to show uniqueness up to diffeomorphisms defined at a certain neighborhood of M in V that equal identity on M and preserve the vector field N.

Let $(\tilde{g}, f)$ be the unique solution of (IV.27) satisfying the already imposed initial conditions and such that $\tilde{g}$ is harmonic with respect to the reference metric $\hat{g}$. Assume that $(\tilde{g}', f')$ is another solution of (IV.27) satisfying the same initial conditions. Let us look for a diffeomorphism Φ of a neighborhood $\mathcal{A}$ of M in V to itself which is identity on M, preserves the transversal field N and such that $\Phi : (\mathcal{A}, \tilde{g}') \longrightarrow (\mathcal{A}, \hat{g})$ is harmonic. In any local chart, this last condition can be expressed as

$$-\tilde{g}^{\alpha\beta}\left(\frac{\partial^2\Phi^\lambda}{\partial x^\alpha \partial x^\beta} - \frac{\partial \Phi^\lambda}{\partial x^\mu}\tilde{\Gamma}^\mu_{\alpha\beta} + \frac{\partial \Phi^\mu}{\partial x^\alpha}\frac{\partial \Phi^\nu}{\partial x^\beta}\hat{\Gamma}^\lambda_{\mu\nu}(\Phi(x))\right) = 0\ .$$

Since the terms containing second-order derivatives are $-\tilde{g}^{\alpha\beta}\frac{\partial^2\Phi^\lambda}{\partial x^\alpha \partial x^\beta}$, the latter system is hyperbolic of the same type we have been studying up to now (those systems that being reduced to order 1 by introducing auxiliary variables lead to quasilinear systems, hyperbolic and symmetric). Therefore it has a unique solution with the given initial conditions on M. Consider the metric $\bar{g}' = (\Phi^{-1})^*\tilde{g}'$. By Proposition IV.2, the mapping $I : (\mathcal{A}, \bar{g}') \longrightarrow (\mathcal{A}, \hat{g})$ is harmonic. Let $\bar{f}' = (\Phi^{-1})^* f'$ (a section of the fiber bundle $(\Phi^{-1})^*(F)$). By uniqueness of the solutions $(\tilde{g}, f)$ of (IV.27) satisfying the already imposed initial conditions and such that $\tilde{g}$ is harmonic with respect to the reference metric, both solutions $(\bar{g}', \bar{f}')$ and $(\tilde{g}, f)$ are equal. This shows that the diffeomorphism Φ^{-1} brings the solution $(\tilde{g}', f')$ to $(\tilde{g}, f)$ as is required to complete the proof. □

Remark IV.10 (*Continuity with respect to the Cauchy data*). The above proof of the fundamental theorem of this section (Theorem IV.7) has consisted in application of Theorem IV.1 to a first-order system obtained in each local chart from system (IV.28). Now, the statement of Theorem IV.1 contains an assertion concerning continuity of the solutions with respect to the Cauchy data, while the statement of the fundamental theorem of this section does not mention this continuity at all. We have done this in order not to make longer the statement of the theorem. However, it is clear that, by Theorem IV.1, solutions of Einstein's equation will depend in continuously on the Cauchy data. An exact statement of this fact is the following. Assume that $\mathcal{P}$ is the neighborhood of the hypersurface M in V (mentioned in the statement of the theorem) where solutions of Einstein's equation exist. Any $x \in \mathcal{P}$ has a neighborhood $\tilde{U}$ with compact closure such that there exists an open set U of M with compact closure, $U \subset \tilde{U}$, such that if the restrictions to U of the Cauchy data (g, k, ω) satisfy $g \in \mathcal{W}^{s+1}(U)_2$, $k \in \mathcal{W}^s(U)_2$, $\omega \in \mathcal{W}^{s-1}(\Gamma(U, F))$, then the restriction of the solution $(\tilde{g}, f)$ to $\tilde{U}$ satisfies $\tilde{g} \in \mathcal{W}^{s+1}(\tilde{U})_2$, $f \in \mathcal{W}^{s-1}(\Gamma(\tilde{U}, F))$, and the mapping $(g, k, \omega)|_U \to (\tilde{g}, f)|_{\tilde{U}}$ is continuous in these Sobolev spaces.

IV.9 An example: the stress-energy tensor of holonomic media

Among all assumptions in Theorem IV.7 there is one which is technical, and by no means intuitive, which requires us to suppose that the equation $\mathrm{div}_{\tilde{g}} T(\tilde{g}, f) = 0$ can be written at a certain neighborhood of each point of M which is the domain of a local chart, as a first-order symmetric, hyperbolic system with respect to f, for which the hypersurface M is nowhere characteristic. In this section, we will give an example of a rather wide admissible class of stress-energy tensors (containing those of perfect fluids) satisfying, under certain constraints, this (technical) assumption in Theorem IV.7.

In [44], Lichnerowicz introduced holonomic media as those continuous media having a stress-energy tensor having the form

$$T = ru \otimes u + \theta \ ,$$

where u is a unitary time-like vector field of velocities, $\tilde{g}(u, u) = -1$, r is a positive scalar (called pseudodensity) and θ is a symmetric 2-tensor whose divergence is equal to the gradient of some function f, that is $\tilde{\nabla}_\alpha \theta^{\alpha\beta} = \tilde{g}^{\alpha\beta} \partial_\alpha f$. Assume that both the pseudodensity r and the function f depend on the same scalar parameter ζ, $r = r(\zeta)$ and $f = f(\zeta)$.

Observe first that perfect fluids mentioned in Section IV.4 are holonomic. Indeed, the stress-energy tensor of these fluids is given by (IV.11), which has the above form with $r = \rho + p$ and $\theta = p\tilde{g}$. Since one can parametrize p by ρ by virtue of the state equation, the parameter ζ used in perfect fluids (when we would like to describe them as holonomic media) will be ρ. Observe that in this case the divergence of $\theta = p\tilde{g}$ is the gradient of p. Therefore the function $f(\zeta)$ will also be p (depending on $\zeta = \rho$).

By means of the tools introduced in Section IV.4, the class of stress-energy tensors of holonomic media can be described as a class of admissible stress-energy tensors in the following way. Consider a 4-dimensional manifold V such that there exists a rank 3 subbundle F' of the tangent bundle $T(V)$ and a vector field N of V such that $T_x(V) = F'_x \oplus \langle N_x \rangle$ at each $x \in V$. For every Lorentzian metric $\tilde{g}$ on V for which N_x is time-like at each point, F'_x is space-like, and for every $v \in \Gamma(F')$, consider the vector field u given at every x by

$$u_x = v_x + zN_x \ ,$$

where z is given as a function of v_x, N_x and $\tilde{g}_x$ by the formula

$$z = \frac{-\tilde{g}(v_x, N_x) - \sqrt{\tilde{g}(v_x, N_x)^2 - \tilde{g}(N_x, N_x)(\tilde{g}(v_x, v_x) + 1)}}{\tilde{g}(N_x, N_x)} .$$

By construction, we will have $\tilde{g}(u, u) = -1$. Let $\mathcal{R}$ be the trivial vector bundle of rank 1 on V given by $\mathcal{R} = V \times \mathbb{R}$. Assume that E is the direct sum $E = S_2(V) \oplus F' \oplus \mathcal{R}$. Let W be an open set of E such that if $w \in W \subset E$ belongs to the fiber E_x, then w has the form $w = (\tilde{g}_x, v_x, \zeta_x)$, where $\tilde{g}_x$ is a Lorentzian metric on the vector space $T_x(V)$ for which N_x is time-like and F'_x is space-like. Suppose that we have a mapping

$$\begin{array}{rccc} \theta: & \Gamma(W, S_2(V) \oplus \mathcal{R}) & \longrightarrow & \Gamma(W, S_2(V)) \\ & (\tilde{g}, \zeta) & \to & \theta(\tilde{g}, \zeta) \end{array} ,$$

such that the value of $\theta(\tilde{g}, \zeta)$ at each $x \in V$ only depends on the values of $\tilde{g}$ and ζ at x (and do not depend on the derivatives either of $\tilde{g}$ or ζ). Moreover, assume that we have a mapping

$$\begin{array}{rccc} T: & \Gamma(W, E) & \longrightarrow & \Gamma(W, S_2(V)) \\ & (\tilde{g}, v, \zeta) & \to & r(\zeta) u \otimes u + \theta(\tilde{g}, \zeta) \end{array} ,$$

such that $u_x = v_x + zN_x$ and that $\mathrm{div}_{\tilde{g}}\theta = \mathrm{grad}_{\tilde{g}} f(\zeta)$ holds for a function $f(\zeta)$ of ζ. In this case, the mapping T will be the admissible class of stress-energy tensors of a holonomic medium.

The objective of this section is to establish the following statement.

Proposition IV.11. *Assume that T is a class of stress-energy tensors of a holonomic medium. Let M be a space-like hypersurface of V with respect to a fixed Lorentzian metric $\tilde{g}$. Then if:*

1. *the function $r(\zeta)$ is positive over its whole domain;*
2. *the functions $r(\zeta)$ and $f(\zeta)$ belong to the class C^1, f' is always positive and* $\frac{r'}{f'} > 2$*;*

then the equation $\mathrm{div}_{\tilde{g}} T = 0$ can be written, in some coordinates adapted to M, as a first-order symmetric, hyperbolic system for which the hypersurface M is nowhere characteristic.

Proof. Let us first check that in some adapted coordinates (t, x^1, x^2, x^3) where M is written as $t = 0$, M is nowhere characteristic of $\mathrm{div}(T) = 0$. To begin with, we have five equations ($\mathrm{div}(T) = 0$ and $\tilde{g}(u, u) = -1$) and five unknowns (u^α and ζ). However, we will deduce another five equations from the initial ones. The equation $\mathrm{div}(T) = 0$ gives

$$\begin{aligned} 0 = \nabla_\alpha T^{\alpha\beta} &= \nabla_\alpha \{ r u^\alpha u^\beta \} + \tilde{g}^{\alpha\beta} \partial_\alpha f \\ &= u(r) u^\beta + r \,\mathrm{div}(u) u^\beta + r (\nabla_u u)^\beta + \tilde{g}^{\alpha\beta} \partial_\alpha f. \end{aligned} \tag{IV.38}$$

Since $u^\alpha u_\alpha = -1$, multiplication by $-u_\beta$ gives

$$u(r) + r\,\mathrm{div}(u) - r\tilde{g}(u, \nabla_u u) - u(f) = 0\ .$$

The third term equals zero since $\tilde{g}(u, \nabla_u u) = (1/2)\nabla_u(\tilde{g}(u,u)) = 0$. Therefore we obtain the equation

$$u(r - f) + r\,\mathrm{div}(u) = 0. \tag{IV.39}$$

Another four equations are obtained by isolating $u(r)$ in (IV.39) and substituting them into (IV.38):

$$u(f)u^\beta + r(\nabla_u u)^\beta + \tilde{g}^{\alpha\beta}\partial_\alpha f = 0. \tag{IV.40}$$

The principal part of the system formed by (IV.39)) and (IV.40) is obtained by replacing $\nabla_\alpha u^\beta$ by $\partial_\alpha u^\beta$. Denoting by B^0 and B^β the terms of lower order and using $\partial_\alpha r = r'(\zeta)\partial_\alpha \zeta$ and $\partial_\alpha f = f'(\zeta)\partial_\alpha f$, the above system becomes

$$\begin{cases} (r'(\zeta) - f'(\zeta))u^\alpha \partial_\alpha \zeta + r\delta^\alpha_\beta \partial_\alpha u^\beta &= B^0 \\ (\tilde{g}^{\lambda\alpha} + u^\lambda u^\alpha) f'(\zeta)\partial_\alpha \zeta + r u^\alpha \delta^\lambda_\beta \partial_\alpha u^\beta &= B^\lambda\ . \end{cases}$$

The Kronecker deltas δ are required in order to write this system in matrix form:

$$\begin{pmatrix} (r'(\zeta) - f'(\zeta))u^\alpha & r\delta^\alpha_\beta \\ f'(\zeta)(\tilde{g}^{\lambda\alpha} + u^\lambda u^\alpha) & r u^\alpha \delta^\lambda_\beta \end{pmatrix} \partial_\alpha \begin{pmatrix} \zeta \\ u^\beta \end{pmatrix} = \begin{pmatrix} B^0 \\ B^\lambda \end{pmatrix}. \tag{IV.41}$$

This is a matrix-valued equation of the form $A^\alpha \partial_\alpha \Psi = B$. M will be characteristic if the determinant of the matrix $A^\alpha \xi_\alpha$ becomes zero, ξ being the normal 1-form of M given in some adapted coordinates by $\xi_\alpha = \delta^0_\alpha$. Therefore $A^\alpha \xi_\alpha = A^\alpha \delta^0_\alpha = A^0$. By the properties of determinants,

$$\begin{aligned} \det(A^0) &= \det \begin{pmatrix} (r' - f')u^0 & r\delta^0_\beta \\ f'(\tilde{g}^{\lambda 0} + u^\lambda u^0) & r u^0 \delta^\lambda_\beta \end{pmatrix} = r^4 \det \begin{pmatrix} (r' - f')u^0 & \delta^0_\beta \\ f'(\tilde{g}^{\lambda 0} + u^\lambda u^0) & u^0 \delta^\lambda_\beta \end{pmatrix} \\ &= r^4 \det \begin{pmatrix} (r' - f')u^0 & 1 \\ f'(\tilde{g}^{00} + (u^0)^2) & u^0 \end{pmatrix} \det(u^0 \delta^i_j) \\ &= r^4 (u^0)^3 [(r' - f')(u^0)^2 - f'(\tilde{g}^{00} + (u^0)^2)] \\ &= r^4 (u^0)^3 [(u^0)^2 (r' - 2f') - \tilde{g}^{00} f']. \end{aligned}$$

The condition $\det(A^0) = 0$ implies that either $r = 0$, or $u^0 = 0$, or

$$[(u^0)^2(r' - 2f') - \tilde{g}^{00} f'] = 0.$$

The first possibility cannot be fulfilled since, by the assumption, r is a positive scalar. Neither can the second possibility occur since u is a time-like vector. The last possibility leads us (by isolating $\tilde{g}^{00}$) to

$$\tilde{g}^{00} = (u^0)^2 \left(\frac{r'}{f'} - 2\right).$$

Since $\tilde{g}^{00}$ is negative, we have $r'/f' < 2$. The assumption, however, states that $r'/f' > 2$. Therefore M cannot be characteristic.

It only remains to find a representation where system of equations (IV.41) would be symmetric and hyperbolic. Instead of working with a 5×5-system, we take the parameter ζ and three spatial components of the velocity vector (ζ, u^1, u^2, u^3) as variables, u^0 being related to u^i's by $\tilde{g}(u,u) = u^\alpha \tilde{g}_{\alpha\beta} u^\beta = -1$ (recall that this equation has been used to obtain the second equation in (IV.41); therefore it can be recovered by means of them). We have

$$\nabla_\alpha u^\alpha \simeq \partial_\alpha u^\alpha = \partial_i u^i + \partial_t u^0.$$

In order to establish a link between $\partial_t u^0$ and u^i, we need only take into account that $u_\alpha u^\alpha = -1$ implies $u_\alpha \partial_\beta u^\alpha \simeq 0$. Hence

$$u_0 \partial_\beta u^0 \simeq -u_i \partial_\beta u^i. \tag{IV.42}$$

In particular, $\partial_t u^0 \simeq -v_i \partial_t u^i$, where we have set $v_i \equiv u_i/u_0$. Then $\partial_\alpha u^\alpha \simeq \partial_i u^i - v_i \partial_t u^i$ which can be written as $(\delta_i^\alpha - v_i \delta_0^\alpha)\partial_\alpha u^i$. By substituting into equation (IV.39), written component-wise as

$$(r'(\zeta) - f'(\zeta))u^\alpha \partial_\alpha \zeta + r\nabla_\alpha u^\alpha = 0,$$

and by dividing by r, we obtain

$$\frac{(r'(\zeta) - f'(\zeta))}{r} u^\alpha \partial_\alpha \zeta + (\delta_i^\alpha - v_i \delta_0^\alpha)\partial_\alpha u^i = A^0. \tag{IV.43}$$

Equation (IV.40), written in intrinsic form, is

$$\nabla f + u(f)u + r\nabla_u u = 0. \tag{IV.44}$$

In order for this equation and (IV.43) to form a symmetric system, the combination $(\delta_i^\alpha - v_i \delta_0^\alpha)\partial_\alpha \zeta = \partial_i \zeta - v_i \partial_t \zeta$ should appear in it.

Therefore we have to handle the three equations of (IV.44) corresponding to $i = 1, 2, 3$ written in contravariant form:

$$r\nabla_u u_i + \partial_i f + u(f)u_i = 0.$$

We recognize the term $\partial_i f = f'\partial_i \zeta$, but $u(f)u_i = u_i(u^0 \partial_t f + u^i \partial_i f)$ gives undesired terms. Instead of expanding the factor $u(f)$ we use the component $\alpha = 0$ in system (IV.44):

$$r\nabla_u u_0 + \nabla_t f + u(f)u_0 = 0.$$

Hence

$$u_i u(f) = -\frac{u_i}{u_0}\{r\nabla_u u_0 + \partial_t f\} = -v_i\{r\nabla_u u_0 + f'\partial_t \zeta\},$$

giving us the missing term $-v_i f'\partial_t\zeta$. It remains to check whether the coefficient of the principal part of $\partial_\alpha u^i$ is symmetric. The principal parts of $\nabla_u u_i = u^\alpha \nabla_\alpha u_i$ and $\nabla_u u_0$ are, by (IV.42),

$$\begin{aligned} u^\alpha \partial_\alpha u_i &= u^\alpha \partial_\alpha (\tilde{g}_{i\beta} u^\beta) \simeq u^\alpha \tilde{g}_{ij} \partial_\alpha u^j + u^\alpha \tilde{g}_{i0} \partial_\alpha u^0 \\ &= \tilde{g}_{ij} u^\alpha \partial_\alpha u^j - v_j \tilde{g}_{i0} u^\alpha \partial_\alpha u^j , \\ u^\alpha \partial_\alpha u_0 &= u^\alpha \partial_\alpha (\tilde{g}_{0\beta} u^\beta) \simeq u^\alpha \tilde{g}_{0j} \partial_\alpha u^j + u^\alpha \tilde{g}_{00} \partial_\alpha u^0 \\ &= \tilde{g}_{0j} u^\alpha \partial_\alpha u^j - v_j \tilde{g}_{00} u^\alpha \partial_\alpha u^j . \end{aligned}$$

Finally the principal part of the three equations $r\nabla_u u_i + \nabla_i f + u(f) u_i = 0$ is

$$\begin{aligned} & r\nabla_u u_i + \nabla_i f + u(f) u_i \\ &\qquad \simeq f'\partial_i \zeta - v_i f' \partial_t \zeta + r\{\tilde{g}_{00} v_i v_j - \tilde{g}_{0j} v_i - \tilde{g}_{0i} v_j + \tilde{g}_{ij}\} u^\alpha \partial_\alpha u^j \\ &\qquad = (\delta_i^\alpha - v_i \delta_0^\alpha) f' \partial_\alpha \zeta + r\{\tilde{g}_{00} v_i v_j - \tilde{g}_{0j} v_i - \tilde{g}_{0i} v_j + \tilde{g}_{ij}\} u^\alpha \partial_\alpha u^j \ . \end{aligned}$$

Summing up, the equations $\operatorname{div}(ru \otimes u + \theta) = 0$ are equivalent to the system

$$\begin{aligned} \frac{(r' - f')u^\alpha}{r} \partial_\alpha \zeta (\delta_i^\alpha - v_i \delta_0^\alpha) \partial_\alpha u^i &= B_0(\tilde{g}, \partial \tilde{g}) \\ (\delta_i^\alpha - v_i \delta_0^\alpha) \partial_\alpha \zeta + \frac{r}{f'} \{\tilde{g}_{00} v_i v_j - \tilde{g}_{0j} v_i - \tilde{g}_{0i} v_j + \tilde{g}_{ij}\} u^\alpha \partial_\alpha u^j &= B_i(\tilde{g}, \partial \tilde{g}) \end{aligned}$$

which is of the form $A^\alpha \partial_\alpha \psi = B$, where $\psi = (\zeta, u^1, u^2, u^3)$ and the symmetric matrices A^α are

$$\begin{pmatrix} \dfrac{(r' - f')u^\alpha}{r} & (\delta_i^\alpha - v_i \delta_0^\alpha) \\ (\delta_i^\alpha - v_i \delta_0^\alpha) & \dfrac{r}{f'} u^\alpha \{\tilde{g}_{00} v_i v_j - \tilde{g}_{0j} v_i - \tilde{g}_{0i} v_j + \tilde{g}_{ij}\} \end{pmatrix}.$$

In order to classify this first-order system as symmetric, hyperbolic, we are led to find the assumptions making the matrix

$$A^\alpha \xi_\alpha = A^\alpha \delta_\alpha^0 = A^0 = \begin{pmatrix} \dfrac{r' - f'}{r} u^0 & -v_i \\ -v_i & \dfrac{r}{f'} u^0 \{\tilde{g}_{00} v_i v_j - \tilde{g}_{0j} v_i - \tilde{g}_{0i} v_j + \tilde{g}_{ij}\} \end{pmatrix}$$

positive definite.

$$\begin{aligned} & (A^0 \Phi, \Phi) \\ &= \frac{r' - f'}{r} u^0 (\Phi^0)^2 - 2 v_i \Phi^i \Phi^0 + \frac{r}{f'} u^0 (\tilde{g}_{00} v_i v_j + \tilde{g}_{ij} - \tilde{g}_{0i} v_j - \tilde{g}_{0j} v_i) \Phi^i \Phi^j \\ &= \frac{r' - f'}{r} u^0 (\Phi^0)^2 - 2 (v_i \Phi^i) \Phi^0 + \frac{r}{f'} u^0 \{\tilde{g}_{00} (v_i \Phi^i)^2 + \tilde{g}_{ij} \Phi^i \Phi^j - 2 \tilde{g}_{0i} \Phi^i (v_j \Phi^j)\}. \end{aligned}$$

At any point belonging to the manifold V, the Lorentzian metric $\tilde{g}$ admits a splitting

$$\tilde{g} = -dt^2 + g_{ij}dx^i \otimes dx^j ,$$

where g is a Riemannian metric. Then $u_0 = \tilde{g}_{0\alpha}u^\alpha = -u^0$, $u_i = \tilde{g}_{i\alpha}u^\alpha = g_{ij}u^j$ and the condition $\tilde{g}(u,u) = -1$ for the velocity field u of the fluid is written as $-u_0^2 + g^{ij}u_iu_j = -1$. By dividing by u_0^2, we obtain

$$v_i g^{ij} v_j = g(v,v) = 1 - \frac{1}{u_0^2} .$$

Since $g(v,v)$ has to be positive, we have $u_0^2 > 1$. Assume, first, that $-u^0 = u_0 < 0$. Then $u_0 < -1$ and $u^0 > 1$. Using the above splitting of $\tilde{g}$, we obtain

$$(A^0\Phi,\Phi) = \frac{r'-f'}{r}u^0(\Phi^0)^2 - 2g(v,\Phi)\Phi^0 + \frac{r}{f'}u^0(-g^2(v,\Phi) + \|\Phi\|^2)$$

where $\|\cdot\|$ stands for the norm with respect to g. If Φ is a tangent vector to M written in a local chart as $\Phi = (0,\Phi^1,\Phi^2,\Phi^3)$, we have

$$(A^0\Phi,\Phi) = \frac{r}{f'}u^0(\|\Phi\|^2 - g^2(v,\Phi)) .$$

Since $g(v,v) = 1 - \frac{1}{(u_0)^2} < 1$ if $u^0 > 1$, the Schwarz inequality implies $g^2(v,\Phi) < \|\Phi\|^2$. Therefore, since the assumptions of the theorem give $r/f' > 0$, we have $(A^0\Phi,\Phi) > 0$.

Now consider the vector $\Phi' = \lambda e_0 + \Phi$ with $e_0 = (1,0,0,0)$ and Φ as before. Then

$$(A^0\Phi',\Phi') = \frac{r'-f'}{r}u^0\lambda^2 - 2g(v,\Phi)\lambda + \frac{r}{f'}u^0(\|\Phi\|^2 - g^2(v,\Phi)) .$$

If the discriminant of this second-order polynomial with respect to λ is negative, then $(A^0\Phi',\Phi') > 0$, $\forall\Phi'$. By using the inequalities $g^2(v,\Phi) < (1-\frac{1}{u_0^2})\|\Phi\|^2$ and $r'/f' > 2$, we arrive at

$$\begin{aligned}
\Delta &= 4g^2(v,\Phi) - 4\left(\frac{r'}{f'}-1\right)u_0^2(\|\Phi\|^2 - g^2(v,\Phi)) \\
&= 4\left(g^2(v,\Phi)\left(1+\left(\frac{r'}{f'}-1\right)u_0^2\right) - \left(\frac{r'}{f'}-1\right)u_0^2\|\Phi\|^2\right) \\
&< 4\left(\left(1-\frac{1}{u_0^2}\right)\left(1+\left(\frac{r'}{f'}-1\right)u_0^2\right) - \left(\frac{r'}{f'}-1\right)u_0^2\right)\|\Phi\|^2 \\
&= -4\left(\frac{1}{u_0^2} + \left(\frac{r'}{f'}-2\right)\right)\|\Phi\|^2 < 0.
\end{aligned}$$

If we had assumed $u_0 > 0$ (and therefore $u^0 < 0$) we would have arrived at A^0 negative definite and, changing the orientation of the normal vector ξ to M, then $A^\alpha\xi_\alpha = -A^0$ would be positive definite. □

If we apply this argument to the particular case of perfect fluids where the stress-energy tensor is given by (IV.11), the parameter ζ is the density ρ, the function r is $r = \rho + p(\rho)$ and if the function $f(\zeta)$ is $p(\rho)$, then the above proposition implies the following corollary.

Corollary IV.12. *Let T be the class of stress-energy tensors of a perfect fluid. Assume that M is a space-like hypersurface of V with respect to a fixed Lorentz metric $\tilde{g}$. Then if:*

1. $\rho + p(\rho) > 0$;
2. *the function $p(\rho)$ belongs to the class C^1 and $0 < p'(\rho) < 1$;*

then the equation $div_{\tilde{g}} T = 0$ can be written, in some coordinates adapted to M, as a first-order symmetric, hyperbolic system for which the hypersurface M is nowhere characteristic.

It is clear that one can arrive at the same conditions without mentioning holonomic media ([11],[37],[56]).

IV.10 The Cauchy problem in the vacuum

What we have done up to now has been related to the Cauchy problem for Einstein's equation with matter. Let us say some words concerning the same problem for Einstein's equation in the vacuum. Observe first that some assumptions in the fundamental theorem IV.7 make no sense in the vacuum. For example, the assumption requiring the equation $\mathrm{div}_{\tilde{g}} T(\tilde{g}, f) = 0$ to be written in a certain neighborhood of each point of M that is the domain of a local chart as a first-order symmetric, hyperbolic system with respect to f, for which the hypersurface M is nowhere characteristic.

However we can follow the lines of the argument used for proving Theorem IV.7 to obtain (in a much simpler way) the following theorem for Einstein's equation in the vacuum.

Theorem IV.13 (Y. Choquet-Bruhat [21]). *Let V be a 4-dimensional manifold. Assume that M is a hypersurface of V admitting a vector field N of V with support on M, transversal to M. The transversal field N is supposed to be given. Let $s > 3$ be an integer. Assume that we have on M:*

- *A pair of quadratic forms (g, k), with $g \in \mathcal{W}^{s+1}_{loc}(M)_2$ and $k \in \mathcal{W}^{s}_{loc}(M)_2$ such that g and k are symmetric and g is positive definite. It is supposed that the data (g, k) on M satisfy the constraint equations*

$$\begin{cases} \gamma(g,k) = 0 \\ \mathcal{H}(g,k) = 0 \,. \end{cases}$$

Then in a neighborhood $\mathcal{P}$ of M in V there exists a Lorentzian metric $\tilde{g} \in \mathcal{W}^{s+1}_{loc}(\mathcal{P})_2$ which is a solution of Einstein's equation in the vacuum satisfying the following conditions:

- *At any point $x \in M$, we have $\tilde{g}(N_x, N_x) = -1$ and N_x is perpendicular to $T_x(M)$ with respect to $\tilde{g}$.*
- *The restriction of M to $\tilde{g}$ is g. Two times the second fundamental quadratic form of the hypersurface M of V with respect to $\tilde{g}$ is k.*

The solution $\tilde{g}$ is unique up to diffeomorphisms defined in a certain neighborhood of M in V equal to the identity on M and preserving the vector field N.

Chapter V

Stability by Linearization of Einstein's Equation, General Concepts

The stability by linearization of Einstein's equation in the vacuum has been widely studied in the literature (refs. [22], [23], [32], [34], [2], [49], [52]). In [50] an interesting bibliography on the subject may be found. The same concept for Einstein's equation in the presence of matter has been studied in [14] and [16] for Robertson-Walker cosmological models. In sections V.1, V.2 and V.3 we will provide the reader with the general concepts concerning stability by linearization of Einstein's equation, in the vacuum as well as in the presence of matter. Section V.5 is devoted to clarifying a technical question concerning Sobolev spaces, and finally Section V.6 covers some calculus needed in the following chapters.

V.1 Classical concept of stability by linearization of Einstein's equation in the vacuum

In this section we will describe the classical concept of stability by linearization of Einstein's equation in the vacuum. We start from an initial Lorentzian metric $\tilde{g}_0$ which is a solution to Einstein's equation in the vacuum, $G(\tilde{g}_0) = 0$. One is interested in metrics $\tilde{g} = \tilde{g}_0 + \tilde{h}$ close to $\tilde{g}_0$ (that is to say, with small $\tilde{h}$) which are also solutions to Einstein's equation. Therefore, $G(\tilde{g}_0 + \tilde{h}) = 0$. The equation obtained from this one, taking exclusively the linear terms in $\tilde{h}$, is called the linearization of the first one at the initial metric $\tilde{g}_0$ and it is expressed as

$$(D_{\tilde{g}_0} G)(\tilde{h}) = 0 \ .$$

An intuitive definition of the concept of stability by linearization of Einstein's equation in the vacuum (without a thorough clarification of the spaces in which we are working) might be the following: Einstein's equation in the vacuum is stable by linearization at the initial metric $\tilde{g}_0$ if for every solution $\tilde{h}$ to the linearized equation, there exists a differentiable curve $\lambda \to \tilde{g}(\lambda)$ of solutions to the initial equation $G(\tilde{g}) = 0$ which for $\lambda = 0$ passes through $\tilde{g}_0$ and such that the derivative of $\tilde{g}(\lambda)$ with respect to λ at point $\lambda = 0$ is $\tilde{h}$.

Assume that the given initial Lorentzian metric $\tilde{g}_0$ is a metric in a four dimensional manifold V. Suppose a hypersurface M in V is given which is space-like with respect to $\tilde{g}_0$. Assume that we have a vector field N in V with support on M which fulfills $\tilde{g}_0(N_x, N_x) = -1$ and that $T_x(M)$ is perpendicular to N_x with respect to $\tilde{g}_0$ for any $x \in M$. From Theorem IV.13 the solutions $\tilde{g}$ to the equation $G(\tilde{g}) = 0$ that satisfy the requirements that $\tilde{g}(N_x, N_x) = -1$ and that $T_x(M)$ be perpendicular to N_x with respect to $\tilde{g}$, are uniquely defined (up to diffeomorphisms) in a neighborhood of M in V by means of the Cauchy data (g, k) on M, which must fulfill the two constraint equations

$$\begin{cases} \gamma(g,k) = 0 \\ \mathcal{H}(g,k) = 0 \,, \end{cases} \tag{V.1}$$

where $\gamma(g, k)$ and $\mathcal{H}(g, k)$ are defined by (IV.23) and (IV.25).

As in a certain neighborhood of M in V, every solution $\tilde{g}$ to Einstein's equation is uniquely determined by the Cauchy data (g, k) on M, the concept of stability by linearization of Einstein's equation in the vacuum can be formulated solely in terms of Cauchy data in the following way. Let (g_0, k_0) be a couple of covariant symmetric 2-tensors on M, with g_0 positive definite (a Riemannian metric) and that satisfy the equations $\gamma(g_0, k_0) = 0$ and $\mathcal{H}(g_0, k_0) = 0$. Einstein's equation is stable by linearization at the initial couple (g_0, k_0) if for each solution (h, K) to the linearized system

$$\begin{cases} (D_{(g_0,k_0)}\gamma)(h,K) = 0 \\ (D_{(g_0,k_0)}\mathcal{H})(h,K) = 0 \,, \end{cases}$$

there exists a smooth curve, $\lambda \to (g(\lambda), k(\lambda))$, of solutions to the system (V.1), which for $\lambda = 0$ passes through (g_0, k_0) and such that the derivative of $(g(\lambda), k(\lambda))$ with respect to λ at point $\lambda = 0$ is (h, K).

Notice that this concept leads us, in a slightly more general framework not linked to relativity, to the following definition.

Definition V.1 (Classical). Let $f : U \to Y$ be a C^1-mapping between an open set U of a Banach space X and another Banach space Y. Let $x_0 \in U$ and $y_0 \in Y$ be such that $f(x_0) = y_0$. The equation $f(x) = y_0$ is stable by linearization at $x = x_0 \in U$ if for every solution h to the linearized equation $(D_{x_0}f)(h) = 0$ there exists a differentiable curve, $\lambda \to x(\lambda)$, of solutions to the original equation $f(x) = y_0$ that is tangent to h at x_0. (In the framework of Einstein's equation in the vacuum we would apply this definition to the map $(g,k) \to (1/\chi)(\mathcal{H}(g,k), \gamma(g,k))$ at point (g_0, k_0), which is mapped to $(0,0)$ by the considered mapping.)

By definition of derivative

$$x(\lambda) = x_0 + h\lambda + E(\lambda) \ , \ \text{with} \ \lim_{\lambda \to 0} \frac{E(\lambda)}{\lambda} = 0 \ .$$

Therefore, if we replace the true solutions $x(\lambda)$ to the original equation by the approximate solutions $\bar{x}(\lambda) = x_0 + h\lambda$ constructed from a linearized solution h, the error $E(\lambda) = x(\lambda) - x_0 - h\lambda$ made by this replacement is $o(\lambda)$ (that is to say, it vanishes faster than λ when λ goes to zero).

The following theorem (an immediate consequence of the inverse function theorem in Banach spaces) provides a sufficient condition for the stability by linearization of an equation $f(x) = y_0$.

Theorem V.2. *Let $f : U \to Y$ be a C^1-mapping between an open set U of a Banach space X and another Banach space Y. Let $x_0 \in U$ and $y_0 \in Y$ be such that $f(x_0) = y_0$. If the tangent linear map $D_{x_0} f$ of f at x_0 is surjective and if the kernel of $D_{x_0} f$ has a topological supplement, then the equation $f(x) = y_0$ is stable by linearization at point x_0, in the sense of Definition* V.1.

Proof. By the inverse function theorem, if both hypothesis of the theorem are satisfied, the set $f^{-1}(y_0)$ is a differentiable manifold in a neighborhood of x_0, with tangent space the kernel of $D_{x_0} f$. Thus, if h satisfies $(D_{x_0} f)(h) = 0$, then $h \in T_{x_0}(f^{-1}(y_0))$ and, by definition of tangent space, there exists a curve $x(\lambda)$ of $f^{-1}(y_0)$ tangent to h. □

V.2 A new concept of stability by linearization of Einstein's equation in the presence of matter

In the former section we talked about Einstein's equation in the vacuum. When we consider Einstein's equation in the presence of matter, $G(\tilde{g}) = \chi T$, the situation is slightly different. Let $\tilde{g}_0$ and T_0 be such that $G(\tilde{g}_0) = \chi T_0$. Now, for any small perturbation of the initial matter we will be interested in the gravitational field created by the perturbed matter. Naively (we will see later on that it is not possible to face the problem in this way) we may consider that for any T close to T_0 such that $T = T_0 + H$ (corresponding to a small perturbation of the initial matter), we are interested in metrics $\tilde{g}$ close to $\tilde{g}_0$ expressed in the form $\tilde{g} = \tilde{g}_0 + \tilde{h}$, which are solutions to the equation $G(\tilde{g}) = \chi T$. The study of the true equation $G(\tilde{g}) = \chi T$ being too complex, we would like to replace the former by the study of the linearized equation $(D_{\tilde{g}_0} G)(\tilde{h}) = \chi H$, always that this procedure makes sense in a certain way. That is to say, that the solutions to the linearized equation would be close to the solutions to the true equation.

In order to account for this new point of view we might notice that in Chapter III we started from the Minkowski metric $\tilde{\eta}$ which is a solution to Einstein's equation in the vacuum, $G(\tilde{\eta}) = 0$, and we studied the metrics $\tilde{g} = \tilde{\eta} + \tilde{h}$ close to $\tilde{\eta}$ which are approximate solutions to Einstein's equation in the presence of matter $G(\tilde{g}) = \chi T$, for small T (notice that we were not limited to the solutions to $G(\tilde{g}) = 0$), and we related then the linearized Einstein's equation (at the initial Minkowski metric) to the classical wave equation.

Notice that in the former section we started from an initial metric g_0 satisfying $G(\tilde{g}_0) = 0$ and we were interested in solutions $\tilde{g} = \tilde{g}_0 + \tilde{h}$ to $G(\tilde{g}) = 0$. In that section the initial T_0 vanished and we did not perturb this initial T_0 because we were just interested in metrics $\tilde{g}$ satisfying Einstein's equation in the vacuum, $G(\tilde{g}) = 0$. However in Chapter III, although the initial stress-energy tensor T_0 vanished, we did perturb T_0.

In order to put forward this new point of view we should slightly modify the classical definition (Definition V.1). Before trying to find such a modification we consider an example in finite dimension. Let $X = \mathbb{R}^2$, $Y = \mathbb{R}$, $x_0 = (0,0) \in X$, and f the mapping $\mathbb{R}^2 \to \mathbb{R}$ given by $f(x_1, x_2) = x_2 - x_1^2 + x_1 x_2$. The linearization of this mapping at the origin is $(D_{(0,0)} f)(x_1, x_2) = x_2$. For any $q \in Y = \mathbb{R}$ denote by H_q the set of solutions to the equation $f(x) = q$, and by L_q the set of those $x \in X = \mathbb{R}^2$ such that $(D_{(0,0)} f)(x - x_0) = q$. We have $H_q = \{(x_1, x_2)$ with $x_2 = (q + x_1^2)/(x_1 + 1)$, for $x_1 \in \mathbb{R}, x_1 \neq -1\}$ and $L_q = \{(x_1, q)$ for $x_1 \in \mathbb{R}\}$. Let $L =$ Ker of $D_{(0,0)} f = L_0$. Let W be the open set of $L = \mathbb{R}$ given by $\{z \in \mathbb{R}, z > -1\}$ and U the open set of $X = \mathbb{R}^2$ given by $\{(x_1, x_2) \in \mathbb{R}^2, x_1 > -1\}$. Consider the map

$$\begin{array}{rccc} \varphi : & Y \times W & \longrightarrow & U \\ & (q, z) & \to & (z, (q + z^2)/(z + 1)) . \end{array}$$

For any $q \in Y$ the mapping $z \to \varphi(q, z)$ maps W into $H_q \cap U$. We have

$$\varphi(q, z) = \varphi(0,0) + (D_{(0,0)} \varphi)(q, z) + E(q, z) = (D_{(0,0)} \varphi)(q, z) + E(q, z) ,$$

where

$$\lim_{(q,z) \to (0,0)} \frac{E(q, z)}{\sqrt{|q|^2 + |z|^2}} = 0 .$$

But here $(D_{(0,0)} \varphi)(q, z) = (z, q)$ so that the error $E(q, z)$ is here the difference $\varphi(q, z) - (z, q)$. And $(z, q) \in L_q$ is a solution to the linearized equation. Thus the difference between the true solution $\varphi(q, z)$ and the solution (z, q) to the linearized equation goes to zero faster then $|q|$ and $|z|$ when these two go to zero.

Remark that L_0 is the tangent space to the manifold H_0 at point $x_0 = 0$. However, for any other $q \in Y$, close to 0, $q \neq 0$, L_q is not tangent to H_q. The following picture shows the level sets H_q and L_q in a neighborhood of the origin for $q = 0, 0.2, 0.4, 0.6$.

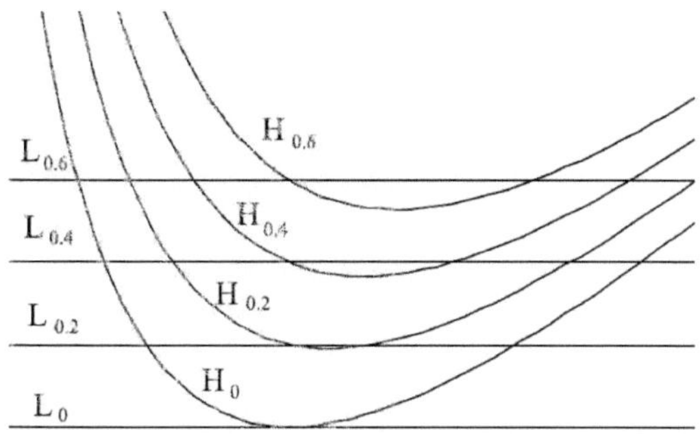

In order to modify the classical definition we note that sometimes it is useful to let q vary in a subspace $\mathcal{K}$ of Y instead of in the whole space Y. The following is a more elaborate version of the definition contained in [14] and [16].

Definition V.3. Let $f : U \to Y$ be a C^1-mapping between an open set U of a Banach space X and a Banach space Y. Let $\mathcal{K}$ be a closed subspace of Y. Let $x_0 \in U$ and $y_0 = f(x_0)$. For any $q \in \mathcal{K}$ denote by H_q and L_q the sets

$$\begin{aligned} H_q &= \{x \in U : \text{ solution to } f(x) = y_0 + q\}, \\ L_q &= \{x \in U : x = x_0 + z, \text{ with } z \text{ solution to } (D_{x_0} f)(z) = q\} \ . \end{aligned}$$

Let $L = \text{Ker } D_{x_0} f = L_0$. Let $H = f^{-1}(y_0 + \mathcal{K}) = \cup_{q \in \mathcal{K}} H_q$. We say that equation $f(x) = y_0 + q$ is stable by linearization at point x_0 in the direction of the subspace $\mathcal{K}$ if there exist a neighborhood V of the origin in $\mathcal{K}$, a neighborhood W of the origin in L, a neighborhood U' of x_0 in X, $U' \subset U$, and a C^1-map φ,

$$\begin{array}{cccc} & \mathcal{K} \times L & & U \\ & \cup & & \cup \\ \varphi : & V \times W & \to & U' \end{array}$$

such that

a) $\varphi(0, 0) = x_0$.

b) $D_{(0,0)}\varphi$ is injective.

c) $(D_{(0,0)}\varphi)(0, z) = z$ for any $z \in L$. Moreover $x = (D_{(0,0)}\varphi)(q, 0)$ is a solution to the equation $(D_{x_0} f)(x) = q$ for any $q \in \mathcal{K}$.

d) φ gives a homeomorphism between $V \times W$ and $H \cap U'$ that maps $\{q\} \times W$ into $H_q \cap U'$ for any $q \in V$.

Concerning this definition we should make some important remarks.

Remark V.4. The map φ in the definition gives a parametrization of $H \cap U'$ by a neighborhood of the origin in $\mathcal{K} \times L$ in such a way that $H \cap U'$ is a differentiable submanifold of X, modeled on the Banach space $\mathcal{K} \times L$. As for any $q \in V$ the map φ maps $\{q\} \times W$ into $H_q \cap U'$, the manifold $H \cap U'$ is endowed with a foliation $\mathcal{H}$ whose leaves are the level sets H_q of f in a neighborhood of x_0. Condition c) ensures that the tangent space to H_0 at x_0 is L and that for any $q \in \mathcal{K}$ there are solutions to the equation $(D_{x_0} f)(x) = q$ belonging to the tangent space $T_{x_0}(H \cap U')$.

The following remark justifies the fact that if Definition V.3 is fulfilled then one can replace the true solutions to $f(x) = q$ by the solutions to the linearized equation.

Remark V.5. Suppose that Definition V.3 is fulfilled. Denote by α the continuous, linear map

$$\begin{array}{rccc} \alpha: & \mathcal{K} & \longrightarrow & X \\ & q & \rightarrow & (D_{(0,0)}\varphi)(q,0). \end{array}$$

By condition c) one has $(D_{x_0} f)(\alpha(q)) = q$. Thus $x_0 + \alpha(q) \in L_q$ and L_q can be described as the affine submanifold of X given by $L_q = x_0 + \alpha(q) + L$. For any $z \in L$ let $\psi_q(z) = x_0 + \alpha(q) + z \in L_q$. Note that ψ_q is an affine parametrization (by the space L) of the affine submanifold L_q. For any $(q, z) \in V \times W \subset \mathcal{K} \times L$ one has

$$\varphi(q, z) = x_0 + (D_{(0,0)}\varphi)(q, z) + E(q, z)$$

and the term $E(q, z)$ fulfils

$$\lim_{(q,z)\to(0,0)} \frac{E(q,z)}{\sqrt{||q||^2 + ||z||^2}} = 0\ .$$

But by virtue of condition c) one has $(D_{(0,0)}\varphi)(q, z) = (D_{(0,0)}\varphi)((q, 0) + (0, z)) = \alpha(q)+z$. Thus $E(q, z) = \varphi(q, z)-(x_0+\alpha(q)+z) = \varphi(q, z)-\psi_q(z)$. This fact justifies the substitution of the true solution $\varphi_q(z)$ to the equation $f(x) = q$ (belonging to H_q) by $\psi_q(z) \in L_q$, because the error made by this substitution goes to zero faster then z and q when these variables go to zero.

Remark V.6. Note that if we take $\mathcal{K} = \{0\}$ in Definition V.3, then that definition only requires that H_0 be a differentiable submanifold of X in a neighborhood of x_0 having L as tangent space at x_0. Thus Definition V.3 with $\mathcal{K} = \{0\}$ coincides with Definition V.1.

Definition V.7. Using the same notation as in Definition V.3, when $f(x) = y_0 + q$ is stable by linearization at the initial point x_0 in the direction of $\mathcal{K} = Y$, we will simply say that it is stable by linearization at that point.

We will now see that Theorem V.2 is still true with respect to the new definition of stability by linearization. Specifically,

Theorem V.8. *Let $f : U \to Y$ be a C^1-mapping between an open set U of a Banach space X and another Banach space Y. Let $x_0 \in U$ and $y_0 \in Y$ be such that $f(x_0) = y_0$. If the linear tangent map $D_{x_0} f$ of f at x_0 is surjective and if the kernel of $D_{x_0} f$ has a topological supplement, then the equation $f(x) = y$ is stable by linearization at point x_0, in the sense of Definition* V.7.

Proof. Denote by L the kernel of $D_{x_0}f$ and by S the topological supplement, so that $X = S \oplus L$. Let us consider the map

$$\begin{array}{rccc} \gamma : & X & \longrightarrow & Y \times L \\ & x_0 + u + z & \rightarrow & (f(x_0 + u + z) - f(x_0), z) \end{array}$$

(defined only in a neighborhood of x_0) with $u \in S$ and $z \in L$.

Since $D_{x_0}f : S \oplus L \to Y$ is surjective, its restriction to S gives an isomorphism between S and Y. Thus

$$\begin{array}{rccc} D_{x_0}\gamma : & S \oplus L = X & \longrightarrow & Y \times L \\ & u + z & \rightarrow & ((D_{x_0}f)(u), z) \end{array}$$

is an isomorphism. Then the inverse function theorem ensures the existence of a neighborhood U' of x_0 in X and a neighborhood $V \times W$ of $\gamma(x_0) = (0,0)$ in $Y \times L$ so that γ gives an isomorphism between U' and $V \times W$. Let φ be the inverse map

$$\begin{array}{rccc} \varphi : & V \times W & \longrightarrow & U' \subset X \\ & (q, z) & \rightarrow & \gamma^{-1}(q, z). \end{array}$$

The map φ fulfils (by construction) the properties a), b), c) and d) of Definition V.7. □

Remark V.9. Concerning the existence of the foliation $\mathcal{H}$ in a neighborhood of x_0 whose leaves are the level sets of f, we should cite Lemma A.2 in [26] and its proof.

V.3 How to apply the definition of stability by linearization of Einstein's equation in the presence of matter

We are now going to apply both Definition V.7 and Theorem V.8 to Einstein's equation in the presence of matter, $G(\tilde{g}) = \chi T$. We first specify which Banach spaces X and Y we must consider in this case and also which map f from an open set U of X into Y we must take. We explained in the beginning of the former section that, given an initial Lorentzian metric $\tilde{g}_0$ and an initial stress-energy tensor T_0 satisfying Einstein's equation $G(\tilde{g}_0) = \chi T_0$, for any stress-energy tensor T close to T_0 we were in principle interested in metrics $\tilde{g}$ close to $\tilde{g}_0$ which are solutions to the equation $G(\tilde{g}) = \chi T$. This may suggest that the most convenient map in order to apply Definition V.7 to this situation is $\tilde{g} \to G(\tilde{g})$. But it is necessary to notice that we can not take any arbitrary T close to T_0 and look for metrics $\tilde{g}$ such that $G(\tilde{g}) = \chi T$ because, on one hand, the expression of T in many cases contains the metric $\tilde{g}$, and, on the other hand, T is not arbitrary, as it must satisfy the supplementary condition $\mathrm{div}_{\tilde{g}} T = 0$. This is why it is preferable, instead of working directly with Einstein's equation $G(\tilde{g}) = \chi T$, to work with Cauchy data on a hypersurface M of V.

Despite the fact that all that we say can be formulated for quite general stress-energy tensor types (specifically, for tensors corresponding to the holonomic media described in Section IV.9), in order to delineate the ideas and make the exposition of this section clearer, we situate ourselves in the class of stress-energy tensors of perfect fluids.

Let us recall which are the Cauchy data for perfect fluids as stated in Theorem IV.7 and Corollary IV.12. In the first place we must assume that a state function is given which relates density ρ and pressure p of perfect fluids, so that it is always possible to express one in terms of the other. We are therefore situated in a manifold V of dimension 4 in which we give a hypersurface M and a vector field N of V with support in M, transverse to M. We can consequently take as Cauchy data on M the four objects $(g, k, \rho|_M, v|_M)$, where g is a Riemannian metric on M, k is a covariant symmetric 2-tensor on M, $\rho|_M$ is a function on M and $v|_M$ a vector field of M. The field $v|_M$ will stand for the tangential component of the vector field u of velocities of the fluid, so that for each $x \in M$ we have $u_x = v_x + zN_x$, with $v_x \in T_x(M)$. The function $\rho|_M$ represents the restriction to M of the density of the perfect fluid. The Cauchy data $(g, k, \rho|_M, v|_M)$ are supposed to satisfy the constraint equations required in Theorem IV.7:

$$\begin{cases} \gamma(g,k) = \chi X \\ \mathcal{H}(g,k) = \chi F \ , \end{cases} \tag{V.2}$$

where X is the 1-form on M given by $X = i(N)T$ and F is the function on M given by $F = -(i(N)T)(N)$. As T (at points of M) depends on g, on $\rho|_M$ and on $v|_M$, both F and X will depend on these variables and, thus, equations (V.2) relate the Cauchy data. If the Cauchy data are taken in the Sobolev spaces specified in Theorem IV.7 and $\rho|_M$ also satisfies the conditions of Corollary IV.12, then Theorem IV.7 ensures that we can find, in a neighborhood of M in V, the Lorentzian metric $\tilde{g}$ and the stress-energy tensor T expressed in the form $T = (\rho+p)u^* \otimes u^* + p\tilde{g}$, where u^* is the 1-form associated by the metric $\tilde{g}$ to the vector field u of velocities that satisfies $\tilde{g}(u,u) = -1$, so that Einstein's equation $G(\tilde{g}) = \chi T$ is fulfilled, and that the following restriction conditions in M are satisfied: (a) At each point $x \in M$ we have that $\tilde{g}(N_x, N_x) = -1$ and N_x is perpendicular to $T_x(M)$ with respect to $\tilde{g}$. (b) The restriction of $\tilde{g}$ to M is g. (c) Twice the second fundamental form of the hypersurface M of V with respect to $\tilde{g}$ is k. (d) At points $x \in M$ the following is satisfied: $\rho|_M = \rho$ and $u_x = (v|_M)_x + zN_x$ with $z > 0$.

Let us now see in further detail which are, in our, case, the function F and the 1-form X which appear in the constraint equations. At each point of M we have $F = -(i(N)T)(N) = -T(N,N) = -(\rho + p)(u^*(N))^2 + p$. But $u^*(N) = \tilde{g}(u,N) = \tilde{g}(v + zN, N) = -z$. However, the equality $\tilde{g}(u,u) = -1$ is written as

$\tilde{g}(v + zN, v + zN) = -1$, which gives $g(v,v) - z^2 = -1$. Therefore (as z must be positive), $z = \sqrt{g(v,v)+1}$. Thus,

$$F = -(\rho|_M + p|_M)(1 + g(v|_M, v|_M)) + p|_M . \tag{V.3}$$

Furthermore, $X = i(N)T$. If x is a point of M and w a vector belonging to $T_x(M)$, at point x we have $X(w) = T(N,w) = (\rho + p)u^*(N)u^*(w) = (\rho + p) \cdot \tilde{g}(u,N)\tilde{g}(u,w) = (\rho+p)\tilde{g}(v+zN,N)\tilde{g}(v+zN,w) = (\rho+p)((-z)g(v,w))$. Taking into account the former expression of z we finally obtain

$$X = -(\rho|_M + p|_M)\sqrt{1 + g(v|_M, v|_M)}(v|_M)^* , \tag{V.4}$$

where $(v|_M)^*$ here denotes the 1-form on M associated through the Riemannian metric g to the vector field $v|_M$ on M.

Let us now specify in which spaces we take the Cauchy data $(g, k, \rho|_M, v|_M)$. We respectively denote by $\mathcal{F}^s(M)$, $\Omega^s(M)$ and $\mathcal{S}^s(M)$ the Sobolev spaces of functions, of 1-forms and of symmetric covariant 2-tensors of regularity degree s. As we have seen in Section I.4, the Sobolev spaces depend (generally speaking) on the Riemannian metric considered. When we consider the Sobolev spaces on the manifold M it will always be with respect to the initial metric g_0. Using the notation adopted in Section I.4, $\mathcal{F}^s(M)$ is nothing but $\mathcal{W}^s(M)_0$ and $\Omega^s(M)$ agrees with $\mathcal{W}^s(M)_1$. We must as well notice that to be able to apply the Sobolev lemma to these spaces we will assume that the Riemannian manifold (M, g_0) *is complete, it has positive injectivity radius and bounded curvature.*

Given an initial C^∞-function f_0 on M, we will denote by $\mathcal{F}^s(f_0)$ the set of functions that can be expressed in the form $f = f_0 + h$, with $h \in \mathcal{F}^s(M)$. We topologize the set $\mathcal{F}^s(f_0)$ by imposing that the map $f \to h = f - f_0$ from $\mathcal{F}^s(f_0)$ to $\mathcal{F}^s(M)$ should be a homeomorphism. In an analogous way, given a C^∞ 1-form ω_0, we denote by $\Omega^s(\omega_0)$ the set of 1-forms ω such that $\omega - \omega_0 \in \Omega^s(M)$, and given a C^∞ symmetric covariant 2-tensor, α_0, we denote by $\mathcal{S}^s(\alpha_0)$ the set of symmetric covariant 2-tensors, α, such that $\alpha - \alpha_0 \in \mathcal{S}^s(M)$. We topologize the sets $\Omega^s(\omega_0)$ and $\mathcal{S}^s(\alpha_0)$ in an analogous manner as $\mathcal{F}^s(f_0)$.

While we are only talking about Cauchy data $(g, k, \rho|_M, v|_M)$ on M and there exists no danger of confusion, we simply write ρ and v instead of $\rho|_M$ and $v|_M$. Given some initial C^∞ Cauchy data (g_0, k_0, ρ_0, v_0), satisfying the constraint equations (V.2) (with X and F given by (V.4) and (V.3)), we will work with 4-tuples (g, k, ρ, v) belonging to $\mathcal{S}^s(g_0) \times \mathcal{S}^{s-1}(k_0) \times \mathcal{F}^{s-2}(\rho_0) \times \Omega^{s-2}(v_0)$, where s will be a positive integer greater than 4. The reason to use spaces $\mathcal{S}^s(g_0)$, $\mathcal{S}^{s-1}(k_0)$, $\Omega^{s-2}(v_0)$ and $\mathcal{F}^{s-2}(\rho_0)$ instead of $\mathcal{S}^s(M)$, $\mathcal{S}^{s-1}(M)$, $\Omega^{s-2}(M)$ and $\mathcal{F}^{s-2}(M)$ is the following: The functions of a Sobolev space belong specifically to $L^2(M)$, but very often the coefficients of the initial metric g_0 are not in $L^2(M)$. This is why, very often, g_0 is not in $\mathcal{S}^s(M)$. The same happens with k_0, v_0 and ρ_0.

We may notice that $\mathcal{S}^s(g_0) \subset \mathcal{S}^s_{loc}(M)$ and an analogous inclusion happens with the other spaces $\mathcal{S}^{s-1}(k_0)$, $\Omega^{s-2}(v_0)$ and $\mathcal{F}^{s-2}(\rho_0)$. Consequently, when we apply Theorem IV.7 with the Cauchy data $(g,k,\rho,v) \in \mathcal{S}^s(g_0) \times \mathcal{S}^{s-1}(k_0) \times \mathcal{F}^{s-2}(\rho_0) \times \Omega^{s-2}(v_0)$ the assumptions of that theorem relative to the Cauchy data belonging to local Sobolev spaces are satisfied.

We now take the hypersurface M of V and the vector field N (vector field of V with support in M) transverse to M in a convenient way, as we explain next. We will take M in such a way that at each point $x \in M$, the vector field u of velocities of the initial fluid corresponding to T_0 (which is a perfect fluid) is perpendicular to M with respect to the initial metric $\tilde{g}_0$ (let us assume that this is possible) and we take the transverse field N to be equal to u at points of M. Thanks to this choice of M and N, the Cauchy data (g_0,k_0,ρ_0,v_0) on M corresponding to the initial metric $\tilde{g}_0$ and the initial tensor T_0, will satisfy $v_0 = 0$.

Let us consider the map

$$\begin{array}{ccc} \mathcal{S}^s(g_0) \times \mathcal{F}^{s-2}(\rho_0) \times \Omega^{s-2}(v_0) & \xrightarrow{H} & \mathcal{S}^s(g_0) \times \mathcal{F}^{s-2}(F_0) \times \Omega^{s-2}(X_0) \\ (g,\rho,v) & \to & (g,F,X)\,, \end{array}$$

where F and X are given as a function of g, ρ and v through the expressions (V.3) and (V.4). Specifically, the expressions of F and X are $F = -(\rho + p(\rho)) \cdot (1 + g(v,v)) + p(\rho)$ and $X = -(\rho + p(\rho))\sqrt{1+g(v,v)}v^*$. Let us now calculate the linear tangent map of H at point (g_0,ρ_0,v_0), that is to say $(DH)_{(g_0,\rho_0,v_0)}$. To do this, we replace g with $g_0 + h$ and ρ with $\rho_0 + f$ on the former expressions of F and X, while we do not change v because $v_0 = 0$. We must bear in mind that, in order to make these substitutions, the component-wise expression of v^* in a local chart are $(v^*)_i = g_{ij}v^j = ((g_0)_{ij} + h_{ij})v^j$. Once these substitutions are made we consider only the linear terms in h, f and v of the resulting expressions. We finally see that the map $(DH)_{(g_0,\rho_0,v_0)}$ is

$$\begin{array}{ccc} \mathcal{S}^s(M) \times \mathcal{F}^{s-2}(M) \times \Omega^{s-2}(M) & \longrightarrow & \mathcal{S}^s(M) \times \mathcal{F}^{s-2}(M) \times \Omega^{s-2}(M) \\ (h,f,v) & \to & (h,-f,-(\rho_0 + p_0)v)\;. \end{array}$$

As $\rho_0 + p_0$ is a positive function on each point of M, the former map is an isomorphism. By the inverse function theorem in Banach spaces, the map H can be inverted in a neighborhood of the initial Cauchy data. Therefore, ρ and v can be expressed (through this inverse function) as functions of g, F and X, in a convenient neighborhood of the initial data (g_0, F_0, X_0). This will allow the following construction:

Given a couple (F,X) close to the initial couple (F_0,X_0), we consider the couple (g,k), close to (g_0,k_0), whose components are solutions to (V.2) (with X and F on the right-hand side of (V.2) being the ones in the given couple (F,X)). We complete each couple (g,k), solution to (V.2) with the only couple (ρ,v) obtained from g, F and X through the expressions we mentioned before (obtained through the inverse H^{-1}). Thus, the data (g,k,ρ,v) satisfies the constraint equations. This determines a Lorentzian metric $\tilde{g}$ and a stress-energy tensor T (of a perfect fluid) in a neighborhood of M in V such that $G(\tilde{g}) = \chi T$, such that the restriction of $\tilde{g}$ to M is g, that twice the second fundamental form of M is k and that we have $F = -(i(N)T)(N)$ and $X = i(N)T$ on M.

In short, given a couple (F, X) close to (F_0, X_0) we may ask which are the Lorentzian metrics $\tilde{g}$ in a neighborhood of M in V that satisfy $G(\tilde{g}) = \chi T$ and $\mathrm{div}_{\tilde{g}} T = 0$, with tensor T such that $(i(N)T)(N) = -F$ and $i(N)T = X$ on M. The answer is that they are those $\tilde{g}$ such that their (g, k) on M are solution to (V.2) with X and F on the right-hand side of (V.2) being the ones in the given couple (F, X).

These considerations lead us to take the map

$$\begin{array}{cccc} & \mathcal{S}^s(g_0) \times \mathcal{S}^{s-1}(k_0) & & \\ & \cup & & \\ \Phi : & \mathcal{U} & \longrightarrow & \mathcal{F}^{s-2}(F_0) \times \Omega^{s-2}(X_0) \\ & (g, k) & \to & (\mathcal{H}(g, k), \gamma(g, k)) \ , \end{array}$$

where $\mathcal{U}$ is a certain neighborhood of (g_0, k_0) in space $\mathcal{S}^s(g_0) \times \mathcal{S}^{s-1}(k_0)$, as the map of Definition V.7 when we apply it to the case of Einstein's equation in the presence of matter. One should see, though, that when (g, k) belongs to $\mathcal{S}^s(g_0) \times \mathcal{S}^{s-1}(k_0)$ then $\mathcal{H}(g, k)$ belongs to $\mathcal{F}^{s-2}(F_0)$ and $\gamma(g, k)$ belongs to $\Omega^{s-2}(X_0)$. Proposition V.10 which we will establish later on ensures that this is true with very general assumptions.

As a summary of all the former, we will now explain **in an intuitive way** (i.e. avoiding technical terms) what it means that Einstein's equation corresponding to a perfect fluid is stable by linearization (or linearization-stable) at an initial metric $\tilde{g}_0$ and an initial stress-energy tensor T_0.

Let us first take a hypersurface M of V in such a way that on each point $x \in M$ the velocity vector field u of the fluid corresponding to T_0 (which is a perfect fluid) will be perpendicular to M with respect to the initial metric $\tilde{g}_0$ (we assume that this is possible) and we take the transverse field N to be equal to u at any point of M. We denote by F_0 and X_0 the function and the 1-form on M given by $F_0 = -(i(N)T_0)(N)$ and $X_0 = i(N)T_0$. For each couple (F, X) close to (F_0, X_0) we consider the set $H_{(F,X)}$ of Lorentzian metrics $\tilde{g}$ on a neighborhood of M in V which are close to $\tilde{g}_0$ and satisfy the equations $G(\tilde{g}) = \chi T$ and $\mathrm{div}_{\tilde{g}} T = 0$ with stress-energy tensor T satisfying $(i(N)T)(N) = -F$ and $i(N)T = X$ on M. We also consider the set $L_{(F,X)}$ of covariant tensors $\tilde{h}$ which are solutions to the linearized Einstein equation $(DG)_{\tilde{g}_0}(\tilde{h}) = \chi(T - T_0)$. The fact that Einstein's equation is linearization-stable at the initial metric $\tilde{g}_0$ means that the following conditions are satisfied:

1. For each (F, X) close to (F_0, X_0) the set $L_{(F,X)}$ is a linear manifold parametrized by a Banach space L (independent of F and X).

2. Each $H_{(F,X)}$ is a differentiable manifold parametrized by the same space L.
3. The initial metric, which belongs to $H_{(F_0,X_0)}$, corresponds to parameter $z = 0 \in L$.
4. The error made in replacing the true solution belonging to $H_{(F,X)}$, corresponding to a parameter $z \in L$, by the solution to the linearized equation belonging to $L_{(F,X)}$ corresponding to the same parameter $z \in L$, vanishes faster than z, than $F - F_0$ and than $X - X_0$, when these three quantities go to zero.

All these properties can be summarized by saying that it is admissible to linearize Einstein's equation at the initial metric and work with the solutions of the linearization instead of the solutions of the true equation. If we wish to specify the meaning of the expression "close to" which we have used a number of times throughout the latter explanation, we must then turn to the different Sobolev spaces used in the definition of the map Φ. For instance, when we talked about couple (F, X) close to (F_0, X_0) we should have said that couples (F, X) are in a neighborhood of (F_0, X_0) in the space $\mathcal{F}^{s-2}(F_0) \times \Omega^{s-2}(X_0)$. Also, when we mentioned the Lorentzian metrics $\tilde{g}$ close to $\tilde{g}_0$ we should have said that every couple (g, k) corresponding to each $\tilde{g}$ (where g is the restriction to M of $\tilde{g}$ and k is twice the second fundamental form of M in V with respect to the metric $\tilde{g}$) is in a neighborhood of (g_0, k_0) in the space $\mathcal{S}^s(g_0) \times \mathcal{S}^{s-1}(k_0)$. Finally, when we said that the error vanishes faster than z, than $F - F_0$ and than $X - X_0$ when these three quantities go to zero, we should have specified that we were referring to the convergence of z, $F - F_0$ and $X - X_0$ in their respective Banach spaces (which are Sobolev spaces).

V.4 Change of notation

As from now on we will make calculations in certain local charts and we will use sub-indexes and super-indexes for the components of tensors, it is convenient to change the expression of the initial couple (g_0, k_0) provided that the sub-indexes of this couple are not be confused with the sub-indexes of the components of these tensors. Consequently from now on, we will denote the initial couple by (g, k) and others couples different from the initial one by (g', k'). For such a couple, we name h and K the differences $h = g' - g$ and $K = k' - k$, in such a way that we have $g' = g + h$, $k' = k + K$. Similarly we denote by T the initial stress-energy tensor and by (F, X) the initial couple given by $F = -(i(N)T)(N)$ and $X = i(N)T$.

V.5 Technical details concerning the map Φ

As much in the case of Einstein's equation in the presence of matter, as for the case of Einstein's equation in the vacuum, the mapping which we consider for the concept of stability by linearization must be the map

$$\begin{array}{lccc} & \mathcal{S}^s(g)\times\mathcal{S}^{s-1}(k) & & \\ & \cup & & \\ \Phi: & \mathcal{U} & \longrightarrow & \mathcal{F}^{s-2}(F)\times\Omega^{s-2}(X) \\ & (g',k') & \to & (\mathcal{H}(g',k'),\gamma(g',k'))\ , \end{array}$$

where $\mathcal{U}$ is a certain neighborhood of (g,k) in the space $\mathcal{S}^s(g)\times\mathcal{S}^{s-1}(k)$ (even though, in the vacuum we have $F=X=0$).

We must see, though, that the map Φ actually maps a certain neighborhood $\mathcal{U}$ of (g,k) in the space $\mathcal{S}^s(g)\times\mathcal{S}^{s-1}(k)$ into $\mathcal{F}^{s-2}(F)\times\Omega^{s-2}(X)$. This is a consequence of the following proposition.

Proposition V.10. *Assume that the initial Riemannian manifold (M,g) (with g of class C^∞) is complete. We denote by $|\ \ |_g$ the norm associated to the inner product $(\ ,\ )$ of p-tensors relative to metric g, which has been introduced in Section I.3. Take a fixed integer $s>4$. We assume an initial C^∞ symmetric 2-tensor k is given on M, such that for each integer l such that $0\le l\le s-1$ the function $|\nabla^{(l)}k|_g$ is bounded by a constant (here $\nabla^{(l)}$ indicates the l-th times iteration of the covariant differentiation ∇). Then there exists a neighborhood $\mathcal{U}$ of (g,k) in the space $\mathcal{S}^s(g)\times\mathcal{S}^{s-1}(k)$ such that for each $(g',k')\in\mathcal{U}$ the difference $\mathcal{H}(g',k')-\mathcal{H}(g,k)$ belongs to $\mathcal{F}^{s-2}(M)$ and the difference $\gamma(g',k')-\gamma(g,k)$ belongs to $\Omega^{s-2}(M)$.*

To prove this proposition we will need the two following lemmas of algebraic character.

Lemma V.11. *Let E be a real vector space of dimension n, endowed with a positive definite inner product g. If A is a covariant p-tensor on E and B is a covariant q-tensor, we denote by $C(A\otimes B)$ the covariant $(p+q-2)$-tensor obtained from $A\otimes B$ by contraction of the first factors of A and B through the metric g. In components with respect to a basis:*

$$(C(A\otimes B))_{i_2\dots i_p j_2\dots j_q}=g^{ij}A_{ii_2\dots i_p}B_{jj_2\dots j_q}=A^j{}_{i_2\dots i_p}B_{jj_2\dots j_q}\ .$$

For each given couple of tensors A and B of respective degrees p and q, the following inequality is satisfied:

$$|C(A\otimes B)|_g\le\mathcal{K}|A|_g|B|_g\ ,$$

where $\mathcal{K}$ is a constant that depends solely on the dimension n of E and on the degrees p and q of the tensors, but not on A and neither on B.

Proof. To prove this, we choose an orthonormal basis of E with respect to the metric g and we use the components of the tensors A and B with respect to this basis. If $(\ ,\ )$ denotes the inner product of $(p+q-2)$-tensors with respect to the metric g, we have

$$(C(A\otimes B),C(A\otimes B)) = \sum_{i_2\dots i_p j_2\dots j_q} \left(C(A\otimes B)_{i_2\dots i_p j_2\dots j_q}\right)^2$$
$$= \sum_{i_2\dots i_p j_2\dots j_q} \left(\sum_i A_{ii_2\dots i_p} B_{ij_2\dots j_q}\right)^2 . \qquad \text{(V.5)}$$

However, if we fix the indexes $i_2\dots i_p, j_2\dots j_q$ and denote by X the vector of components $X_i = A_{ii_2\dots i_p}$ and by Y the vector of components $Y_i = B_{ij_2\dots j_q}$, we have, owing to Schwarz inequality,

$$\sum_i A_{ii_2\dots i_p} B_{ij_2\dots j_q} = g(X,Y) \le |X|_g\,|Y|_g = \sqrt{\sum_i (A_{ii_2\dots i_p})^2}\,\sqrt{\sum_j (B_{jj_2\dots j_q})^2}\,.$$

But, obviously,

$$\sum_i (A_{ii_2\dots i_p})^2 \le \sum_{i_2\dots i_p}\sum_i (A_{ii_2\dots i_p})^2 = |A|_g^2$$

and similarly,

$$\sum_j (B_{jj_2\dots j_q})^2 \le |B|_g^2\,.$$

Thus,

$$\sum_i A_{ii_2\dots i_p} B_{ij_2\dots j_q} \le |A|_g\,|B|_g\,.$$

Replacing this in (V.5) we may obtain the desired inequality. □

Lemma V.12. *Let E be a real vector space of dimension n, endowed with a positive definite inner product g.*

1. *Let $\{e_1\dots e_n\}$ be a basis of E. Let α be a symmetric covariant 2-tensor on E such that its components α_{ij} in the former basis form an invertible matrix. Let $Inv(\alpha)$ be the contravariant 2-tensor whose components in the basis $\{e_i\}$ are the elements α^{ij} of the inverse matrix of (α_{ij}). The tensor $Inv(\alpha)$ does not depend on the chosen basis (it is intrinsic).*

2. *There exist two constants C and C' such that for each symmetric covariant 2-tensor h such that $|h|_g < C$, the contravariant tensor $Inv(g+h)$ is well defined (see former section) and satisfies $|Inv(g+h) - Inv(g)|_g < C'|h|_g$.*

Proof. The first assertion is immediate using the component-wise formulas for the change of basis of covariant and contravariant tensors. We will prove the second. Let us consider the general linear group $GL(n,\mathbb{R})$ as a open subset of $\mathbb{R}^{n^2}$. Let C be a positive constant such that for every matrix $h \in \mathbb{R}^{n^2}$ such that $|h| < C$ we have $Id + h \in GL(n,\mathbb{R})$ (here $|\ |$ indicates the norm of the usual inner product in $\mathbb{R}^{n^2}$). As the mapping which maps every matrix of $GL(n,\mathbb{R})$ to its inverse is of class C^∞, the mean value theorem in each one of the n^2 coordinates of $GL(n,\mathbb{R})$ ensures that there exists a constant C' such that, for each matrix h with $|h| < C$, one has $|(Id+h)^{-1} - Id| < C'|h|$. Then, for each symmetric covariant 2-tensor h on E such that $|h|_g < C$, the contravariant 2-tensor $Inv(g+h)$ will satisfy $|Inv(g+h) - Inv(g)|_g < C'|h|_g$. Indeed, it is sufficient to use components referred to any orthonormal basis with respect to g and apply to them the inequality obtained in the general linear group. □

Proof of Proposition V.10. Let us recall the explicit expressions of $\mathcal{H}(g',k')$ and $\gamma(g',k')$ given by (IV.25) and (IV.23), which are:

$$\begin{cases} \mathcal{H}(g',k') = \dfrac{1}{8}(k' \cdot_{g'} k' - 4R' - (\mathrm{tr}_{g'}k')^2) \\ \gamma(g',k')_i = \dfrac{1}{2}\left\{g'^{rs}\nabla'_r k'_{is} - \partial_i(\mathrm{tr}_{g'}k')\right\} \ , \end{cases} \tag{V.6}$$

where $\gamma(g',k')_i$ indicates the i-th component of the 1-form $\gamma(g',k')$ in a local chart of M of coordinates (x^1,x^2,x^3), ∇' indicates the covariant differentiation with respect to the metric g' and ∂_i indicates, as usual, $\partial/\partial x^i$.

Let us see that for couples (g',k') in a certain neighborhood of (g,k) in $\mathcal{S}^s(g)\times\mathcal{S}^{s-1}(k)$ the difference $\gamma(g',k')-\gamma(g,k)$ belongs to $\Omega^{s-2}(M)$. As we have assumed that the Riemannian manifold (M,g) is complete, the Sobolev space $\Omega^{s-2}(M)$ consists of those 1-forms which are in L^2 and which have weak derivatives in L^2 till order $s-2$ (see Section I.4). Consequently, we must prove that the difference $\gamma(g',k')-\gamma(g,k)$ is in L^2 and has derivatives in L^2 till order $s-2$. Recall that owing to the notation adopted in Section I.4, when we say that the difference $\gamma(g',k')-\gamma(g,k)$ is in L^2 we mean exactly that the function $|\gamma(g',k')-\gamma(g,k)|_g$ belongs to $L^2(d\mu)$, where $d\mu$ is the measure given by the volume element of g. The terms $g'^{rs}\nabla'_r k'_{is}$ and $-\partial_i(\mathrm{tr}_{g'}k')$ from the expression of $\gamma(g',k')_i$ given in (V.6) correspond to global 1-forms, despite the fact that there, they are given in components with respect to a local chart. We denote by $T(g',k')$ the first term. The second term is precisely $-d(\mathrm{tr}_{g'}k')$. We will then have $2\gamma(g',k') = T(g',k') - d(\mathrm{tr}_{g'}k')$. To prove that the difference $\gamma(g',k')-\gamma(g,k)$ is in L^2 we will show that each of the differences $T(g',k')-T(g,k)$ and $d(\mathrm{tr}_{g'}k') - d(\mathrm{tr}_g k)$ are in L^2. We shall start with $T(g',k')-T(g,k)$.

In Section III.1, we introduced the two-times covariant and one-time contravariant tensor Q defined by $Q(X,Y) = \nabla'_X Y - \nabla_X Y$. In expression $g'^{rs}\nabla'_r k'_{is}$ of $T(g',k')_i$ we write the covariant differentiation ∇' as a function of ∇ and of Q. To do this, we must bear in mind that if X, Y, Z are vector fields,

$$(\nabla'_X k')(Y,Z) = X(k'(Y,Z)) - k'(\nabla'_X Y, Z) - k'(Y, \nabla'_X Z) \ ,$$

and we replace here $\nabla'_X Y$ by $\nabla_X Y + Q(X,Y)$ and we do the same with $\nabla'_X Z$. By doing all this we obtain

$$g'^{rs}\nabla'_r k'_{is} = g'^{rs}(\nabla_r k'_{is} - Q^l_{ri}k'_{ls} - Q^l_{rs}k'_{il}) \ . \tag{V.7}$$

We now replace k' with $k+K$. We obtain

$$g'^{rs}\nabla'_r k'_{is} = g'^{rs}(\nabla_r(k+K)_{is} - Q^l_{ri}(k+K)_{ls} - Q^l_{rs}(k+K)_{il}) \,. \tag{V.8}$$

Let us recall the expression of Q obtained in (III.2):

$$Q^i_{jk} = \frac{1}{2} g'^{il}(\nabla_j h_{kl} + \nabla_k h_{lj} - \nabla_l h_{jk}) \,. \tag{V.9}$$

The components g'^{rs} and g'^{il} that appear in both expressions mean the corresponding elements of the inverse matrix of g'. Owing to Lemma V.12 these elements define in the tangent space at each point an intrinsic contravariant tensor which we named $Inv(g')$. Replacing now g' by $g+h$, the two expressions (V.8) and (V.9) become

$$\begin{aligned} g'^{rs}\nabla'_r k'_{is} &= Inv(g+h)^{rs}\big(\nabla_r(k+K)_{is} - Q^l_{ri}(k+K)_{ls} - Q^l_{rs}(k+K)_{il}\big) \,, \\ Q^i_{jk} &= \frac{1}{2} Inv(g+h)^{il}(\nabla_j h_{kl} + \nabla_k h_{lj} - \nabla_l h_{jk}) \,. \end{aligned} \tag{V.10}$$

We replace $Inv(g+h)$ by $Inv(g) + (Inv(g+h) - Inv(g))$ on the right-hand side of the first equality of (V.10). Then

$$\begin{aligned} (T(g',k') - T(g,k))_i = {} & Inv(g)^{sr}\{\nabla_r K_{is} - Q^l_{ri}(k+K)_{ls} \\ & - Q^l_{rs}(k+K)_{il}\} + (Inv(g+h) - Inv(g))^{rs}\{\nabla_r(k+K)_{is} \\ & - Q^l_{ri}(k+K)_{ls} - Q^l_{rs}(k+K)_{il}\} \,. \end{aligned} \tag{V.11}$$

All the terms in this equality correspond to globally defined 1-forms and the terms on the right-hand side are obtained by convenient contractions of tensorial products. We denote by A the 1-form given by

$$A_i = Inv(g)^{sr}\{\nabla_r K_{is} - Q^l_{ri}(k+K)_{ls} - Q^l_{rs}(k+K)_{il}\} \,,$$

and by B the 1-form given by

$$B_i = (Inv(g+h) - Inv(g))^{rs}\{\nabla_r(k+K)_{is} - Q^l_{ri}(k+K)_{ls} - Q^l_{rs}(k+K)_{il}\} \,.$$

On one hand (by definition of $|\ |_g$) we have $|Inv(g)_x|_g < C$ for each $x \in M$. On the other hand, Q is obtained by contractions of $Inv(g+h)$ and ∇h. Owing to Lemma V.12, there exist constants M_1 and M_2 such that if $|h|_g < M_1$ then $|Inv(g+h) - Inv(g)|_g < M_2|h|_g$. Owing to Lemma V.11 we see that if $|h|_g < C'$, then $|Q|_g < C''|h|_g|\nabla h|_g$, where C'' is a certain constant. Thus

$$|A|_g < C(|\nabla K|_g + C''|h|_g|\nabla h|_g(|k|_g + |K|_g)) \,.$$

Similarly, we have

$$|B|_g < C_1|h|_g\{|\nabla k|_g + |\nabla K|_g + C_2|h|_g|\nabla h|_g(|k|_g + |K|_g)\} \,.$$

From expression (V.11) we have $|T(g',k') - T(g,k)|_g < |A|_g + |B|_g$. As $|k|_g$ and $|\nabla k|_g$ are functions on M bounded by a constant, and as $|h|_g$, $|\nabla h|_g$, $|K|_g$ and $|\nabla K|_g$ are functions in L^2, it is clear from the obtained bounds for $|A|_g$ and $|B|_g$ that $|T(g',k') - T(g,k)|_g$ is in L^2.

It only remains to see that the successive covariant differentiations of the difference $T(g',k') - T(g,k)$ till order $s-2$ are also in L^2. If we covariantly differentiate expression (V.11), covariant differentiations of the contravariant tensor $Inv(g+h) - Inv(g)$ appear which we need to calculate. Let us place in a local chart. As (g'^{ij}) is the inverse matrix of (g'_{ij}), we have:

$$\nabla_r g'^{is} = -g'^{ij}(\nabla_r g'_{jk})g'^{ks} \ .$$

This shows that tensor $\nabla Inv(g')$ is obtained by convenient contractions of the tensor product $-g' \otimes \nabla g' \otimes g'$. Provided that $g' = g+h$ and $\nabla g = 0$, we can say that tensor $\nabla Inv(g+h)$ is obtained by contractions of product $-(g+h) \otimes \nabla h \otimes (g+h)$. Thus,

$$|\nabla(Inv(g+h) - Inv(g))|_g = |\nabla Inv(g+h)|_g < C(C_1 + C_2|h|_g + C_3|h|_g^2)|\nabla h|_g \ .$$

If we now take covariant differentiation in (V.11), using this inequality and proceeding as before, we see that $\nabla(T(g',k') - T(g,k))$ is in L^2. Similarly, successive derivatives till order $s-2$ will be in L^2.

The proof that the difference $d(\mathrm{tr}_{g'}k') - d(\mathrm{tr}_g k)$ and all of its covariant differentiations are in L^2 can be done in an analogous way, but is much easier in its calculations.

Neither will the proof that the difference $\mathcal{H}(g',k') - \mathcal{H}(g,k)$ and all of its covariant differentiations till order $s-2$ are in L^2 be made explicitly. We just mention that as in the expression of $\mathcal{H}(g',k')$ appears the scalar curvature of g', one should use equation (III.1) giving the expression of Ricci tensor of g' as a function of Ricci tensor of g and of tensor Q. By contraction of expression (III.1) we obtain an expression of the scalar curvature of g', using which, one can copy almost exactly the reasoning we made to see that the difference $T(g',k') - T(g,k)$ and all its covariant differentiations till order $s-2$ were in L^2. □

V.6 Tangent linear map of Φ

For use in subsequent chapters we will give now the explicit expression of $D_{(g,k)}\Phi$ (tangent linear map of Φ at the initial couple (g,k)). As $\Phi(g',k')$ is the couple $(\mathcal{H}(g',k'), \gamma(g',k'))$, we will separately calculate the tangent linear maps of each element of the couple. We start with $D_{(g,k)}\gamma$. To calculate its tangent linear map we replace g' by $g+h$ and k' by $k+K$ in the expression of $\gamma(g',k')$ and we keep solely the linear part in h and K of the resulting expression. We use symbol $\approx$ to denote equality of these two expressions, up to terms of order > 1 in h or K.

We start from the expression in coordinates of $\gamma(g',k')$ given by (V.6). Concerning the factor g'^{rs} which appears in the term $g'^{rs}\nabla'_r k'_{is}$ of the expression of $\gamma(g',k')_i$, notice that equation (III.7) shows that

$$g'^{rs} = (g+h)^{rs} \approx g^{rs} - g^{ri}h_{ij}g^{js} = g^{rs} - h^{rs} \ .$$

In the former expression the meaning of the super-indexes is unclear. While g'^{rs} stands for the term corresponding to the inverse matrix of (g'_{ij}), h^{rs} stands for the component corresponding to the contravariant 2-tensor obtained from covariant tensor h through the initial metric. From equation (V.7) of $g'^{rs}\nabla'_r k'_{is}$ we find

$$g'^{rs}\nabla'_r k'_{is} \approx (g^{rs} - h^{rs})(\nabla_r(k+K)_{is} - Q^l_{ri}(k+K)_{ls} - Q^l_{rs}(k+K)_{il}) \ ,$$

and therefore,

$$g'^{rs}\nabla'_r k'_{is} \approx g^{rs}\nabla_r k_{is} + g^{rs}\nabla_r K_{is} - h^{rs}\nabla_r k_{is} - g^{rs}Q^l_{ri}k_{ls} - g^{rs}Q^l_{rs}k_{il} \ . \tag{V.12}$$

Let us calculate the last two terms. Bearing in mind expression (V.10) of Q, we have

$$\begin{aligned} g^{rs}Q^l_{ri}k_{ls} &\approx \frac{1}{2}g^{rs}g^{lm}(\nabla_r h_{im} + \nabla_i h_{mr} - \nabla_m h_{ri})k_{ls} \\ &= \frac{1}{2}(\nabla^s h_{im})k^m_s + \frac{1}{2}(\nabla_i h_{mr})k^{mr} - \frac{1}{2}(\nabla^l h_{ri})k^r_l \ . \end{aligned}$$

But the first and last term on the right-hand side of the former equality are equal, but opposite in sign, so they cancel. Thus,

$$g^{rs}Q^l_{ri}k_{ls} \approx \frac{1}{2}(\nabla_i h_{mr})k^{mr} \ .$$

Similarly, we obtain

$$g^{rs}Q^l_{rs}k_{il} \approx (\nabla^s h_{sm})k^m_i - \frac{1}{2}(\nabla^l \text{tr } h)k_{il} \ .$$

Replacing all of this into (V.12) we find

$$\begin{aligned} g'^{rs}\nabla'_r k'_{is} &\approx g^{rs}\nabla_r k_{is} + g^{rs}\nabla_r K_{is} - h^{rs}\nabla_r k_{is} \\ &\quad - \frac{1}{2}(\nabla_i h_{mr})k^{mr} - (\nabla^s h_{sm})k^m_i + \frac{1}{2}(\nabla^l \text{tr } h)k_{il} \ . \end{aligned} \tag{V.13}$$

We need to calculate now the term $\partial_i(\text{tr}_{g'}k')$ of the expression of γ in (V.6). We replace g' by $g+h$ and k' by $k+K$, and we leave out terms of order greater than 1 in h and K. As

$$\text{tr}_{g'}k' \approx (g^{rs} - h^{rs})(k_{rs} + K_{rs}) \approx \text{tr}_g k + \text{tr}_g K - h^{rs}k_{rs} \ ,$$

we have

$$\begin{aligned}\partial_i(\mathrm{tr}_{g'}k') &\approx \partial_i(\mathrm{tr}_g k + \mathrm{tr}_g K) - \partial_i(h^{rs}k_{rs}) \\ &= \partial_i(\mathrm{tr}_g k + \mathrm{tr}_g K) - (\nabla_i h^{rs})k_{rs} - h^{rs}\nabla_i k_{rs} \ . \end{aligned} \tag{V.14}$$

Replacing in the expression of γ in (V.6) the expressions (V.13) and (V.14), we obtain

$$\gamma(g',k')_i - \gamma(g,k)_i \approx \frac{1}{2}\left\{\nabla^s K_{is} - h^{rs}\nabla_r k_{is} - \frac{1}{2}(\nabla_i h_{mr})k^{mr}\right.$$
$$\left. -(\nabla^s h_{sm})k_i^m + \frac{1}{2}(\nabla^l \mathrm{tr}_g h)k_{il} - \partial_i(\mathrm{tr}_g K) + (\nabla_i h^{rs})k_{rs} + h^{rs}\nabla_i k_{rs}\right\} .$$

However, the third and seventh terms on the right-hand side can be grouped (with coefficient 1/2), and we finally see that:

$$\begin{aligned}\gamma(g',k')_i - \gamma(g,k)_i \approx \frac{1}{2}\{&\nabla^s K_{is} - h^{rs}\nabla_r k_{is} + \\ &+ \frac{1}{2}(\nabla_i h_{mr})k^{mr} - (\nabla^s h_{sm})k_i^m + \frac{1}{2}(\nabla^l \mathrm{tr}_g h)k_{il} \\ &-\partial_i(\mathrm{tr}_g K) + h^{rs}\nabla_i k_{rs}\} \ . \end{aligned} \tag{V.15}$$

To calculate the tangent linear map of $\mathcal{H}$ at point (g,k) we start from the expression of $\mathcal{H}(g',k')$ in (V.6) and replace g' by $g+h$ and k' by $k+K$. We obtain

$$\begin{aligned}\mathcal{H}(g',k') &= \frac{1}{8}\left\{g'^{ir}g'^{js}k'_{ij}k'_{rs} - (g'^{ij}k'_{ij})^2 - 4R(g')\right\} \\ &\approx \frac{1}{8}\left\{(g^{ir}-h^{ir})(g^{js}-h^{js})(k_{ij}+K_{ij})(k_{rs}+K_{rs})\right. \\ &\quad \left. - \left\{(g^{ij}-h^{ij})(k_{ij}+K_{ij})\right\}^2 - 4R(g')\right\} \ . \end{aligned} \tag{V.16}$$

Recall now equation (III.6) of the Ricci tensor of g'. By contraction of that expression with respect to the metric g' we easily find:

$$R(g') \approx R(g) - h^{ij}R_{ij} + \nabla^i\nabla^s h_{is} - \nabla^s\nabla_s \mathrm{tr}_g h \ . \tag{V.17}$$

The last term but one in (V.16) can be calculated by:

$$\begin{aligned}\left\{(g^{ij}-h^{ij})(k_{ij}+K_{ij})\right\}^2 &= (g.k - h.k + g.K - h.K)^2 \\ &\approx (g.k)^2 - 2(g.k)(h.k) + 2(g.k)(g.K) \\ &= (\mathrm{tr}\ k)^2 - 2(\mathrm{tr}\ k)(h.k) + 2(\mathrm{tr}\ k)(\mathrm{tr}\ K) \ . \end{aligned} \tag{V.18}$$

Replacing (V.17) and (V.18) in (V.16) we finally obtain

$$\begin{aligned}\mathcal{H}(g',k') - \mathcal{H}(g,k) \approx \frac{1}{8}\{&-2h.(k\times k) + 2K.k \\ &+ 2(\mathrm{tr}_g k)(h.k) - 2(\mathrm{tr}_g k)(\mathrm{tr}_g K) \\ &+4h.Ric(g) - 4\nabla^i\nabla^s h_{is} + 4\nabla^s\nabla_s \mathrm{tr}_g h\} \ . \end{aligned} \tag{V.19}$$

All calculations made in this section can be summarized in the following proposition.

Proposition V.13. *Let us assume that the initial Riemannian manifold* (M,g) *(with* g *of class* C^∞*) is complete. We take a fixed integer* $s > 4$*. Suppose that the initial* C^∞ *symmetric 2-tensor* k *on* M *is such that for each integer* l *such that* $0 \le l \le s-1$ *the function* $|\nabla^{(l)}k|_g$ *on* M *is bounded by a constant (these are exactly the hypotheses of Proposition* V.10*). Then, the tangent linear map of* Φ *at the initial couple* (g,k) *is the map*

$$\begin{array}{llll} D_{(g,k)}\Phi : & \mathcal{S}^s(M) \times \mathcal{S}^{s-1}(M) & \longrightarrow & \mathcal{F}^{s-2}(M) \times \Omega^{s-2}(M) \\ & (h,K) & \rightarrow & ((D_{(g,k)}\mathcal{H})(h,K)\ ,\ (D_{(g,k)}\gamma)(h,K))\ , \end{array}$$

where $(D_{(g,k)}\mathcal{H})(h,K)$ *is given by the right-hand side of* (V.19) *and* $(D_{(g,k)}\gamma)(h,K)$ *is given by the right-hand side of* (V.15).

Chapter VI

General Results on Stability by Linearization when the Submanifold M of V is Compact

We continue with the situation from the previous chapter. We have a 4-dimensional manifold V with an initial Lorentzian metric $\tilde{g}$ and an initial stress-energy tensor T corresponding to a perfect fluid and which together fulfil Einstein's equation $G(\tilde{g}) = \chi T$. Let us consider a hypersurface M of V such that at each point $x \in M$ the velocity vector u of the perfect fluid is perpendicular to M with respect to the initial metric $\tilde{g}$. As observed in the previous chapter, study of the linearization stability of Einstein's equation at the initial metric leads us to the study of the linearization stability of the mapping

$$\begin{array}{rccc} & \mathcal{S}^s(g) \times \mathcal{S}^{s-1}(k) & & \\ & \cup & & \\ \Phi : & \mathcal{U} & \longrightarrow & \mathcal{F}^{s-2}(F) \times \Omega^{s-2}(X) \\ & (g', k') & \to & (\mathcal{H}(g', k'), \gamma(g', k')) \end{array}$$

at the initial pair (g, k).

In this chapter we study this situation in the case where the hypersurface M of V is compact, a hypothesis that throughout the chapter we assume to be verified. It must be pointed out that in this case it is not necessary to introduce the additional technical assumptions of Propositions V.10 and V.13, since the compactness of M implies those hypotheses.

VI.1 Adjoint of $D_{(g,k)}\Phi$

Let us recall that in the Riemannian manifold (M, g) we have the local inner product $(\ ,\ )$ for p-tensors and the global inner product $\langle \alpha, \beta \rangle = \int_M (\alpha, \beta) d\mu$

introduced in Section I.3.

We now introduce the following local inner product for elements $(h, K) \in \mathcal{S}^s(M) \times \mathcal{S}^{s-1}(M)$:

$$((h, K), (h', K')) = (h, h') + (K, K') = g^{ir} g^{js} (h_{ir} h'_{js} + K_{ir} K'_{js}) ,$$

and the corresponding global inner product:

$$\langle (h, K), (h', K') \rangle = \int_M ((h, k), (h', K')) d\mu .$$

We also introduce the global inner product for pairs $(f, Y) \in \mathcal{F}^{s-2}(M) \times \Omega^{s-2}(M)$ by

$$\langle (f, Y), (f', Y') \rangle = 4 \int_M f f' d\mu + 4 \int_M g(Y, Y') d\mu$$

(the coefficients 4 are included in order to facilitate the calculations).

We now consider the adjoint $(D_{(g,k)}\Phi)^*$ of $D_{(g,k)}\Phi$ with respect to the inner products introduced. The action of this adjoint is first defined on the pairs $(f, Y) \in \mathcal{F}(M) \times \Omega(M)$ belonging to the class C^∞, and it is characterized by the equality

$$\begin{aligned} &\langle (h, K), (D_{(g,k)}\Phi)^*(f, Y) \rangle = \langle (D_{(g,k)}\Phi)(h, K), (f, Y) \rangle \\ &= 4 \int_M (D_{(g,k)}\mathcal{H})(h, K) f d\mu + 4 \int_M g((D_{(g,k)}\gamma)(h, K), Y) d\mu . \qquad \text{(VI.1)} \end{aligned}$$

The definition of $(D_{(g,k)}\Phi)^*$ is then extended by a standard procedure to the corresponding Sobolev spaces. We wish to calculate in this section the explicit expression for $(D_{(g,k)}\Phi)^*$. To this end, it is necessary to calculate the two integrals that appear in the previous expression separately. Let us first consider the inner product that appears in the second of these integrals.

By virtue of the expression for $D_{(g,k)}\gamma$ given in Proposition V.13, we have

$$\begin{aligned} 2g((D_{(g,k)}\gamma)(h, K), Y) = {} & (\nabla^s K_{is}) Y^i - (\nabla_i \mathrm{tr}\ K) Y^i \\ & - h^{rs} (\nabla_r k_{is}) Y^i + h^{rs} (\nabla_i h_{rs}) Y^i + \frac{1}{2} (\nabla_i h^{rs}) k_{rs} Y^i \\ & - (\nabla^s h_{sm}) k_i^m Y^i + \frac{1}{2} (\nabla^l \mathrm{tr}\ h) k_{il} Y^i . \qquad \text{(VI.2)} \end{aligned}$$

The term $(\nabla^s K_{is}) Y^i$ from the previous expression can be written as

$$\nabla^s (K_{is} Y^i) - K_{is} \nabla^s Y^i .$$

However, the first term is a divergence, so we have $(\nabla^s K_{is})Y^i = -K_{is}\nabla^s Y^i + \text{div}$. If we do the same with some of the terms in (VI.2), this expression is written as

$$\begin{aligned}2g((D_{(g,k)}\gamma)(h,K),Y) &= -K_{is}\nabla^s Y^i + (\text{tr } K)(\text{div } Y)\\ &- h^{rs}(\nabla_r k_{is})Y^i + h^{rs}(\nabla_i k_{rs})Y^i - \frac{1}{2}h^{rs}(\nabla_i k_{rs})Y^i\\ &- \frac{1}{2}h^{rs}k_{rs}\text{div } Y + h_{sm}(\nabla^s k_i^m)Y^i + h_{sm}k_i^m(\nabla^s Y^i)\\ &- \frac{1}{2}(\text{tr } h)(\nabla^l k_{il})Y^i - \frac{1}{2}(\text{tr } h)k_{il}\nabla^l Y^i + \text{div} \ . \qquad \text{(VI.3)}\end{aligned}$$

The third and seventh terms on the right-hand side are now simplified, while the fourth and the fifth are combined into a single term that is written $\frac{1}{2}h^{rs}(\nabla_i k_{rs})Y^i$. Furthermore, taking into account Proposition I.7 and Corollary I.8, the term $-K_{is}\nabla^s Y^i$ of (VI.3) can be written $-\frac{1}{2}K.L_Y g$ and the sum $\frac{1}{2}h^{rs}(\nabla_i k_{rs})Y^i + h_{sm}k_i^m(\nabla^s Y^i)$ can be written $\frac{1}{2}h.L_Y k$. Analogously, the term $-\frac{1}{2}(\text{tr } h)k_{il}\nabla^l Y^i$ is written $-\frac{1}{4}(\text{tr } h)(k.L_Y g)$. Having done this, the expression (VI.3) becomes

$$\begin{aligned}2g((D_{(g,k)}\gamma)(h,K),Y) = &-\frac{1}{2}K.L_Y g + (\text{tr } K)(\text{div } Y)\\ &+ \frac{1}{2}h.L_Y k - \frac{1}{2}(h.k)\text{div } Y - \frac{1}{2}(\text{tr } h)(\nabla^l k_{il})Y^i\\ &\frac{1}{4}(\text{tr } h)(k.L_Y g) + \text{div} \ . \qquad \text{(VI.4)}\end{aligned}$$

The term $-\frac{1}{2}(\text{tr } h)(\nabla^l k_{il})Y^i$ may be written as $-\frac{1}{2}(\text{tr } h)(\text{div } k)(Y)$ (bearing in mind that k is a 2-tensor, div k is a 1-form, and this 1-form is made to act on the field Y).

Let us now consider the first integral on the right-hand side of (VI.1). From the expression of $(D_{(g,k)}\mathcal{H})(h,K)$ given in Proposition V.13, we have

$$\begin{aligned}8(D_{(g,k)}\mathcal{H})(h,K)f = &-2h.(k\times k)f + 2K.kf\\ &+ 2(\text{tr } k)(h.k)f - 2(\text{tr } k)(\text{tr } K)f + 4(h.Ric(g))f\\ &- 4(\nabla^i\nabla^s h_{is})f + 4(\nabla^s\nabla_s \text{tr } h)f \ . \qquad \text{(VI.5)}\end{aligned}$$

In the last two terms of this expression we move the covariant differentiations of the first factor to the factor containing f by means of adding a divergence. Thus for example

$$(\nabla^i\nabla^s h_{is})f = -(\nabla^s h_{is})\nabla^i f + \text{div} = h_{is}(\nabla^s\nabla^i f) + \text{div} = h.\text{Hess}(f) + \text{div} \ ,$$

where Hess(f) indicates the Hessian of f. By doing the same with the last factor of (VI.5), we obtain:

$$\begin{aligned}8(D_{(g,k)}\mathcal{H})(h,K)f = &-2h.(k\times k)f + 2K.kf\\ &+ 2(\text{tr } k)(h.k)f - 2(\text{tr } k)(\text{tr } K)f + 4(h.Ric(g))f\\ &- 4h.\text{Hess}(f) - 4(\text{tr } h)\Delta f + \text{div} \ , \qquad \text{(VI.6)}\end{aligned}$$

where Δf is the Laplacian of f associated to the Riemannian metric. The adjoint $(D_{(g,k)}\Phi)^*$ we wish to find, applied to an element (f, Y) of class C^∞ (we have already stated that the adjoint is first defined on pairs C^∞), will be a pair $(A, B) \in \mathcal{S}_2(M) \times \mathcal{S}_2(M)$, which by virtue of the adjoint definition (VI.1) will fulfil

$$\int_M (h, A)d\mu + \int_M (K, B)d\mu = \langle\langle (h, K) \,, (A, B)\rangle\rangle = \langle\langle (D_{(g,k)}\Phi)(h, K) \,, (f, Y)\rangle\rangle$$

$$= 4\int_M (D_{(g,k)}\mathcal{H})(h, K)f d\mu + 4\int_M g((D_{(g,k)}\gamma)(h, K), Y)d\mu$$

for any pair (h, K). However, we have calculated the two expressions that appear in the integrals on the right-hand side, and they are given by the formulas (VI.6) and (VI.4). As the integral on M of any divergence vanishes, we may deduce that the expressions of A and B will be

$$\begin{aligned} A &= -f(k \times k) + f(\text{tr } k)k + 2f\,Ric(g) - 2\text{Hess}(f) \\ &\quad - 2(\Delta f)g + L_Y k - (\text{div } Y)k - \{(\text{div } k)(Y)\}\, g \\ &\quad - \frac{1}{2}(k.L_Y g)g \ , \\ B &= fk - f(\text{tr } k)g - L_Y g + 2(\text{div } Y)g \ . \end{aligned} \tag{VI.7}$$

(In these expressions of A and B we consider Y as a vector field. Actually, Y is a 1-form. However, by the metric we may identify the 1-forms and the vector fields whenever it is convenient to do so.)

Let us now summarize all the above in the following proposition.

Proposition VI.1. $(D_{(g,k)}\Phi)^*(f, Y) = (A, B)$, *where A and B are given in terms of f and Y by* (VI.7).

Remark VI.2. When the initial stress-energy tensor is zero (in the vacuum), the expression of A given in (VI.7) is usually written in a different manner by using the constraint condition $\gamma(g, k) = 0$. This constraint equation is written in components $\nabla^s k_{is} = \partial_i(\text{tr}_g k)$, which is equivalent (intrinsically) to div $k = d\,\text{tr}_g k$. Therefore, the term $-\{(\text{div } k)(Y)\}g$ in the expression of A may be written as $-\{Y(\text{tr}_g k)\}g$.

Remark VI.3. If we now consider the expressions of A and B, we will see that A is a linear differential operator of order 2 in the variable f (which is the first variable), and of order 1 in the variable Y (which is the second one). Furthermore, B is a linear differential operator of order 0 in the first variable, while in the second it is of order 1. Therefore we have:

$$(D_{(g,k)}\Phi)^* : \ \mathcal{F}^s(M) \times \Omega^{s-1}(M) \longrightarrow \mathcal{S}^{s-2}(M) \times \mathcal{S}^{s-2}(M) \ .$$

VI.2 Results by A. Fischer and J. E. Marsden

From Theorem V.8, if the mapping $D_{(g,k)}\Phi$ is surjective, and if the subspace $\text{Ker}(D_{(g,k)}\Phi)$ (subspace of $\mathcal{S}^s(M) \times \mathcal{S}^{s-1}(M)$) has a topological supplement, then

Φ is stable by linearization at the initial pair (g, k). But Theorem V.8 was stated in the context of Banach spaces, while here we apply it to Sobolev spaces based on L^2, which are in particular Hilbert spaces. Therefore the hypothesis on the existence of a topological supplement of $\operatorname{Ker}(D_{(g,k)}\Phi)$ is not necessary because it is fulfilled automatically, since in a Hilbert space every closed subspace has a topological supplement (its orthogonal). In order to study the surjectivity of $D_{(g,k)}\Phi$, we employ the theory of elliptic operators in compact manifolds, a summary of which can be found in the Appendix to this Chapter (Section VI.4). Let us consider the linear differential operator

$$D_{g,k)}\Phi : \mathcal{S}^s(M) \times \mathcal{S}^{s-1}(M) \longrightarrow \mathcal{F}^{s-2}(M) \times \Omega^{s-2}(M) .$$

If the symbol of $(D_{(g,k)}\Phi)^*$ were injective, then Theorem VI.11 in the above-mentioned Appendix would yield a splitting of $\mathcal{F}^{s-2}(M) \times \Omega^{s-2}(M)$ into a direct topological sum $\operatorname{Im}(D_{(g,k)}\Phi) \oplus \operatorname{Ker}((D_{(g,k)}\Phi)^*)$, and in order to study the surjectivity of $D_{(g,k)}\Phi$ it would suffice to study whether $\operatorname{Ker}((D_{(g,k)}\Phi)^*$ vanishes or not. In order to pursue this scheme, we begin by proving:

Proposition VI.4. *The symbol of $(D_{(g,k)}\Phi)^*$ is injective.*

Proof. By virtue of expression (VI.7) of A and B, the symbol of $(D_{(g,k)}\Phi)^*$ associates to each $\xi \in T_x(M)^*$ the linear mapping from $\mathbb{R} \times T_x(M)^*$ to $\mathcal{S}_x \times \mathcal{S}_x$ that applies each $(f, Y) \in \mathbb{R} \times T_x(M)^*$ into the pair $(\alpha, \beta) \in \mathcal{S}_x \times \mathcal{S}_x$ given by

$$\begin{cases} \alpha_{ij} = -2\xi_i\xi_j f + 2(g^{rs}\xi_r\xi_s) f g_{ij} + Y^r\xi_r k_{ij} + \xi_i Y^r k_{rj} \\ \qquad\quad +\xi_j Y^r k_{ir} - (\xi_r Y^r)k_{ij} - (Y^r\xi^l k_{rl})g_{ij} - (k_{rs}\xi^s Y^r)g_{ij} \\ \beta_{ij} = -\xi_i Y_j - \xi_j Y_i + 2(\xi_r Y^r)g_{ij} . \end{cases}$$

It is necessary to ensure that the symbol associated to a given $x \in M$ and to a given $\xi \in T_x(M)^*$, $\xi \neq 0$, is injective. Let us take a system of coordinates for which the matrix (g_{ij}) at the point x is the identity and such that $\xi = (\xi_1, 0, 0)$ with $\xi_1 \neq 0$. Let us assume that $\alpha = \beta = 0$. We wish to prove that $f = 0$ and $Y = 0$. In these coordinates we have

$$0 = \beta_{22} = \underbrace{-2\xi_2 Y_2}_{=0} + 2\xi_1 Y_1 .$$

Therefore, $Y_1 = 0$. Then $\beta_{21} = 0$ implies $Y_2 = 0$, and $\beta_{31} = 0$ implies that $Y_3 = 0$. Thus $Y = 0$. Now by substituting the field $Y = 0$ in the expression of α_{ij}, we see that $\alpha_{22} = 0$ implies that $f = 0$. □

As a corollary to this construction, we may state the following result.

Theorem VI.5. *If the initial Lorentzian metric $\tilde{g}$ of V and the compact submanifold M of V are such that the pair (g, k) fulfils that $\operatorname{Ker}((D_{(g,k)}\Phi)^*) = \{0\}$, then Einstein's equation is stable by linearization at the initial metric $\tilde{g}$.*

This construction is due largely to A. Fischer and J. E. Marsden [34], and therefore the above-mentioned theorem should be attributed to these authors, although they are working in the case of the vacuum and gave geometric conditions to ensure that $\mathrm{Ker}((D_{(g,k)}\Phi)^*$ vanishes. In view of the historical interest of this approach, we now give a proof of the original result by A. Fischer and J. E. Marsden for the case of the vacuum:

Corollary VI.6 (Fischer and Marsden). *In the case of the vacuum (initial stress-energy tensor zero), if the submanifold M of V is compact, and if the pair (g,k) corresponding to the initial metric $\tilde{g}$ fulfils the following conditions:*

(1) *k is not identically zero and $\mathrm{tr}_g k$ is constant,*

(2) *there is no non-vanishing vector field Y on M that fulfils the two conditions $L_Y g = 0$, $L_Y k = 0$,*

then Einstein's equation is stable by linearization at the initial metric $\tilde{g}$.

Proof. We must see that if these conditions are fulfilled, then $\mathrm{Ker}((D_{(g,k)}\Phi)^*$ vanishes. Let us take $(f,Y) \in \mathrm{Ker}((D_{(g,k)}\Phi)^*$. From Proposition VI.1, one has $(D_{(g,k)}\Phi)^*(f,Y) = (A,B)$, where A and B are given in terms of f and Y by (VI.7). By hypothesis we have $A=0$, $B=0$. If we take traces in the expressions of A and B, then we have

$$0 = \mathrm{tr}_g B = -2f\mathrm{tr}\ k + 4\,\mathrm{div}\ Y$$

(when we take the trace of B, it is necessary to take into account that, from Corollary I.8, $\mathrm{tr}\ L_Y g = 2\,\mathrm{div}\ Y$). Therefore, $\mathrm{div}\ Y = \frac{1}{2} f\ \mathrm{tr}\ k$. If this is substituted in the expression of B, from $B=0$ we obtain $L_Y g = fk$.

Let us now take traces to the expression of A. Before doing so, however, let us recall that since we are dealing with the case of the vacuum, by virtue of Remark VI.2, the term $-\{(\mathrm{div}\ k)(Y)\}g$ of the expression of A is written $-\{Y(\mathrm{tr}_g k)\}g$, and since we assume that $\mathrm{tr}\ k$ is constant, this term is null. Before taking traces to the expression of A, it is also necessary to know how to calculate $\mathrm{tr}\ L_Y k$. From Proposition I.7 we have, in coordinates,

$$(L_Y k)_{ij} = Y^r \nabla_r k_{ij} + \nabla_i Y^r k_{rj} + \nabla_j Y^r k_{ir}\ .$$

Taking traces to this expression, we obtain

$$\mathrm{tr}\ L_Y k = Y(\mathrm{tr}\ k) + k.L_Y g = k.L_Y g$$

(since $\mathrm{tr}\ k$ is constant). Observe that if f is a function on M, then $\mathrm{tr}_g \mathrm{Hess}(f) = -\Delta f$, where Δ is De Rham's Laplacian associated to the Riemannian metric g. In fact, from expression (I.14) of the Laplacian we obtain that $\Delta f = -g^{ij}\nabla_i\nabla_j f = -\mathrm{tr}_g\mathrm{Hess}(f)$ (here f is a function). Now we are ready to take traces to the expression of A given in (VI.7). Bearing in mind all that thus far we have obtained

($\{(\text{div } k)(Y)\}g$=0, $L_Y g = fk$, div $Y = \frac{1}{2} f \text{ tr } k$), we have

$$0 = \text{tr } A = -\frac{3}{2} fk.k + \frac{1}{2} f(\text{tr } k)^2 + 2fR - 4\Delta f \ .$$

Now, from the constraint equation $\mathcal{H}(g,k) = 0$ (recalling once again that we are dealing with the case of the vacuum), we obtain $R = \frac{1}{4}k.k - \frac{1}{4}(\text{tr } k)^2$. Substituting this in the previous expression of tr A, we have

$$0 = \text{tr } A = -fk.k - 4\Delta f \ .$$

Take the global inner product or this equality by f (this inner product was defined in (I.11)). We have

$$4\langle \Delta f, f\rangle = -\langle fk.k, f\rangle = -\int_M f^2 k.k d\mu \ .$$

However, the left-hand side on this equality is ≥ 0, since $\langle \Delta f, f\rangle = \langle \delta df, f\rangle = \langle df, df\rangle$. Moreover, the right-hand side is ≤ 0. Thus both sides vanish. Therefore $df = 0$ and f is constant. Hence

$$0 = \int_M f^2 k.k d\mu = f^2 \int_M k.k d\mu \ .$$

Since k is not identically zero, f must vanish. Then we have $L_Y g = fk = 0$ and div $Y = \frac{1}{2} f \text{ tr } k = 0$. From the expression of A given in (VI.7), we obtain that $0 = A = L_Y k$. Therefore Y fulfils the two conditions $L_Y g = 0$ and $L_Y k = 0$. Thus $Y = 0$. □

VI.3 A result by V. Moncrief

Corollary VI.6 provides sufficient conditions for the kernel of $(D_{(g,k)}\Phi)^*$ to vanish in the case of the vacuum. However, V. Moncrief [52] characterized this kernel (also in the case of the vacuum) by providing conditions on the metric $\tilde{g}$ of the Lorentzian manifold rather than on the pair (g,k) of M. The following technical lemma, which may be employed for the proof of Moncrief's result, can be stated without the assumption that the stress-energy tensor is zero. We therefore deal with a case more general than is strictly necessary.

Lemma VI.7. *For each pair (f,Y) formed by a function f and a 1-form Y on M, let A' and B' be the symmetric covariant 2-tensors defined by*

$$\begin{aligned} A' &= -L_Y k + f\, k \times k + 2\, Hess(f) - 2f\, Ric(g) \\ &\quad - \frac{1}{2} f(tr\ k)k - \chi g(X,Y)g - 2\chi f F g \\ B' &= fk - L_Y g \ , \end{aligned}$$

where X and F as usual represent $F = \frac{1}{\chi}\mathcal{H}(g,k)$ and $X = \frac{1}{\chi}\gamma(g,k)$. The simultaneous vanishing of A' and B' is equivalent to the simultaneous vanishing of A and B given by (VI.7).

Proof. From the definition of B' we deduce that $\text{tr } B' = f\text{tr } k - 2\,\text{div } Y$. If we now observe the expression of B given in (VI.7), we see that

$$B = B' - (\text{tr } B')g \ . \tag{VI.8}$$

From the expression of A and the definition of A', it follows that

$$\begin{aligned} A = &-A' + \frac{1}{2}f(\text{tr } k)k - \chi g(X,Y)g - 2\chi f F g \\ &- 2(\Delta f)g - (\text{div } Y)k - \{(\text{div } k)(Y)\}g - \frac{1}{2}(k.L_Y g)g \ . \end{aligned} \tag{VI.9}$$

Taking traces to the expression of A', and taking into account that as seen in Section VI.2 $\text{tr } L_Y k = Y(\text{tr } k) + k.L_Y g$, we have

$$\begin{aligned} \text{tr } A' = &-Y(\text{tr } k) - k.L_Y g + fk.k - 2\Delta f \\ &- 2f\,R(g) - \frac{1}{2}f(\text{tr } k)^2 - 3\chi g(X,Y) - 6\chi f F \ . \end{aligned} \tag{VI.10}$$

We now employ the constraint equation $k.k - 4R - (\text{tr } k)^2 = 8\chi F$. Hence we obtain $-2R = 4\chi F - \frac{1}{2}k.k + \frac{1}{2}(\text{tr } k)^2$, which when substituted in the expression (VI.10) of $\text{tr } A'$ gives

$$\begin{aligned} \text{tr } A' = &-Y(\text{tr } k) - k.L_Y g + \frac{1}{2}fk.k - 2\Delta f \\ &- 3\chi g(X,Y) - 2\chi f F \ . \end{aligned} \tag{VI.11}$$

The constraint equation

$$\nabla^s k_{is} - \partial_i(\text{tr } k) = 2\chi X_i$$

implies $(\text{div } k)(Y) = Y(\text{tr } k) + 2\chi g(X,Y)$, which when substituted in the expression (VI.9) of A yields

$$\begin{aligned} A = &-A' + \frac{1}{2}f(\text{tr } k)k - 3\chi g(X,Y)g - 2\chi f F g \\ &- 2(\Delta f)g - (\text{div } Y)k - Y(\text{tr } k)g - \frac{1}{2}(k.L_Y g)g \ . \end{aligned}$$

Bearing in mind the expression (VI.11) of $\text{tr } A'$, this may now be written

$$\begin{aligned} A = &-A' + (\text{tr } A')g + \frac{1}{2}(k.L_Y g)g \\ &- \frac{1}{2}f(k.k)g + \frac{1}{2}f(\text{tr } k)k - (\text{div } Y)k \ . \end{aligned} \tag{VI.12}$$

Now, from the expression of B' it follows that $k.B' = fk.k - k.L_Y g$. Therefore, $fk.k = k.B' + k.L_Y g$, which substituted in (VI.12) gives

$$A = -A' + (\mathrm{tr}\ A')g - \frac{1}{2}(k.B')g + \frac{1}{2}f(\mathrm{tr}\ k)k - (\mathrm{div}\ Y)k\ .$$

However, from the expression of B' it follows that $\mathrm{tr}\ B' = f\mathrm{tr}\ k - 2\,\mathrm{div}\ Y$. Therefore, the term $\frac{1}{2}f\mathrm{tr}\ k$ in the previous expression of A can be written as $\frac{1}{2}\mathrm{tr}\ B' + \mathrm{div}\ Y$, whereby the expression of A is written

$$A = -A' + (\mathrm{tr}\ A')g - \frac{1}{2}(k.B')g + \frac{1}{2}(\mathrm{tr}\ B')k\ . \tag{VI.13}$$

The expressions (VI.8) and (VI.13) which give A and B in terms of A' and B' show that if $A' = 0$ and $B' = 0$, then $A = 0$ and $B = 0$. In order to prove the converse, it is necessary to obtain the expressions of A' and B' in terms of A and B, which as we now see is very easy. From the expression (VI.8) of B it follows that $\mathrm{tr}\ B = -2\,\mathrm{tr}\ B'$, thus

$$B' = B - \frac{1}{2}(\mathrm{tr}\ B)g\ . \tag{VI.14}$$

From expression (VI.13) of A it follows that

$$\mathrm{tr}\ A = 2\,\mathrm{tr}\ A' - \frac{3}{2}k.B' + \frac{1}{2}(\mathrm{tr}\ B')(\mathrm{tr}\ k)\ .$$

Therefore,

$$\mathrm{tr}\ A' = \frac{1}{2}\mathrm{tr}\ A + \frac{3}{4}k.B' - \frac{1}{4}(\mathrm{tr}\ B')(\mathrm{tr}\ k)\ ,$$

and replacing here B' by its expression, we obtain

$$\mathrm{tr}\ A' = \frac{1}{2}\mathrm{tr}\ A + \frac{3}{4}k.B - \frac{1}{4}(\mathrm{tr}\ B)(\mathrm{tr}\ k)\ .$$

If we now substitute this expression and (VI.14) in (VI.13), we finally arrive at

$$A' = -A + \frac{1}{2}(\mathrm{tr}\ A)g + \frac{1}{4}(k.B)g - \frac{1}{4}(\mathrm{tr}\ B)k\ . \tag{VI.15}$$

The expressions (VI.14) and (VI.15) show that when $A = 0$ and $B = 0$, then $A' = 0$ and $B' = 0$. □

We are now ready to prove Moncrief's following result.

Theorem VI.8 (V. Moncrief). *In the case of the vacuum (initial stress-energy tensor zero), if the hypersurface M of V is compact, and if any vector field Y defined in a neighborhood of M in V and fulfilling $L_Y\tilde{g} = 0$ must necessarily be zero, then Einstein's equation is stable by linearization at the initial metric $\tilde{g}$.*

Proof. In order to describe a certain neighborhood of the hypersurface M in V, we use the Gauss representation already employed in Section IV.6. It is assumed that M has a unitary transverse field N. Let us recall that for each $x \in M$ we take the geodesic $\gamma_x(t)$ passing through x and having N_x as a vector tangent at x. We denote by U' the neighborhood of $M \times \{0\}$ in $M \times \mathbb{R}$ formed by (x,t) such that $\gamma_x(t)$ remains defined. The mapping

$$\begin{array}{cccc} \gamma: & U' & \longrightarrow & V \\ & (x,t) & \rightarrow & \gamma_x(t) \end{array}$$

gives a diffeomorphism between U' (open subset of $M \times \mathbb{R}$) and a certain neighborhood U of M in V. In the case with which we are concerned, since M is compact we may select a neighborhood of the form $M \times (-\varepsilon, \varepsilon)$ contained in U'. Once again we denote by U the image of $M \times (-\varepsilon, \varepsilon)$ by γ. For every $t \in (-\varepsilon, \varepsilon)$, we denote by M_t the image by γ of $M \times \{t\}$ (which is a hypersurface of V). The vector field $\frac{\partial}{\partial t}$ of $M \times (-\varepsilon, \varepsilon)$ gives rise by the diffeomorphism γ to a field on $U \subset V$ (which we continue to denote by $\frac{\partial}{\partial t}$), and which for $t = 0$ coincides with the transverse field N given on M. Since the field $\frac{\partial}{\partial t}$ is geodesic and for $t = 0$ fulfils $\tilde{g}(\frac{\partial}{\partial t}, \frac{\partial}{\partial t}) = -1$, it fulfils this equality for every t. Furthermore, Gauss's well-known lemma ensures that the field $\frac{\partial}{\partial t}$ is perpendicular to all the hypersurfaces M_t.

A local chart of M of coordinates (x^1, x^2, x^3) always gives rise by the diffeomorphism γ to a local chart of U of coordinates (x^1, x^2, x^3, t) (which we call coordinates adapted to Gauss's representation). Let us recall that in such a system of coordinates we have

$$\begin{aligned} \partial_t g_{ij} &= k_{ij}, \\ \partial_t k_{ij} &= (k \times k)_{ij} - 2R_{ij} - \frac{1}{2}(\operatorname{tr} k)k_{ij} \,, \end{aligned} \tag{VI.16}$$

where R_{ij} indicates the components of the Ricci tensor of g. The first of these two inequalities is none other than (IV.20), while the second arises from (IV.17), taking into account that as we are in the vacuum the Ricci tensor of $\tilde{g}$ is zero. Equations (VI.16) are called evolution equations for the pair (g,k) in the vacuum. For evolution equations for perfect fluids we must cite [7].

The proof of the theorem will consist in assigning to each pair (f, Y) of the kernel of $(D_{(g,k)}\Phi)^*$ a vector field $\tilde{Y}$ defined in a neighborhood of M in V, and such that $L_{\tilde{Y}}\tilde{g} = 0$ and that the mapping $(f, Y) \rightarrow \tilde{Y}$ is injective. Then the hypothesis regarding the Killing vector fields of $\tilde{g}$ imply that $\tilde{Y} = 0$, and thus $(f, Y) = (0, 0)$. We then have that $\operatorname{Ker}((D_{(g,k)}\Phi)^*) = \{0\}$, and Theorem VI.5 ensures the linearization stability of Einstein's equation.

We first extend the function f, which is only defined on M, to a function on the neighborhood U of M, a function that we continue to denote by f. The way in which this extension is performed will be detailed later. We also extend the vector field Y, which is only defined on M, to a field Y on the neighborhood U, a field that we continue to denote by Y and which we wish to be tangent at each point to the hypersurface M_t that passes through that point. The way in which this extension is performed will likewise be detailed later. In every adapted local chart of coordinates (x^1, x^2, x^3, t), the extension f of the function f will be a function $f(x^1, x^2, x^3, t)$, and the extension Y of Y, since it is tangent to the M_t, will be a field of the form

$$Y = \sum_{i=1}^{3} Y^i(x^1, x^2, x^3, t) \frac{\partial}{\partial x^i} .$$

Assuming both f and Y to be extended to the whole neighborhood U of M, we denote by $\tilde{Y}$ the vector field

$$\tilde{Y} = f \frac{\partial}{\partial t} - Y .$$

In a system of adapted coordinates (x^1, x^2, x^3, t), the bracket $[\tilde{Y}, \frac{\partial}{\partial t}]$ will be

$$\left[\tilde{Y}, \frac{\partial}{\partial t}\right] = \left[f \frac{\partial}{\partial t} - Y^k \frac{\partial}{\partial x^k}, \frac{\partial}{\partial t}\right] = -\partial_t f \frac{\partial}{\partial t} + \partial_t Y^k \frac{\partial}{\partial x^k} ,$$

and the bracket $[\tilde{Y}, \frac{\partial}{\partial x^i}]$ will be

$$\left[\tilde{Y}, \frac{\partial}{\partial x^i}\right] = \left[f \frac{\partial}{\partial t} - Y^k \frac{\partial}{\partial x^k}, \frac{\partial}{\partial x^i}\right] = -\partial_i f \frac{\partial}{\partial t} + \partial_i Y^k \frac{\partial}{\partial x^k} .$$

We now calculate the Lie derivative $L_{\tilde{Y}}\tilde{g}$ in a system of adapted coordinates (x^1, x^2, x^3, t). Then we have

$$(L_{\tilde{Y}}\tilde{g}) \left(\frac{\partial}{\partial t}, \frac{\partial}{\partial t}\right) = \tilde{Y}\left(\tilde{g}\left(\frac{\partial}{\partial t}, \frac{\partial}{\partial t}\right)\right) - 2\tilde{g}\left(\left[\tilde{Y}, \frac{\partial}{\partial t}\right], \frac{\partial}{\partial t}\right) .$$

Since $\tilde{g}(\frac{\partial}{\partial t}, \frac{\partial}{\partial t}) = -1$, the first term on the right-hand side of the previous expression is zero. From the expression of $[\tilde{Y}, \frac{\partial}{\partial t}]$, previously calculated, it finally follows that

$$(L_{\tilde{Y}}\tilde{g}) \left(\frac{\partial}{\partial t}, \frac{\partial}{\partial t}\right) = -2\partial_t f . \tag{VI.17}$$

But

$$(L_{\tilde{Y}}\tilde{g}) \left(\frac{\partial}{\partial t}, \frac{\partial}{\partial x^i}\right) = \tilde{Y}\left(\tilde{g}\left(\frac{\partial}{\partial t}, \frac{\partial}{\partial x^i}\right)\right) - \tilde{g}\left(\left[\tilde{Y}, \frac{\partial}{\partial t}\right], \frac{\partial}{\partial x^i}\right) - \tilde{g}\left(\frac{\partial}{\partial t}, \left[\tilde{Y}, \frac{\partial}{\partial x^i}\right]\right) .$$

Since $\tilde{g}(\frac{\partial}{\partial t}, \frac{\partial}{\partial x^i}) = 0$, from the previous expressions of $[\tilde{Y}, \frac{\partial}{\partial t}]$ and of $[\tilde{Y}, \frac{\partial}{\partial x^i}]$ we obtain

$$(L_{\tilde{Y}}\tilde{g}) \left(\frac{\partial}{\partial t}, \frac{\partial}{\partial x^i}\right) = -\partial_t Y^j g_{ji} - \partial_i f . \tag{VI.18}$$

We now calculate $(L_{\tilde{Y}}\tilde{g})(\frac{\partial}{\partial x^i}, \frac{\partial}{\partial x^j})$. We have

$$\begin{aligned}
&(L_{\tilde{Y}}\tilde{g})\left(\frac{\partial}{\partial x^i}, \frac{\partial}{\partial x^j}\right) \\
&= \tilde{Y}(g_{ij}) - \tilde{g}\left(\left[\tilde{Y}, \frac{\partial}{\partial x^i}\right], \frac{\partial}{\partial x^j}\right) - \tilde{g}\left(\frac{\partial}{\partial x^i}, \left[\tilde{Y}, \frac{\partial}{\partial x^j}\right]\right) \\
&= f\partial_t g_{ij} - Y(g_{ij}) - \partial_i Y^k g_{kj} - \partial_j Y^k g_{ik} .
\end{aligned}$$

Now, from (VI.16), we are able to replace $\partial_t g_{ij}$ by k_{ij} in this equality, and moreover the expression $-Y(g_{ij}) - \partial_i Y^k g_{kj} - \partial_j Y^k g_{ik}$ is equal to $-(L_Y g)(\frac{\partial}{\partial x^i}, \frac{\partial}{\partial x^j})$. Thus

$$(L_{\tilde{Y}}\tilde{g})\left(\frac{\partial}{\partial x^i}, \frac{\partial}{\partial x^j}\right) = fk_{ij} - (L_Y g)_{ij} . \tag{VI.19}$$

Bearing in mind the formulas (VI.17), (VI.18) and (VI.19), we now explain how we take the extensions of f and Y in a neighborhood U of M in V. We take the extension of f in such a way that it does not depend on the coordinate t, so $\partial_t f = 0$, and then $(L_{\tilde{Y}}\tilde{g})(\frac{\partial}{\partial t}, \frac{\partial}{\partial t}) = 0$. In order for $(L_{\tilde{Y}}\tilde{g})(\frac{\partial}{\partial t}, \frac{\partial}{\partial x^i}) = 0$ also to be fulfilled, it is necessary to take the extension Y in such a way that $-\partial_t Y^j g_{ji} - \partial_i f = 0$ is fulfilled (f is now defined on the whole U). It is also necessary to point out that this expression, despite having been obtained by fixing a local chart (x^1, x^2, x^3) of M, is a global and intrinsic expression. In fact, it can be written

$$-\frac{\partial}{\partial t}Y - \sharp d_M f = 0 , \tag{VI.20}$$

where $d_M f$ denotes the exterior derivative of f in M, and $\sharp$ denotes the isomorphism between 1-forms and vector fields given by the metric g of M. Thus we take the extension of Y to the whole U as the only solution of (VI.20), which for $t = 0$ coincides with the Y given on M. In this way we arrive at $(L_{\tilde{Y}}\tilde{g})(\frac{\partial}{\partial t}, \frac{\partial}{\partial x^i}) = 0$.

We now wish to check that if the pair (f, Y) belongs to the kernel of the map $(D_{(g,k)}\Phi)^*$, then the vector field $\tilde{Y}$ constructed from the pair (f, Y) in the manner explained fulfils the condition $L_{\tilde{Y}}\tilde{g} = 0$. In the interests of brevity we write

$$a = L_{\tilde{Y}}\tilde{g} .$$

In every adapted local chart (x^1, x^2, x^3, t) we have $a_{0i} = a_{00} = 0$, and it must only be checked that $a_{ij} = 0$. By virtue of (VI.19) we have

$$a = fk - L_Y g .$$

(Recall that both f and Y have already been extended to the neighborhood U of M, and that g and k are also considered to have been extended to the whole U in such a way that g is the restriction of $\tilde{g}$ to M_t for every t, and that k is twice the second fundamental form of M_t). We now wish to check that if (f, Y) belongs to the kernel of $(D_{(g,k)}\Phi)^*$, then a is identically zero. By virtue of Lemma VI.7, a vanishes for $t=0$; that is, on the hypersurface M. Let us first see if the derivative $\partial_t a$ also vanishes for $t=0$. From derivation of a we have that $\partial_t a = (\partial_t f)k + f\partial_t k - \partial_t L_Y g$. Now, bearing in mind (VI.16) as well as that $\partial_t f = 0$, we obtain

$$\partial_t a = f(k\times k) - 2f Ric(g) - \frac{1}{2}f(\mathrm{tr}\ k)k - \partial_t L_Y g\ . \tag{VI.21}$$

It is convenient in the last term to invert the order of the operators ∂_t and L_Y. In order to see how this may be achieved, observe that if Z is any vector field, we have:

$$\begin{aligned}[\partial_t, L_Y](Z) &= \partial_t L_Y Z - L_Y\partial_t Z\\ &= \partial_t[Y,Z] - [Y,\partial_t Z] = [\partial_t Y, Z] = L_{\partial_t Y} Z\ .\end{aligned}$$

Furthermore, if h is any function, we have analogously

$$\begin{aligned}[\partial_t, L_Y](h) &= \partial_t L_Y h - L_Y\partial_t h\\ &= \partial_t(Y(h)) - Y(\partial_t h)\ .\end{aligned}$$

Since in any local chart $Y=\sum_k Y^k\partial_k$ one has $\partial_t(Y(h)) = (\partial_t Y)(h) + Y(\partial_t h)$, the last equality becomes

$$[\partial_t, L_Y](h) = L_{\partial_t Y} h\ .$$

The two above equalities concerning $[\partial_t, L_Y]$ show that the equality $[\partial_t, L_Y] = L_{\partial_t Y}$ is fulfilled when the two sides are applied to a vector field, and when they are applied to a function. It will thus also be fulfilled for any tensor. Therefore, we have

$$[\partial_t, L_Y](g) = \partial_t L_Y g - L_Y\partial_t g = L_{\partial_t Y} g\ .$$

This enables us to invert the order of the operators ∂_t and L_Y that appear in the last term of (VI.21), by adding therein one more term. We then have

$$\partial_t a = f(k\times k) - 2f Ric(g) - \frac{1}{2}f(\mathrm{tr}\ k)k - L_Y\partial_t g - L_{\partial_t Y} g\ .$$

Since $\partial_t g = k$, and since, by construction, $\partial_t Y = -\sharp d_M f$, we arrive at

$$\partial_t a = f(k\times k) - 2f Ric(g) - \frac{1}{2}f(\mathrm{tr}\ k)k - L_Y k + L_{\sharp d_M f} g\ . \tag{VI.22}$$

We now calculate the Lie derivative of g with respect to the field $\sharp d_M f$. By Corollary I.8, we have

$$L_{\sharp d_M f} g\left(\frac{\partial}{\partial x^i}, \frac{\partial}{\partial x^j}\right) = g\left(\nabla_{\frac{\partial}{\partial x^i}}\sharp df\,,\ \frac{\partial}{\partial x^j}\right) + g\left(\frac{\partial}{\partial x^i}\,,\ \nabla_{\frac{\partial}{\partial x^j}}\sharp df\right)\ .$$

But, $(\sharp df)^k = g^{kj}\partial_j f = g^{kj}\nabla_j f = \nabla^k f$. Therefore, the k-th component of $\nabla_{(\partial/\partial x^i)}\sharp df$ will be $\nabla_i\nabla^k f$. Therefore, the foregoing equality is written as

$$(L_{\sharp d_M f} g)_{ij} = 2\nabla_i\nabla_j f\ .$$

Then, (VI.22) is written

$$\partial_t a = f(k \times k) - 2f Ric(g) - \frac{1}{2} f(\mathrm{tr}\ k)k - L_Y k + 2\,\mathrm{Hess}(f)\ . \tag{VI.23}$$

This formula is valid for any t. But, for $t = 0$, since the pair (f, Y) belongs to the kernel of $(D_{(g,k)}\Phi)^*$, by virtue of the Lemma VI.7, $A' = 0$. Since we are dealing with the case of the vacuum, in the expression of A' given by the lemma, X and F vanish, and therefore the expression of A' agrees exactly with the expression of $\partial_t a$ from (VI.23). Thus, $(\partial_t a)_{t=0} = 0$.

To conclude the proof of the theorem, it still remains to be seen that a is identically zero. Let ψ_t be the one-parameter group corresponding to the field $\tilde{Y}$. Since we are in the case of the vacuum, from the equality $\widetilde{Ric}(\tilde{g}) = 0$ we deduce that $\psi_t^*(\widetilde{Ric}(\tilde{g})) = 0$, which can be written $\widetilde{Ric}(\psi_t^*(\tilde{g})) = 0$. Indeed, $\tilde{g}$ and $\psi_t^*(\tilde{g})$ are two isometric metrics on V through ψ_t (for any fixed t). As the Ricci tensor of a metric depends only on the metric, the Ricci tensors of these two isometric metrics must be mapped by ψ_t one into the other. That is, $\psi_t^*(\widetilde{Ric}(\tilde{g})) = \widetilde{Ric}(\psi_t^*(\tilde{g}))$.

By deriving with respect to t the equality $\widetilde{Ric}(\psi_t^*(\tilde{g})) = 0$, we have

$$0 = \frac{d}{dt}\widetilde{Ric}(\psi_t^*\tilde{g})\ |_{t=0} = (D_{\tilde{g}}\widetilde{Ric})(L_{\tilde{Y}} g)\ .$$

Derivative $D_{\tilde{g}}\widetilde{Ric}$ was already calculated in Chapter III for any pseudo-Riemannian manifold, and the expression (III.6) was found, which we reproduce here, adapted to current notation:

$$(D_{\tilde{g}}\widetilde{Ric})(a)_{\nu\mu} = \frac{1}{2}\tilde{g}^{\rho\alpha}(\tilde{\nabla}_\rho\tilde{\nabla}_\mu a_{\nu\alpha} + \tilde{\nabla}_\rho\tilde{\nabla}_\nu a_{\alpha\mu} - \tilde{\nabla}_\rho\tilde{\nabla}_\alpha a_{\mu\nu}) - \frac{1}{2}\tilde{\nabla}_\mu\tilde{\nabla}_\nu \mathrm{tr}_{\tilde{g}} a\ .$$

In the coordinates adapted to Gauss's representation, taking only the spatial components, we have

$$0 = (D_{\tilde{g}}\widetilde{Ric})(a)_{ij} = \frac{1}{2}\tilde{g}^{\rho\alpha}(\tilde{\nabla}_\rho\tilde{\nabla}_j a_{i\alpha} + \tilde{\nabla}_\rho\tilde{\nabla}_i a_{\alpha j} - \tilde{\nabla}_\rho\tilde{\nabla}_\alpha a_{ji}) - \frac{1}{2}\tilde{\nabla}_j\tilde{\nabla}_i \mathrm{tr}\ a\ .$$

Bearing in mind that in Gauss's representation $\tilde{g}_{0i} = 0$ and $\tilde{g}_{00} = -1$, the foregoing equation can now be written as

$$\partial_t^2 a_{ij} + T(a)_{ij} = 0\ ,$$

where $T(a)$ denotes the sum of terms containing the factors a and $\partial_t a$, but not $\partial_t^2 a$. The only solution for such an equation fulfilling the initial conditions $a = 0$ for $t = 0$ and $(\partial_t a)_{t=0} = 0$ is the identically zero solution, whereby we conclude the proof. □

Remark VI.9. Theorem VI.5 and Lemma VI.7 (on which the proof of Moncrief's Theorem VI.8 is based) are valid for the case with matter (not necessarily null stress-energy tensor). However, the hypothesis of being in the vacuum has been used on several occasions in the preceding proof of Theorem VI.8. Thus it would be desirable to find a generalization of Theorem VI.8 for the case with matter.

Remark VI.10. As already mentioned, Moncrief's theorem states that if $\tilde{g}$ has no Killing vector fields, then vacuum Einstein's equation is stable by linearization at the initial metric $\tilde{g}$. J. M. Arms and J. E. Marsden [2] proved that if $\tilde{g}$ has Killing vector fields $\neq 0$ then vacuum Einstein's equation is not stable by linearization at the initial metric $\tilde{g}$. Soon afterwards J. M. Arms, A. Fischer, J. E. Marsden and V. Moncrief studied in detail the cases where $\tilde{g}$ has constant mean curvature and one or several Killing vector fields, [36] and [3].

VI.4 Appendix: general results on elliptic operators in compact manifolds

In this section, we do not wish (neither are we able) to give the theory of elliptic operators on compact manifolds. We merely recall the fundamental concepts.

Recall that in Section IV.2, we defined the concept of a quasi-linear differential operator in vector bundles; in this section, we require the concept of a linear differential operator, which is a particular case of the former. In the former case, however, we were working with operators of order 1 and 2; here it is necessary to work with operators of arbitrary order k. Let M be a differentiable manifold of dimension n and let $\pi : E \to M$ and $\pi' : F \to M$ be two vector bundles whose fibers are $\mathbb{R}$-vector spaces of dimension m (the same dimension for both bundles). We denote by $\Gamma(E)$ and by $\Gamma(F)$ the sets of C^∞-sections of E and F, respectively. A mapping $D : \Gamma(E) \to \Gamma(F)$ is a linear differential operator of order k if for each domain of a local chart U of M over which E and F trivialize, $E \cong U \times \mathbb{R}^m$, $F \cong U \times \mathbb{R}^m$, and for any section $f \in \Gamma(E)$ expressed in the trivialization $E|_U \cong U \times \mathbb{R}^m$ by $f = (f^1 \dots f^m)$, $D(f)$ is expressed on U, in the corresponding trivialization of $F|_U$, by $D(f) = (D(f)^1 \dots D(f)^m)$, with the components $D(f)^A$ of the form

$$D(f)^A(x) = a^A_B(f^B) + F^A \ .$$

Here each coefficient a^A_B is a differential operator of order k on U of the form

$$a^A_B = \sum_{|\alpha|=k} a^A_{B\alpha}(x) D^\alpha_x \ ,$$

with

$$D^\alpha_x = \frac{\partial^{|\alpha|}}{\partial x_1^{\alpha_1} \dots \partial x_n^{\alpha_n}} \ ,$$

the $a^A_{B\alpha}(x)$ being C^∞ functions, and where the F^A are C^∞ functions of x and f whose values at x only depend on the derivatives of f up to order $k-1$ (see Section IV.1 for the meaning of this concept).

Such a differential operator $D : \Gamma(E) \longrightarrow \Gamma(F)$ is called *elliptic* if for any $x \in M$ and $\xi \in T_x(M)^*$, $\xi \neq 0$, the leading symbol $\sigma_L(D)(x,\xi) : E_x \longrightarrow F_x$ is an isomorphism. D gives rise for any s to a mapping between the Sobolev spaces $\mathcal{W}^s(\Gamma(E))$ and $\mathcal{W}^{s-k}(\Gamma(F))$,

$$D_s : \mathcal{W}^s(\Gamma(E)) \longrightarrow \mathcal{W}^{s-k}(\Gamma(F)) .$$

Assume now that the manifold M is Riemannian and oriented, and that the bundles E and F are endowed with inner products. The operator $D^* : \Gamma(F) \to \Gamma(E)$, adjoint of D with respect to the inner products, induces for any s the mappings

$$D^*_s : \mathcal{W}^s(\Gamma(F)) \longrightarrow \mathcal{W}^{s-k}(\Gamma(E)) .$$

The fundamental theorem for elliptic operators states:

Theorem VI.11. *Let M be a compact Riemannian oriented manifold. Let $\pi : E \to M$ and $\pi' : F \to M$ be two vector bundles over M having each of them $\mathbb{R}$-vector spaces of the same dimension as typical fibers, and endowed with inner products. Let $D : \Gamma(E) \to \Gamma(F)$ be a linear differential operator of order k, elliptic and self-adjoint. Then for any $s \geq k$ we have the following splitting into direct orthogonal topological sums:*

$$\mathcal{W}^{s-k}(\Gamma(F)) = D_s(\mathcal{W}^s(\Gamma(E)) \oplus \textit{Kernel of } D_{s-k} .$$

Furthermore, the kernel of D_{s-k} has finite dimension. In addition, if for a certain $s > k$ a certain $f \in L^2(E)$ satisfies $D(f) \in \mathcal{W}^{s-k}(\Gamma(F))$, then $f \in \mathcal{W}^s(\Gamma(E))$ (regularization). In particular, if $f \in L^2(E)$ is a solution of $D(f) = g$, where g is C^∞, then f is C^∞ ($\Gamma(F) = \cap_s \mathcal{W}^s(\Gamma(F))$ and $\Gamma(E) = \cap_s \mathcal{W}^s(\Gamma(E))$).

Hodge's theorem is a particular case of this theorem when the bundles $E = F = \wedge^p T(M)^*$ are taken for bundles E and F, and the De Rham Laplacian Δ is taken for the operator D.

From Theorem VI.11 (whose proof is not given) the following result is easily obtained:

Theorem VI.12. *Let M be a compact Riemannian oriented manifold. Let E and F be vector bundles over M endowed with inner products. Let D be a linear differential operator, $D : \Gamma(E) \longrightarrow \Gamma(F)$, of order k (here, D is not assumed to be self-adjoint), and let D^* be its adjoint. If one of the two operators D or D^* has an injective symbol, then we have a splitting into direct orthogonal topological sums,*

$$\mathcal{W}^{s-k}(\Gamma(F)) = D_s(\mathcal{W}^s(\Gamma(E))) \oplus \textit{Kernel of } D^*_{s-k} .$$

Moreover, if the operator having the injective symbol is D^**, the kernel of* D^*_{s-k} *is finite-dimensional. Furthermore, we also have an analogous splitting for the space of* C^∞ *sections:*

$$\Gamma(F) = D(\Gamma(E)) \oplus \textit{Kernel of } D^* \ .$$

This theorem, which is a consequence of Theorem VI.11, was proven by M. Berger and D. Ebin [9] for the case where D is elliptic. In order to provide a proof based on the Theorem VI.11, we require the following lemma (of Banach spaces):

Lemma VI.13. *Let* X *and* Y *be two Banach spaces and* $T : X \longrightarrow Y$ *a bounded linear mapping (that is,* $\|T(x)\| \leq K\|x\|$ *for a certain constant* K*). If* $T(X)$ *admits an algebraic supplement* C *that is closed, then the splitting* $Y = T(X) \oplus C$ *is topological.*

Proof of the lemma. The mapping $T : X \longrightarrow Y$ factorizes:

$$X \xrightarrow{\pi} X/\text{Ker } T \xrightarrow{\overline{T}} Y \ ,$$

where $\overline{T}$ is injective. $\overline{T}$ is continuous, since T is also injective. Furthermore, $T(X) = \overline{T}(X/\text{Ker } T)$. Therefore, C is an algebraic supplement of $\overline{T}(X/\text{Ker } T)$ in Y. Thus the mapping

$$\begin{array}{rccccc} \varphi : & X/\text{Ker } T & \times & C & \longrightarrow & Y \\ & (\alpha & , & c) & \longrightarrow & \overline{T}(\alpha) + c \end{array}$$

is linear, continuous and bijective, and by virtue of the open mapping theorem it is bicontinuous, thereby proving that the splitting $Y = \overline{T}(X/\text{Ker } T) \oplus C$ is topological. □

Proof of the theorem. The two cases are addressed separately. We first assume that the symbol of D^* is injective. Since the symbol of the product is the product of symbols, and the symbol of the adjoint of an operator is the adjoint mapping of the operator symbol, then DD^* has an injective symbol. Indeed, if $\sigma_L(D)(\sigma_L(D))^*\gamma = 0$, then $\langle \sigma_L(D)^*\gamma, \sigma_L(D)^*\gamma\rangle = \langle \sigma_L(D)(\sigma_L(D))^*\gamma, \gamma\rangle = 0$, which implies that $\sigma_L(D)^*\gamma = 0$, which in turn implies that $\gamma = 0$. Since the symbol of DD^* (for x and $\xi \neq 0$ fixed) goes from F_x to F_x (the same space of finite dimension), the symbol of DD^* is bijective, and therefore DD^* is elliptic. Furthermore, DD^* is self-adjoint, since $\langle DD^*x, y\rangle = \langle D^*x, D^*y\rangle = \langle x, DD^*y\rangle$. We now apply Theorem VI.11 to DD^*,

$$DD^* : \quad \mathcal{W}^{s+k}(\Gamma(F)) \xrightarrow{D^*_{s+k}} \mathcal{W}^{s}(\Gamma(E)) \xrightarrow{D_s} \mathcal{W}^{s-k}(\Gamma(F)) \ .$$

We have that $\mathcal{W}^{s-k}(\Gamma(F)) =$ Image of $D_sD^*_{s+k} \oplus A$, where A is the kernel of $D_{s-2k}D^*_{s-2}$, which has finite dimension. We denote the canonical projections associated to this splitting by π_1 and π_2. Let $B = \pi_2(\text{Im of } D_s))$. B is a subspace of A (which has finite dimension). Remark that $B \subset$ Im of D_s, since if

$b \in B$, $b = \pi_2(x)$ with $x \in$ Im of D, then $b = \pi_2(x) = x - \pi_1(x) \in$ Im $D +$ Im $DD^* \subset$ Im D. Let C be the orthogonal supplement of B in A, $A = B \oplus C$. We now have $\mathcal{W}^{s-k}(\Gamma(F)) = D_s(\mathcal{W}^s(\Gamma(E)) \oplus C$ (algebraic splitting). Indeed, if $x \in \mathcal{W}^{s-2}(\Gamma(F))$, $x = \pi_1(x) + \pi_2(x) = \pi_1(x) + p_1\pi_2(x) + p_2\pi_2(x)$, where p_1 and p_2 are the canonical projections of A on B and C, respectively. Since $p_1\pi_2(x) \in B \subset$ Im D, $\pi_1(x) + p_1\pi_2(x) \in$Im D and $p_2\pi_2(x) \in C$. Now in order to see that the intersection is empty, it suffices to see that C is orthogonal to Im D. In fact, if $x \in$ Im D and $c \in C$, $x = \pi_1(s) + \pi_2(x)$ with $\pi_2(x) \in B$. Therefore $\langle x, c\rangle = \langle \pi_1(x)+\pi_2(x), c\rangle = \langle \pi_1(x), c\rangle = 0$ (since $c \in A \perp \operatorname{Ker}(DD^*)$). By virtue of the lemma, the splitting $\mathcal{W}^{s-k}(\Gamma(F)) = D_s(\mathcal{W}^s(\Gamma(E)) \oplus C$ is topological. Finally, we see that $C = \operatorname{Ker}(D^*_{s-k})$. Indeed, if $c \in C$, $0 = \langle c, Dx\rangle\ \forall x \in \mathcal{W}^s(\Gamma(E))$. This is written $0 = \langle D^*c, x\rangle$; that is, $D^*c = 0$. Reciprocally, $y \in \operatorname{Ker}(D^*)$, $D^*y = 0$. Then $\langle y, Dx\rangle = \langle D^*y, x\rangle = 0\ \forall x \in \mathcal{W}^s(\Gamma(E))$. Therefore $y \in C$ (which is the orthogonal supplement of $D_s(\mathcal{W}^s(\Gamma(E)))$.

The last part of the theorem, giving a splitting of the C^∞ sections of F, is obtained using the part of regularization from Theorem VI.11. Indeed, for that theorem we have the splitting $\Gamma(F) = \operatorname{Im}(DD^*) \oplus A$, where A is now the kernel of DD^* that has finite dimension. All that has been performed for the Sobolev s-spaces can now be repeated in the C^∞ case, and we arrive at the desired splitting.

Now assume that it is the symbol of D which is injective. In this case, D^*D will be elliptic.

$$\mathcal{W}^s(\Gamma(E)) \xrightarrow{D_s} \mathcal{W}^{s-k}(\Gamma(F)) \xrightarrow{D^*_{s-k}} \mathcal{W}^{s-2k}(\Gamma(E)) \ .$$

By Theorem VI.11 we have the splitting

$$\mathcal{W}^{s-k}(\Gamma(E)) = \operatorname{Im}(D^*_{s-k}D_s) \oplus \operatorname{Ker}(D^*_{s-3k}D_{s-2k}) \ .$$

In order to prove the splitting $\mathcal{W}^{s-k}(\Gamma(F)) = \operatorname{Im}(D_s) \oplus \operatorname{Ker}(D^*_{s-k})$, it suffices by virtue of the lemma to prove it algebraically, since $\operatorname{Ker}(D^*_{s-k})$ is closed. Since both subspaces are mutually orthogonal, it suffices to prove that any x of $\mathcal{W}^{s-k}(\Gamma(F))$ can be written as a sum of an element of the image of D_s and an element of the kernel of D^*_{s-2}. Thus let $x \in \mathcal{W}^{s-k}(\Gamma(F))$. Then $D^*_{s-k}x$ splits into sum $D^*_{s-k}x = D^*_{s-k}D_s a + b$, with $b \in \operatorname{Ker}(D^*_{s-3k}D_{s-2k})$. This b belongs to $\operatorname{Ker}(D_{s-2k})$, since $\langle Db, Db\rangle = \langle b, D^*Db\rangle = 0$. We now apply D_{s-2k} to $D^*_{s-k}x$ and obtain

$$D_{s-2k}D^*_{s-k}x = D_{s-2k}D^*_{s-k}D_s a \ .$$

We then see that x and $D_s a$ have the same image by $D_{s-2k}D^*_{s-k}$. They thus differ by one element of the kernel. In other words, $x = D_s a + c$, with $c \in \operatorname{Ker}(D_{s-2k}D^*_{s-k})$. Now we see that $c \in \operatorname{Ker}(D^*_{s-2})$. Indeed, $0 = \langle DD^*c, c\rangle = \langle D^*c, D^*c\rangle$.

As regards the last part of the theorem giving the splitting of $\Gamma(F)$, since both summands are mutually orthogonal, it suffices to prove that any element f

of the space $\Gamma(F)$ splits into sum $Dg + h$, with $g \in \Gamma(E)$ and $D^*h = 0$. But we already know that $f = D_s g + h$ with $g \in \mathcal{W}^s(\Gamma(E))$. Therefore, $D^*f = D^*Dg$. However, since if f is C^∞, then so is D^*f, and by regularity of D^*D (which is elliptic), g is C^∞ and so is $h = f - Dg$. □

Chapter VII

Stability by Linearization of Einstein's Equation at Minkowski's Initial Metric

Let us recall that in Chapter III we studied the case of a gravitational field created by matter with a small stress-energy tensor, and where we saw (by linearization at Minkowski's initial metric) how in this case Einstein's equation was related to the wave equation. However, this naturally raises the question of whether or not it is admissible to linearize Einstein's equation at Minkowski's initial metric. In other words, whether or not Einstein's equation is stable by linearization at Minkowski's initial metric.

The affirmative response was obtained by Y. Choquet-Bruhat and S. Deser, [22]. This chapter is fundamentally devoted to this problem. Obviously, we present Y. Choquet-Bruhat and S. Deser's result in the context of the new perspective of the concept of stability by linearization of Einstein's equation introduced in Chapter V.

For the resolution of their problem, Y. Choquet-Bruhat and S. Deser had recourse to Sobolev spaces with weights, which were introduced by M. Cantor, [17], [18]. In the exposition set out in this chapter, we place great emphasis on showing why the usual Sobolev spaces are insufficient for this problem, and why new types of spaces are required. In order to make this need intuitively clear, in the proof of the Laplacian isomorphisms in Section VII.3, we adopt a completely different point of view from the original approach taken by M. Cantor.

Shortly after Y. Choquet-Bruhat and S. Deser established their result, Y. Choquet-Bruhat, A. Fischer and J. E. Marsden, [23], performed an important generalization of the same result by proving the stability by linearization of Einstein's equation in the vacuum when the initial metric (instead of Minkowski's metric) tends asymptotically to Minkowski's metric when $x \in \mathbb{R}^4$ tends to infinity in any space-like direction. For recent results we should cite [8], [26].

In Section VII.5, we set forth Y. Choquet-Bruhat, A. Fischer and J. E. Marsden's result in detail and adapt it to the case where matter is present. The new version we give of it generalizes the original result even in the vacuum (which is the case considered by these authors).

VII.1 A further expression of $D_{(g,k)}\Phi$

In many cases, in order to prove the stability by linearization of Einstein's equation, Theorem V.8 is applied, which leads to a proof of the surjectivity of the map $D_{(g,k)}\Phi$. The expression of $D_{(g,k)}\Phi = (D_{(g,k)}\mathcal{H}, D_{(g,k)}\gamma)$ obtained at the conclusion of Chapter V is as follows:

$$
\begin{aligned}
(D_{(g,k)}\mathcal{H})(h,K) &= \frac{1}{8}[-2h\cdot(k\times k) + 2k\cdot K - 2\,\mathrm{tr}(k)\,\mathrm{tr}(K) + 2h\cdot k\ \mathrm{tr}(k) \\
&\qquad + 4h\cdot Ric(g) - 4\nabla^i\nabla^j h_{ij} + 4\nabla^i\nabla_i\ \mathrm{tr}(h)\,], \\
(D_{(g,k)}\gamma)(h,K)_i &= \frac{1}{2}[\,\nabla^j K_{ij} - \partial_i\ \mathrm{tr}(K) - h^{jl}\nabla_j k_{il} + h^{jl}\nabla_i k_{jl} + \frac{1}{2}k^{jl}\nabla_i h_{jl} \\
&\qquad - k_i^l\nabla^j h_{jl} + \frac{1}{2}k_{ij}\nabla^j\ \mathrm{tr}(h)\,]\ .
\end{aligned}
$$

In accordance with an idea from [22], we express $D_{(g,k)}\Phi$ in terms of two other variables (τ, Y), where τ is a function and Y a vector field, such that when calculating the prior expressions of $D_{(g,k)}\mathcal{H}$ and of $D_{(g,k)}\gamma$ in terms of the new variables, known operators appear (if possible, Laplacians).

Since we not only employ the expression of $D_{(g,k)}\Phi$ in the case of Minkowski's metric (in the vacuum), but also in the Robertson-Walker models (with matter), which we address in the following chapter, in order to make use of calculations in similar situations we will work here in the general framework of a not necessarily null stress-energy tensor.

From the two components of the map $D\Phi$, it is in $D\gamma$ where most differential operators appear, while in $D\mathcal{H}$ the majority are algebraic operators. Let us begin then with $D\gamma$. If we set $K = L_Y g$, with Y being any vector field ($L_Y g$ is symmetric, since $(L_Y g)_{ij} = \nabla_i Y_j + \nabla_j Y_i$), the term $\nabla^j K_{ij}$ from the expression of $D\gamma$ would be $-\nabla^* L_Y g$, an operator which in the Euclidean case is directly related to the Laplacian. However, since $\mathrm{tr}(K) = -2\nabla^* Y$, the term $\partial_i\ \mathrm{tr}(K)$ would cause $-2\partial_i\nabla^* Y$ to appear. If we wished $D_{(g,k)}\gamma(h,K)$ to be equal only to $\nabla^* L_Y g$, it would be necessary to cause terms to appear in $\partial_i\nabla^* Y$ in order to be able to cancel the term $-2\partial_i\nabla^* Y$. To this end, it would be necessary to add to the term $L_Y g$ a certain scalar multiple of $g\nabla^* Y$. In other words, we would write $K = bL_Y g + cg\nabla^* Y$, where b and c would be real numbers to be determined later.

The part of $D\mathcal{H}$ containing second derivatives is formed by its last two terms (in which only h appears). A good choice for h would be to take h proportional to a product of the metric g by any function τ, since the afore-mentioned terms $\nabla^i\nabla^j h_{ij}$ and $\nabla^i\nabla_i\ \mathrm{tr}(h)$ of $D\mathcal{H}$ would then become Laplacians (since $\nabla^i\nabla^j g_{ij}\tau = -\Delta\tau$ and $\nabla^i\nabla_i\ \mathrm{tr}(g\tau) = -3\Delta\tau$). Returning to the expression of $D\gamma$, on substituting $h = \tau g$ in the remaining five terms of $D\gamma$ (which are contractions of ∇h with k), we obtain that the first and last terms are of order 0 in τ, while the other three are of order 1. In order to counteract these three terms, it would be necessary to add to the expression of K the terms $k\tau$ and $g\tau\ \mathrm{tr}(k)$.

In short, in order to determine the most appropriate coefficients, we write

$$\begin{aligned} h(\tau,Y) &= a\tau g \\ K(\tau,Y) &= bL_Y g + cg\nabla^* Y + dk\tau + eg\tau\ \mathrm{tr}(k)\ , \end{aligned}$$

where a, b, c, d and e are real numbers to be determined. Then $\mathrm{tr}(h) = 3a\tau$ and $\mathrm{tr}(K) = (3c-2b)\nabla^* Y + (3e+d)\tau\ \mathrm{tr}(k)$. With these substitutions, the previous expressions of $D_{(g,k)}\mathcal{H}$ and $D_{(g,k)}\gamma$ become:

$$\begin{aligned} (D_{(g,k)}\mathcal{H})(Y,\tau) &= \frac{1}{8}\big[(2d-2a)\tau k\cdot k + 2b\ k\cdot L_Y g + (4b-4c)\ \mathrm{tr}(k)\nabla^* Y \\ &\quad +(2a-2d-4e)\tau\ \mathrm{tr}^2(k) + 4a\ \tau R - 8a\ \Delta\tau\big]\ , \\ (D_{(g,k)}\gamma)(Y,\tau)_i &= \frac{1}{2}\big[-b(\nabla^* L_Y g)_i + (2b-2c)\nabla_i\nabla^* Y + (d-a)\tau\nabla^j k_{ij} \\ &\quad +(1/2\ a+d)k_{ij}\nabla^j\tau + (1/2\ a - 2e - d)\ \mathrm{tr}(k)\nabla_i\tau \\ &\quad +(a-2e-d)\tau\nabla_i\ \mathrm{tr}(k)\big]\ . \end{aligned}$$

It is thus clear that it is impossible to reduce $(D_{(g,k)}\gamma)(Y,\tau)$ to the term $\nabla^* L_Y g$ by algebraic methods, since the vanishing of all the other terms gives rise to the condition $b=c$ and to a linear and homogeneous system of four equations for the coefficients a, d and e, the only solution to which is the trivial one. By choosing these three coefficients so that the terms of order 1 in τ are removed, we arrive at the conclusion that $b=c$, $d=-1/2\ a$ and $e=1/2\ a$. By setting $a=1$ and $b=-2$, we deduce that $c=-2$, $d=-1/2$ and $e=1/2$, thus obtaining

$$\begin{aligned} (D_{(g,k)}\mathcal{H})(Y,\tau) &= \frac{1}{8}\big[-3\tau k\cdot k - 4\ k\cdot L_Y g + \tau\ \mathrm{tr}^2(k) + 4\tau R - 8\Delta\tau\big]\ , \\ (D_{(g,k)}\gamma)(Y,\tau)_i &= \frac{1}{2}\left[2(\nabla^* L_Y g)_i - \frac{3}{2}\tau\nabla^j k_{ij} + \frac{1}{2}\tau\nabla_i\ \mathrm{tr}(k)\right]\ . \end{aligned}$$

Recall that with these values for the coefficients a, b, c, d and e, the change of variables used is

$$\begin{aligned} h(\tau,Y) &= \tau g, \\ K(\tau,Y) &= -2L_Y g - 2g\nabla^* Y - \frac{1}{2}\,k\tau + \frac{1}{2}\,g\tau\ \mathrm{tr}(k)\ . \end{aligned} \tag{VII.1}$$

The foregoing expressions of $D_{(g,k)}\mathcal{H}$ and $D_{(g,k)}\gamma$ can be simplified a little further if we take into account the constraint equations $\mathcal{H}(g,k) = \chi F$ and $\gamma(g,k) = \chi X$, which explicitly are $k \cdot k - 4R - \operatorname{tr}^2(k) = 8\chi F$ and $\nabla^j k_{ij} - \partial_i \operatorname{tr}(k) = 2\chi X_i$:

$$\begin{aligned}(D_{(g,k)}\mathcal{H})(Y,\tau) &= \frac{1}{8}\left[-2\tau k \cdot k - 4\, k \cdot L_Y g - 8\chi F \tau - 8\Delta\tau\right],\\ (D_{(g,k)}\gamma)(Y,\tau)_i &= \frac{1}{2}\left[2(\nabla^* L_Y g)_i - \tau\nabla_i \operatorname{tr}(k) - 3\chi\tau X_i\right] .\end{aligned} \tag{VII.2}$$

VII.2 The relation between Euclidean Laplacian and stability by linearization at the initial Minkowski metric

Let us now take the case where the manifold V is $\mathbb{R}^4$, the initial metric is Minkowski's metric, the initial stress-energy tensor T is zero, and the submanifold M of V is $\{0\} \times \mathbb{R}^3$. Then the initial pair (g,k) is formed by the Euclidean metric e of $\mathbb{R}^3$ and by the tensor $k = 0$. Since we are dealing with the vacuum, we have that $F = 0$ and $X = 0$. Therefore, the previous expressions of $D_{(g,k)}\mathcal{H}$ and $D_{(g,k)}\gamma$ become

$$\begin{aligned}(D_{(e,0)}\mathcal{H})(Y,\tau) &= -\Delta\tau\\ (D_{(e,0)}\gamma)(Y,\tau) &= \nabla^* L_Y e .\end{aligned}$$

Subsequently, we study the operators $Y \to \nabla^* L_Y e$ and $\tau \to \Delta\tau$, and we will see what conditions enable us to state that they are isomorphisms. For the moment, let us assume that they are isomorphisms and see that in this case $D\Phi$ is surjective and its kernel has a topological supplement. Indeed:

- As regards surjectivity: given $(f,Z) \in \mathcal{F}^{s-2}(M) \times \Omega^{s-2}(M)$ there exists by hypothesis a single $Y \in \mathcal{X}^s$ such that $\nabla^* L_Y e = Z$. For this $Y \in \mathcal{X}^s$, $(D_{(e,0)}\mathcal{H})(Y,\tau) = f$ implies that

$$\Delta\tau = -f .$$

  Recall that, by hypothesis, Δ is an isomorphism. Thus, there exists a sole function τ such that $(D_{(e,0)}\mathcal{H})(Y,\tau) = f$. With this function τ and this field Y, the tensors h and K are built according to the expressions $h = e\tau$ and $K = -2L_Y e - 2e\nabla^* Y$. By construction, $(D_{(e,0)}\Phi)(h,K) = (f,Z)$ and the pair (h,K) is the inverse image of (f,Z).

- As regards the existence of the topological supplement: consider for any element $(h,K) \in \mathcal{S}^s \times \mathcal{S}^{s-1}$ the pair (h_1,K_1) of the form

$$\begin{cases} h_1 = e\tau \\ K_1 = -2L_Y e - 2e\nabla^* Y \end{cases}$$

such that $(D_{(e,0)}\Phi)(h_1, K_1) = (D_{(e,0)}\Phi)(h, K)$ (unique, since $D_{(e,0)}\Phi$ is an isomorphism restricted to the pairs (h_1, K_1)). If we write

$$(h, K) = (h_1, K_1) + (h_2, K_2),$$

then (h_2, K_2) belongs to the kernel of $D\Phi$. The set formed by the pairs (h_1, K_1) is the topological supplement sought for $\ker D\Phi$.

The operators that, without specifying anything, we assume to be topological isomorphisms are $Y \to \nabla^* L_Y e$ and the Laplacian. The Euclidean Laplacian is an operator that has been widely studied, [17] and [18], and is known to be a topological isomorphism between certain Sobolev spaces which will be introduced in the following section. We will also see that the same result easily implies that the operator $Y \to \nabla^* L_Y e$ is an isomorphism.

VII.3 Some proofs on topological isomorphisms in $\mathbb{R}^n$

In Section I.4, we introduced the Sobolev spaces $\mathcal{W}^{p,s}(M)_r$ of the r-tensors with coefficients in L^p and with covariant derivatives up to the order s in L^p. When $r = 0$, (that is, in the case of functions) these spaces are denoted by $\mathcal{F}^{p,s}(M)$ (Section V.3). It is well known that in the case of $M = \mathbb{R}^n$ with the Euclidean metric, the Laplacian Δ does not give an isomorphism between the spaces $\mathcal{F}^{p,s}$ and $\mathcal{F}^{p,s-2}$. However, by conveniently replacing the Sobolev spaces by analogous spaces with *weights*, it is possible to make Δ yield a topological isomorphism, [17], [18]. In this section, these Sobolev spaces with weights are defined precisely, and an intuitive proof is given of the fact that Δ yields an isomorphism between these spaces; this proof differs from that found in the papers referred to.

We frequently work with functions of the type $|x|^\alpha = (x_1^2 + \ldots + x_n^2)^{\frac{\alpha}{2}}$, where $\alpha \in \mathbb{R}$. We now recall the following highly elementary property regarding integrals of these functions in $\mathbb{R}^n$.

Lemma VII.1. *Let B be a ball centered at the origin of $\mathbb{R}^n$, and let ${}^cB$ be its complement in $\mathbb{R}^n$. Let J_1 and J_2 be the two integrals*

$$J_1 = \int_B |x|^\alpha dx \ , \ J_2 = \int_{{}^cB} |x|^\alpha dx \ ,$$

where dx denotes $dx_1 \ldots dx_n$. The integral J_1 is finite if and only if $\alpha > -n$. The integral J_2 is finite if and only if $\alpha < -n$.

Proof. We assign the number $r = |x|$ and the unit vector $\omega = \frac{x}{r} \in S^{n-1}$ to every $x \in \mathbb{R}^n$, $x \neq 0$. The point x can be recovered from the pair (r, ω) (spherical coordinates). The volume element dx of $\mathbb{R}^n$ is expressed in spherical coordinates by $r^{n-1} d\omega dr$, where $d\omega$ is the volume element (or area element) of the unit sphere

S^{n-1}. If R is the radius of the ball B, and if $\alpha + n \neq 0$, the integral J_1, calculated in spherical coordinates, will be

$$J_1 = \int_{\omega \in S^{n-1}} \int_{r \in [0,R]} r^{\alpha+n-1} d\omega dr = C \left[\frac{r^{\alpha+n}}{\alpha+n} \right]_{r=0}^{r=R} ,$$

where C denotes the volume of the sphere S^{n-1}. In the case $\alpha+n=0$, a logarithm will appear. In any case, the integral will be finite if and only if $\alpha + n > 0$. An analogous reasoning is followed for the integral J_2. □

We now study the Laplacian operator on functions. In order to find the inverse of any operator it is necessary to determine its fundamental solution. Indeed, it is well known that the function $v(x) = v(x_1, \ldots, x_n) = |x|^{2-n} = (x_1^2 + \ldots + x_n^2)^{\frac{2-n}{2}}$ defined in $\mathbb{R}^n$ $(n \geq 3)$ for $x \neq 0$ is a solution to the Laplacian equation $\Delta v = 0$. In fact, a short calculation shows that the second derivative with respect to any variable x_i is

$$\frac{\partial^2 v}{\partial x_i^2} = (n-2)(x_1^2 + \ldots + x_n^2)^{(-n/2-1)} [n x_i^2 - (x_1^2 + \ldots + x_n^2)] ,$$

and on adding,

$$\sum_i \frac{\partial^2 v}{\partial x_i^2} = (n-2)(\sum_i x_i^2)^{(-n/2-1)} \left[n \sum_i x_i^2 - n \sum_i x_i^2 \right] = 0 .$$

Let us apply Green's formula

$$\int_U (u\Delta v - v\Delta u) dx = - \int_{\partial U} \left(u \frac{\partial v}{\partial n} - v \frac{\partial u}{\partial n} \right) d\sigma \tag{VII.3}$$

when the region $U = B_R - B_\epsilon$ is limited by the spheres centered at the origin with respective radii R and ϵ, when $v(x) = |x|^{2-n}$ and $u \in C^2(\mathbb{R}^n)$ and has compact support. Here $dx = dx_1 \ldots dx_n$ is the volume element of $\mathbb{R}^n$, and $d\sigma$ the area element of ∂U. The direction of the normal n is that which goes towards the exterior of ∂U. Green's formula as stated here differs from that usually found in the literature by the sign on the right-hand side. This difference is due to the fact that here the Riemannian Laplacian (in accordance with the definitions in Chapter I) is $\Delta = -\sum \partial^2/\partial x_i^2$.

We choose R sufficiently large such that the support of u is included in the ball B_R, and therefore the integral $\int_{\partial B_R} (u\frac{\partial v}{\partial n} - v\frac{\partial u}{\partial n}) d\sigma$ vanishes. Since $\frac{\partial v}{\partial n} = C|x|^{1-n}$ and the measure of ∂B_ϵ is proportional to ϵ^{n-1}, then

$$\int_{\partial B_\epsilon} u \frac{\partial v}{\partial n} d\sigma = C\epsilon^{1-n} \int_{\partial B_\epsilon} u d\sigma = C \frac{1}{\sigma(\partial B_\epsilon)} \int_{\partial B_\epsilon} u d\sigma,$$

$$\int_{\partial B_\epsilon} v \frac{\partial u}{\partial n} d\sigma = C\epsilon^{2-n} \int_{\partial B_\epsilon} \frac{\partial u}{\partial n} d\sigma = C \frac{\epsilon}{\sigma(\partial B_\epsilon)} \int_{\partial B_\epsilon} \frac{\partial u}{\partial n} d\sigma.$$

When $\epsilon \to 0$, by Lebesgue's theorem, the first integral tends to $u(0)$ and the second to zero. Finally, making $R \to \infty$ and using $\Delta v = 0$, from (VII.3) we obtain

$$u(0) = C \int_{\mathbb{R}^n} |y|^{2-n} \Delta u(y) dy \ .$$

Here we have replaced the previous constant C by another constant we also refer to as C.

Note that since u is continuous with compact support and $2 - n > -n$, the part of Lemma VII.1 referring to the integral J_1 ensures that the previous integral is finite. Let τ_x be the transformation $\tau_x(y) = x - y$ (a symmetry with respect to the origin followed by a translation). By applying the previous identity to $(u \circ \tau_x)(y) = u(x - y)$, we arrive at

$$\begin{aligned}(u \circ \tau_x)(0) = u(x) &= C \int_{\mathbb{R}^n} |y|^{2-n} \Delta u(x-y) dy \\ &= C \int_{\mathbb{R}^n} |x-y|^{2-n} \Delta u(y) dy \ .\end{aligned}$$

Here we have used $\Delta(u \circ \tau) = \Delta u \circ \tau$, since the transformation τ_x is an isometry of the Euclidean metric and Δ is a metric operator. We thus recover u from Δu by the convolution product of Δu with the so-called *Riesz potential*, $I_2(x) = C|x|^{2-n}$, and the inverse operator of $u \to \Delta u$ is $f \to I_2 * f$, where

$$(I_2 * f)(x) = C \int_{\mathbb{R}^n} |x-y|^{2-n} f(y) dy \ .$$

Note that the integral is finite by virtue of that part of Lemma VII.1 referring to the integral J_1.

The question now is to find appropriate Banach spaces in which the operator $u \to \Delta u$ is a topological isomorphism with inverse $f \to I_2 * f$. Henceforth, we denote the operator $f \to I_2 * f$ by I. Since Δ is a second-order linear differential operator, if we choose as Banach spaces the Sobolev spaces $\mathcal{F}^{p,s}(\mathbb{R}^n)$ to which we referred at the beginning of this section, then Δ leads to a map $\Delta : \mathcal{F}^{p,s}(\mathbb{R}^n) \to \mathcal{F}^{p,s-2}(\mathbb{R}^n)$, which will automatically be continuous because $\|\Delta u\|_{p,s-2} \leq K\|u\|_{p,s}$, K being a positive constant, since on the left-hand side there are fewer derivatives than on the right-hand side. However, in these Sobolev spaces the inverse operator is not continuous, because it does not fulfill $\|I(f)\|_{p,s} \leq K'\|f\|_{p,s-2}$. Finding among which functional spaces the previous inequality is true constitutes the crux of the question, since if it were true, for example, that $\|I(f)\|_{p,s} \leq K'\|f\|_{p,s-2}$, then operator Δ would be an isomorphism: on the one hand, any $f \in \mathcal{F}^{p,s}(\mathbb{R}^n)$ has an inverse image given by $I(f)$, and by making $f = \Delta u$ in $\|I(f)\|_{p,s} \leq K'\|f\|_{p,s-2}$ we would obtain $\|u\|_{p,s} \leq K'\|\Delta u\|_{p,s-2}$, whereby $\Delta u = 0$ would imply that $u = 0$.

The reason why the Sobolev spaces $\mathcal{F}^{p,s}(\mathbb{R}^n)$ are inappropriate is that the homogeneity is incorrect. Indeed, assume that

$$\|I(f)\|_{p,s} \leq K\|f\|_{p,s-2}\,, \qquad f \in \mathcal{F}^{p,s-2}(\mathbb{R}^n),$$

is fulfilled, and we will arrive at a contradiction. The above inequality for $s = 2$, $\|I(f)\|_{p,2} \leq K\|f\|_{p,0}$ implies that

$$\|\nabla^{(k)}(I(f))\|_p \leq K\|f\|_p\,, \quad k \leq 2.$$

In particular, for $k = 0$ it is necessary that $\|I(f)\|_p \leq K\|f\|_p$. In order to check that this is impossible, we consider the *dilated functions* $f_\lambda(x) = \lambda^{n/p}f(\lambda x)$, $\lambda > 0$. The L^p-norms of f and of f_λ are equal because

$$\begin{aligned}(\|f_\lambda\|_p)^p = \int_{\mathbb{R}^n} |f_\lambda^p(x)|dx &= \int_{\mathbb{R}^n} \lambda^n|f^p(\lambda x)|dx \\ &= \int_{\mathbb{R}^n} \lambda^n|f^p(x)|\frac{1}{\lambda^n}dx = (\|f\|_p)^p\,. \end{aligned} \tag{VII.4}$$

Furthermore,

$$\begin{aligned} I(f)_\lambda(x) &= \lambda^{n/p}\int_{\mathbb{R}^n} |\lambda x - y|^{2-n}f(y)dy \\ &= \lambda^{n/p}\int_{\mathbb{R}^n} |\lambda x - \lambda y|^{2-n}f(\lambda y)\lambda^n dy \\ &= \lambda^{n/p}\lambda^2\int_{\mathbb{R}^n} |x-y|^{2-n}f(\lambda y)dy = \lambda^2 I(f_\lambda)(x)\,. \end{aligned}$$

Thus, if $\|I(f)\|_p \leq K\|f\|_p$ were true, we could write

$$\lambda^{-2}\|I(f)\|_p = \lambda^{-2}\|I(f)_\lambda\|_p = \|I(f_\lambda)\|_p \leq K\|f_\lambda\|_p = K\|f\|_p,$$

for any λ, an inequality that is false when $\lambda \to 0$.

In the case of the first derivatives, since

$$\frac{\partial}{\partial x_i}I(f)(x) = (2-n)C\int \frac{x_i - y_i}{|x-y|^n}f(y)dy\,,$$

we would obtain

$$\lambda^{-1}\|\nabla(I(f)\|_p \leq K\|f\|_p\,,$$

with the same contradiction as before. Thus the operators $f \to I_2 * f$ and $f \to \nabla(I_2 * f)$ are not bounded in L^p. However, the second derivative does have the correct homogeneity, because

$$\frac{\partial^2}{\partial x_i \partial x_j}I(f)(x) = (2-n)\int\left[\frac{\delta_{ij}}{|x-y|^n} - n\frac{(x_i-y_i)(x_j-y_j)}{|x-y|^{n+2}}\right]f(y)dy$$

has a behaviour equivalent to $|x|^{-n}$. Obviously, this does not mean that the operator $f \to \nabla^2(I_2 * f)$ is automatically bounded in L^p, but it can be proven that it is bounded if $p > 1$.

In order to define the appropriate Sobolev spaces for this problem, we use for the derivation operators the following notation: if k is a multi-index, $k = (k_1 \ldots k_n)$, $|k|$ will indicate the sum $|k| = k_1 + \ldots + k_n$ and D^k will indicate the derivation operator

$$D^k = \frac{\partial^{|k|}}{\partial x_1^{k_1} \ldots \partial x_n^{k_n}} .$$

Let $\mathcal{D}(\mathbb{R}^n)$ be the space of C^∞-functions on $\mathbb{R}^n$, with compact support. For every $s \in \mathbb{Z}^+ \cup \{0\}$, $p \in \mathbb{R}^+$ and $\delta \in \mathbb{R}$ we define the Sobolev space of functions $\mathcal{F}_\delta^{p,s}(\mathbb{R}^n)$, weighted by δ, as the completion of the space $\mathcal{D}(\mathbb{R}^n)$ by the norm $\| \ \|_{p,s,\delta}$ defined as

$$\|u\|_{p,s,\delta} = \sum_{|k| \leq s} \left\| (D^k u)(x) \left(\sqrt{1+|x|^2}\right)^{|k|+\delta} \right\|_p .$$

Note that the weights $\left(\sqrt{1+|x|^2}\right)^{|k|+\delta}$ only influence the convergence of the integrals close to infinity. Therefore, if $\Phi \in \mathcal{D}$, then $\Phi\, u \in \mathcal{F}_\delta^{p,s}$ if and only if $u \in \mathcal{F}^{p,s}$ (Sobolev space without weights). In other words, the elements $u \in \mathcal{F}_\delta^{p,s}$ are locally functions of $\mathcal{F}^{p,s}$. It is not difficult to prove that $\mathcal{F}_\delta^{p,s}$ coincides with the space of those functions $u \in \mathcal{F}_{loc}^{p,s}(\mathbb{R}^n)$ whose derivatives (in the sense of distributions) fulfill $(D^k u)(x) \left(\sqrt{1+|x|^2}\right)^{|k|+\delta} \in L^p(\mathbb{R}^n)$, $|k| \leq s$.

For example, the functions u of class C^∞ that outside a certain ball B are of the type $|x|^\lambda$ with $\lambda \in \mathbb{R}$ belong to $\mathcal{F}_\delta^{p,s}(\mathbb{R}^n)$ if $\delta < -\frac{n}{p} - \lambda$. Indeed, for any multi-index k we have

$$\left(\left\| (D^k u)(x) \left(\sqrt{1+|x|^2}\right)^{|k|+\delta} \right\|_p \right)^p = \int_B |(D^k u)(x)|^p \left(\sqrt{1+|x|^2}\right)^{p(|k|+\delta)} dx$$
$$+ \int_{{}^cB} |D^k |x|^\lambda|^p (1+|x|^2)^{\frac{p(|k|+\delta)}{2}} dx .$$

Obviously, the first integral is finite. The second is also finite, since it can be bounded above by an integral of the type

$$\int_{{}^cB} |x|^{-p|k|+p\lambda} (1+|x|^2)^{\frac{p(|k|+\delta)}{2}} dx ,$$

which is finite providing that $p(\lambda + \delta) < -n$, according to Lemma VII.1. For instance, if u outside a certain ball has the same behavior as $|x|^{-\delta-1}$, then for $p > n$ it will be in the space $\mathcal{F}_\delta^{p,s}(\mathbb{R}^n)$, because in this case $\lambda = -\delta - 1$ and the inequality $\delta < -\frac{n}{p} - \lambda$ is fulfilled when $p > n$.

If $\delta \geq 0$, we have $\| \ \|_{p,s} \leq \| \ \|_{p,s,\delta}$, which implies that $\mathcal{F}_{\delta}^{p,s}(\mathbb{R}^n) \subset \mathcal{F}^{p,s}(\mathbb{R}^n)$. Therefore, Sobolev's lemma remains true: $\mathcal{F}_{\delta}^{p,s}(\mathbb{R}^n) \subset C^k(\mathbb{R}^n)$ if $\delta \geq 0$ and $s > \frac{n}{p} + k$.

Proposition VII.2. *The operator $I : \mathcal{F}_{\delta+2}^{p,s-2} \longrightarrow \mathcal{F}_{\delta}^{p,s}$ is bounded when $-n/p < \delta < n/q - 2$, where p and q are conjugate exponents; that is $\frac{1}{p} + \frac{1}{q} = 1$.*

Proof. To facilitate the proof, we take the case $s = 2$. We begin with $f \in \mathcal{F}_{\delta+2}^{p,0}$; that is, we assume that $f(x)(1+|x|^2)^{\frac{2+\delta}{2}} \in L^p(\mathbb{R}^n)$ and wish to arrive at $I(f) \in \mathcal{F}_{\delta}^{p,2}$. It is necessary to prove therefore that $I(f)(x)(1+|x|^2)^{\frac{\delta}{2}}$, $\left(\frac{\partial}{\partial x_i} I(f)(x)\right)(1+|x|^2)^{\frac{1+\delta}{2}}$ and $\left(\frac{\partial^2}{\partial x_i \partial x_j} I(f)(x)\right)(1+|x|^2)^{\frac{2+\delta}{2}}$ belong to L^p. We start by proving the first of these; that is, that $I(f)(x)(1+|x|^2)^{\frac{\delta}{2}} \in L^p$.

Recall that $I(f)(x) = C \int_{\mathbb{R}^n} |x-y|^{2-n} f(y)dy$. By making the change $g(x) = f(x)(1+|x|^2)^{1+\delta/2}$, it suffices to prove that the operator T defined by

$$\begin{aligned}(Tg)(x) &= I(f)(x)(1+|x|^2)^{\delta/2} \\ &= C \int_{\mathbb{R}^n} g(y)(1+|y|^2)^{-1-\delta/2}|x-y|^{2-n}(1+|x|^2)^{\delta/2}dy\end{aligned}$$

is a bounded operator from $L^p(\mathbb{R}^n)$ to $L^p(\mathbb{R}^n)$, for $p > 1$. In order to check this, we use the following lemma (see for example [28]):

Lemma VII.3. *An integral operator with a positive kernel $K(x,y)$ is bounded from $L^p(\mathbb{R}^n)$ to $L^p(\mathbb{R}^n)$ if and only if there exists a function $h \geq 0$ such that*

$$\int K(x,y)h(y)^{-q}dy = O(h(x)^{-q}) \quad , \int K(x,y)h(x)^{-p}dx = O(h(y)^{-p}),$$

where q is the conjugate exponent of p, $\frac{1}{p} + \frac{1}{q} = 1$, $p > 1$.

In our case, the kernel is

$$K(x,y) = (1+|y|^2)^{-1-\delta/2}|x-y|^{2-n}(1+|x|^2)^{\delta/2}.$$

We take $h(y) = 1 + |y|^\beta$, and calculate

$$\begin{aligned}T(h^{-q})(x) &= \int (1+|y|^\beta)^{-q}(1+|y|^2)^{-1-\delta/2}|x-y|^{2-n}(1+|x|^2)^{\delta/2}dy \\ &= (1+|x|^2)^{\delta/2} \int (1+|y|^\beta)^{-q}(1+|y|^2)^{-1-\delta/2}|x-y|^{2-n}dy\end{aligned}$$

in order to check that it is of order $(1+|x|^\beta)^{-q}$. Note first of all that $T(h^{-q})$ is continuous, since in the region $|x-y| \sim 0$ the integral is finite according to Lemma VII.1. Therefore, to check if $T(h^{-q})(x)$ is of order $(1+|x|^\beta)^{-q}$, we may assume that $|x| \gg 1$.

We divide the integration domain of $T(h^{-q})$ into the three following regions:

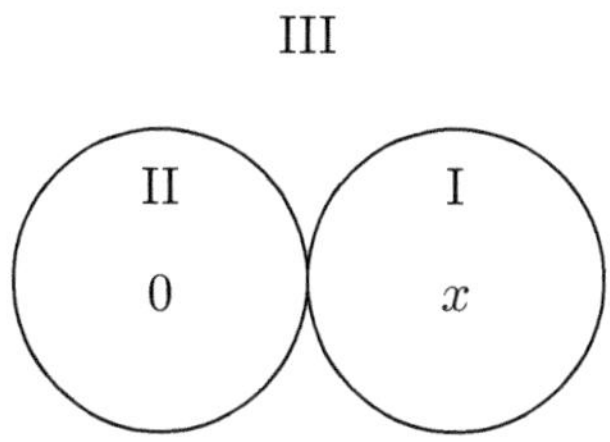

Region I: Here, $|x-y| \leq |x|/2$; that is, $|y| \approx |x| \gg 1$, and the integral is of the type

$$|x|^{-\beta q-2} \int_{|x-y|\leq |x|/2} |x-y|^{2-n} dy = |x|^{-\beta q-2} O(|x|^2) = O(|x|^{-\beta q})$$
$$\cong O((1+|x|^{\beta})^{-q})$$

since $|x|$ is large.

Region II: Now $|y| \leq |x|/2$, which implies $|x-y| \simeq |x|$, and the integral defining $T(h^{-q})(x)$ is in this region of the type

$$|x|^{2-n+\delta} \int_{|y|\leq|x|/2} (1+|y|)^{-2-\beta q-\delta} dy \ .$$

The integral appearing here is convergent (by virtue of Lemma VII.1) if $n < 2 + \beta q + \delta$, and therefore it is finite (of order 1 in $|x|$). If $n = 2+\beta q+\delta$, it is $O(\log(|x|))$, and if $n > 2+\beta q+\delta$, it is $O(|x|^{-2-\beta q-\delta+n})$. Therefore, when $n > 2+\beta q+\delta$ the contribution we are dealing with is $O(|x|^{-\beta q})$, which is $O((1+|x|^{\beta})^{-q})$, because $|x|$ is large.

Region III: In the exterior of the two balls in the preceding figure, $|y| \approx |x-y|$, and the integral defining $T(h^{-q})(x)$ in this region becomes

$$(1+|x|^2)^{\delta/2} \int_{|y|\geq|x|/2} |y|^{-\beta q-2-\delta+2-n} dy = O(|x|^{-\beta q}) \ ,$$

when $\beta q + \delta > 0$, which is the convergence condition of the preceding integral.

In conclusion, if $n > 2+\beta q+\delta$ and $\beta q+\delta > 0$, we have that

$$\int K(x,y) h(y)^{-q} dy = O(h(x)^{-q})$$

when $h(y) = 1 + |y|^{\beta}$.

Now it is necessary to study when

$$\begin{aligned} T(h^{-p})(y) &= \int K(x,y)h^{-p}(x)dx \\ &= \int (1+|y|^2)^{-1-\delta/2}|x-y|^{2-n}(1+|x|^\beta)^{-p}(1+|x|^2)^{\delta/2}dx \\ &= (1+|y|^2)^{-1-\delta/2}\int |x-y|^{2-n}(1+|x|^\beta)^{-p}(1+|x|^2)^{\delta/2}dx \end{aligned}$$

is of order $(1+|y|^\beta)^{-p}$. In the same way as before, we can consider $|y| \gg 1$.

Region I: $|x-y| \leq |y|/2$, $|x| \approx |y| \gg 1$. The integral defining $T(h^{-p})(y)$ in this region becomes

$$\begin{aligned} &(1+|y|^2)^{-1-\delta/2}(1+|y|^\beta)^{-p}(1+|y|^2)^{\delta/2}\int_{|x-y|\leq|y|/2}|x-y|^{2-n}dx \\ &\quad = O(|y|^{-2-\delta})O(|y|^{-p\beta})O(|y|^\delta)O(|y|^2) = O(|y|^{-p\beta}) = O((1+|y|^\beta)^{-p})\ , \end{aligned}$$

since $|y|$ is large.

Region II: $|x| \leq |y|/2$, $|y-x| \simeq |y|$. The integral defining $T(h^{-p})(y)$ in this region becomes

$$|y|^{-\delta-n}\int_{|x|<\frac{|y|}{2}}(1+|x|)^{-\beta p+\delta}dx\ .$$

The integral is finite and of order $O(|y|^{n-\beta p+\delta)})$ when $n > \beta p - \delta$. With this assumption, the global contribution in this region is $O(|y|^{-\beta p})$.

Region III: $|x| \approx |x-y|$ and the integral defining $T(h^{-p})(y)$ is in this region of the type

$$(1+|y|^2)^{-1-\delta/2}\int_{|x|\geq|y|/2}|x|^{2-n-\beta p+\delta}dx = O(|y|^{-\beta p})$$

if $\beta p - \delta - 2 > 0$ (which is the condition of finitude of the previous integral).

The conditions appearing so far on $\beta \in \mathbb{R}$ are $\beta q + \delta > 0$, $n > 2 + \beta q + \delta$, $\beta p - \delta > 2$ and $n > \beta p - \delta$; that is,

$$\max\left(\frac{2+\delta}{p}, \frac{-\delta}{q}\right) < \beta < \min\left(\frac{n+\delta}{p}, \frac{n-2-\delta}{q}\right)\ .$$

The only inequalities imposing conditions on δ are $\frac{2+\delta}{p} < \frac{n-2-\delta}{q}$ and $\frac{-\delta}{q} < \frac{n+\delta}{p}$. The first is equivalent to $\delta < n/q - 2$, and the second is equivalent to $\delta > -n/p$.

Now it is necessary to check that $\left(\frac{\partial}{\partial x_i}I(f)(x)\right)(1+|x|^2)^{\frac{1+\delta}{2}} \in L^p(\mathbb{R}^n)$. We have

$$\frac{\partial}{\partial x_i}I(f)(x) = C\int \frac{x_i-y_i}{|x-y|^n}f(y)dy\ .$$

As before, by taking $g(x) = f(x)(1+|x|^2)^{1+\delta/2} \in L^p(\mathbb{R}^n)$, we define the operator

$$Tg(x) = \int_{\mathbb{R}^n} g(y)(1+|y|^2)^{-1-\delta/2}|x-y|^{1-n}(1+|x|^2)^{(1+\delta)/2}dy,$$

and repeat all the previous analysis with the kernel

$$K(x,y) = (1+|y|^2)^{-1-\delta/2}|x-y|^{1-n}(1+|x|^2)^{(1+\delta)/2}.$$

The result is as follows:

$$\int K(x,y)h(y)^{-q}dy = O(h(x)^{-q}) \quad , \int K(x,y)h(x)^{-p}dx = O(h(y)^{-p}),$$

where $h(x) = 1+|x|^\beta$, when

$$\max\left(\frac{2+\delta}{p}, \frac{-1-\delta}{q}\right) < \beta < \min\left(\frac{n+1+\delta}{p}, \frac{n-2-\delta}{q}\right) ,$$

and we arrive at the same conditions on δ as before: $\delta < n/q - 2$ and $\delta > -n/p$.

Lastly, it is necessary to verify that $\left(\frac{\partial^2}{\partial x_i \partial x_j} I(f)(x)\right)(1+|x|^2)^{1+\frac{\delta}{2}} \in L^p(\mathbb{R}^n)$. Instead of calculating $\frac{\partial^2}{\partial x_i \partial x_j} I(f)(x)$ explicitly, we use the fact referred to above that the operator $f \to \frac{\partial^2}{\partial x_i \partial x_j} I(f)$ is bounded in $L^p(\mathbb{R}^n)$ if $p > 1$.

First of all, setting $u = I(f)$, we have that

$$\begin{aligned}\frac{\partial^2}{\partial x_i \partial x_j}[u(x)(1+|x|^2)^{1+\frac{\delta}{2}}] &= \frac{\partial^2 u}{\partial x_i \partial x_j}(1+|x|^2)^{1+\frac{\delta}{2}} \\ &+ 2x_j \frac{\partial u}{\partial x_i}(1+\delta/2)(1+|x|^2)^{\frac{\delta}{2}} + 2x_i \frac{\partial u}{\partial x_j}(1+\delta/2)(1+|x|^2)^{\frac{\delta}{2}} \\ &+ 4x_i x_j (1+\delta/2)(\delta/2)u(x)(1+|x|^2)^{\frac{\delta}{2}-1} + 2\delta_{ij}u(x)(1+\delta/2)(1+|x|^2)^{\frac{\delta}{2}} .\end{aligned}$$

Since we already know that $u(x)(1+|x|^2)^{\frac{\delta}{2}}$, $\frac{\partial u}{\partial x_i}(1+|x|^2)^{\frac{1+\delta}{2}} \in L^p(\mathbb{R}^n)$, and furthermore, since the norm in L^p of $\frac{\partial u(x)}{\partial x_i} x_j(1+|x|^2)^{\frac{\delta}{2}}$ is comparable to that of $\frac{\partial u(x)}{\partial x_i}(1+|x|^2)^{\frac{1+\delta}{2}}$, and the norm in L^p of $x_i x_j u(x)(1+|x|^2)^{\frac{\delta}{2}-1}$ is comparable to the norm of $u(x)(1+|x|^2)^{\frac{\delta}{2}}$, we have that $\frac{\partial^2 u(x)}{\partial x_i \partial x_j}(1+|x|^2)^{\frac{1+\delta}{2}} \in L^p(\mathbb{R}^n)$ if and only if $\frac{\partial^2}{\partial x_i \partial x_j}[u(x)(1+|x|^2)^{1+\frac{\delta}{2}}] \in L^p(\mathbb{R}^n)$. However, the fact that $\frac{\partial^2}{\partial x_i \partial x_j} I$ is bounded in $L^p(\mathbb{R}^n)$ means that $\|\frac{\partial^2}{\partial x_i \partial x_j} I(\Delta v)\|_p \le C\|\Delta v\|_p$; that is, $\|\frac{\partial^2 v}{\partial x_i \partial x_j}\|_p \le C\|\Delta v\|_p$. Let us write $v(x) = u(x)(1+|x|^2)^{1+\frac{\delta}{2}}$. If we prove that $\|\Delta v\|_p$ is finite (in other words, that Δv belongs to L^p), the proof is thereby concluded. We have

$$\begin{aligned}\Delta v &= (\Delta u)(1+|x|^2)^{1+\delta/2} + 4(1+\delta/2)(\nabla u \cdot x)(1+|x|^2)^{\delta/2} \\ &+ 2n\, u(x)(1+\delta/2)(1+|x|^2)^{\delta/2} + 4(1+\delta/2)(\delta/2)u(x)(1+|x|^2)^{\frac{\delta}{2}-1}|x|^2 .\end{aligned}$$

Recall that $u = I(f)$, and that $f(x)(1+|x|^2)^{1+\delta/2} \in L^p(\mathbb{R}^n)$. Therefore, $\Delta u(x)(1+|x|^2)^{1+\delta/2} = f(x)(1+|x|^2)^{1+\delta/2} \in L^p(\mathbb{R}^n)$. All the other terms in the previous expression of Δv belong to L^p. For example, $(\nabla u \cdot x)(1+|x|^2)^{\delta/2}$ belongs to L^p because its norm in L^p is comparable to that of a linear combination of $\frac{\partial u}{\partial x_i}(1+|x|^2)^{\frac{1+\delta}{2}})$, whose terms we know belong to L^p. In short, $\|\Delta v\|_p$ is finite, which is what we wished to prove. □

As stated above, the proposition just proven implies the following important corollary:

Corollary VII.4. *The Laplacian Δ is a topological isomorphism between the spaces $\mathcal{F}^{p,s}_\delta$ and $\mathcal{F}^{p,s-2}_{\delta+2}$ if*

$$\frac{-n}{p} < \delta < \frac{n}{q} - 2 .$$

Proof. Although the proof of this corollary is immediate and is already included in previous paragraphs (specifically, those following the introduction to the Riesz potential), we recall it here in the interests of clarity. Since Δ is a second-order differential operator, the mapping that it induces between the Sobolev spaces $\mathcal{F}^{p,s}_\delta$ and $\mathcal{F}^{p,s-2}_{\delta+2}$ is continuous. Furthermore, Proposition VII.2 is equivalent to the existence of a positive constant K such that $\|I(f)\|_{p,s,\delta} \leq K\|f\|_{p,s-2,\delta+2}$ for every $f \in \mathcal{F}^{p,s-2}_{\delta+2}$. By taking $f = \Delta u$, since I is the inverse of Δ, we have $\|u\|_{p,s,\delta} \leq K\|\Delta u\|_{p,s-2,\delta+2}$. Thus, $\Delta u = 0$ implies that $u = 0$. Consequently, Δ is injective, continuous, with its inverse I being continuous. □

Remark VII.5. M. Cantor proves in [17] our Corollary VII.4 with the supplementary condition $p > \frac{n}{n-2}$, which is equivalent to stating that 0 belongs to the interval $(-n/p\,, n/q-2)$. Indeed, $-n/p < 0 < n/q-2$ is equivalent to $0 < n/p < n-2$, which is written $p > \frac{n}{n-2}$. It is easy to check that for $\delta = 0$ the operator

$$(Tg)(x) = (I_2 * f)(x) = C\int_{\mathbb{R}^n} g(y)(1+|y|^2)^{-1-\delta/2}|x-y|^{2-n}(1+|x|^2)^{\delta/2}dy$$

is not bounded in $L^p(\mathbb{R}^n)$ if $p \leq \frac{n}{n-2}$. Indeed, the space of the C^∞ functions with compact support is dense in $\mathcal{F}^{p,s-2}_{\delta+2}(\mathbb{R}^n)$. If g has compact support, $(Tg)(x) \sim |x|^{2-n}$ for large $|x|$ and $\delta = 0$, and $(Tg)^p$ is not integrable at infinity (see Lemma VII.1) if $p(2-n) \geq -n$; that is, $p \leq \frac{n}{n-2}$.

VII.4 Stability of the Minkowski metric: Y. Choquet-Bruhat and S. Deser's result

In Section VII.2, we considered the space $\mathbb{R}^3$ endowed with the Euclidean metric e, and were interested in finding suitable Banach spaces for which the operators $\tau \to \Delta\tau$ and $Y \to \nabla^* L_Y e$ gave isomorphisms (where τ was a function and Y a vector field). In the foregoing section, we introduced the Sobolev spaces $\mathcal{F}^{p,s}_\delta$, and saw

(Corollary VII.4) how the Laplacian Δ gave an isomorphism between these spaces, $\Delta : \mathcal{F}^{p,s}_{\delta}(\mathbb{R}^3) \to \mathcal{F}^{p,s-2}_{\delta+2}(\mathbb{R}^3)$, when δ fulfilled the condition $-\frac{3}{p} < \delta < \frac{3}{q} - 2$. Now let $\mathcal{X}^{p,s}_{\delta}(\mathbb{R}^3)$ be the space of vector fields on $\mathbb{R}^3$ whose components are functions of $\mathcal{F}^{p,s}_{\delta}(\mathbb{R}^3)$ (that is, $\mathcal{X}^{p,s}_{\delta}(\mathbb{R}^3) = \left(\mathcal{F}^{p,s}_{\delta}(\mathbb{R}^3)\right)^3$). The following proposition (which is a consequence of Corollary VII.4) solves the problem of the operator $Y \to \nabla^* L_Y e$.

Proposition VII.6. *The operator* $Y \to \nabla^* L_Y e$ *gives a topological isomorphism* $\mathcal{X}^{p,s}_{\delta}(\mathbb{R}^3) \to \mathcal{X}^{p,s-2}_{\delta+2}(\mathbb{R}^3)$ *if* $-\frac{3}{p} < \delta < \frac{3}{q} - 2$.

Proof. We first recall (Proposition I.7) that for any Riemannian metric g we have $(L_Y g)_{ij} = \nabla_i Y_j + \nabla_j Y_i$.

Then,

$$(\nabla^* L_Y g)_j = -\nabla^i (L_Y g)_{ij} = -\nabla^i \nabla_i Y_j - \nabla^i \nabla_j Y_i \,.$$

Here we are interested in the case where g is the Euclidean metric e of $\mathbb{R}^3$. The first term is the Laplacian, and we have that

$$(\nabla^* L_Y e)_i = \Delta Y_i - \sum_j \partial_j \partial_i Y_j \,.$$

Now we check that if $\nabla^* L_Y e = Z$, the inverse image of Z is explicitly given by $Y_j = \Delta^{-1}(Z_j + \sum_k \frac{1}{2}\partial_k \partial_j \Delta^{-1} Z_k)$, and then the isomorphism of Δ implies the result. Indeed, by replacing Y_j in the term $\sum_j \partial_j \partial_i Y_j$ by the previous expression, we obtain (bearing in mind that the derivatives commute and that $\sum_j \partial_j \partial_j = -\Delta$),

$$\begin{aligned}
\sum_j \partial_j \partial_i Y_j &= \sum_j \partial_j \partial_i \Delta^{-1}(Z_j + \sum_k \frac{1}{2}\partial_k \partial_j \Delta^{-1} Z_k) \\
&= \sum_j \partial_j \partial_i \Delta^{-1} Z_j + \frac{1}{2}\sum_{j,k} \partial_j \partial_j \Delta^{-1} \partial_i \partial_k \Delta^{-1} Z_k \\
&= \sum_j \partial_j \partial_i \Delta^{-1} Z_j - \frac{1}{2}\sum_k \partial_i \partial_k \Delta^{-1} Z_k = \frac{1}{2}\sum_j \partial_i \partial_j \Delta^{-1} Z_j \,.
\end{aligned}$$

Finally, replacing Y_i by $Y_i = \Delta^{-1}(Z_i + \sum_j \frac{1}{2}\partial_j \partial_i \Delta^{-1} Z_j)$ in the previous expression of $(\nabla^* L_Y e)_i$, we obtain

$$(\nabla^* L_Y e)_i = \Delta Y_i - \sum_j \partial_j \partial_i Y_j = Z_i + \frac{1}{2}\sum_j \partial_j \partial_i \Delta^{-1} Z_j - \frac{1}{2}\sum_j \partial_j \partial_i \Delta^{-1} Z_j = Z_i.$$

□

In accordance with what was stated in Section VII.2, we are now ready to state further that, when using Sobolev spaces with weights introduced in the foregoing section instead of the usual Sobolev spaces introduced in Chapter I, the mapping $D\Phi$ is surjective, and its kernel has a topological supplement, which entails stability by linearization of Einstein's equation with Minkowski's initial metric. In order to state this result precisely, we repeat the concepts introduced in Chapter V, Section V.3, conveniently changing the Sobolev spaces.

We denote by $\mathcal{S}^{p,s}_{\delta}(\mathbb{R}^3)$ the space of symmetric, covariant, second-order tensors on $\mathbb{R}^3$, whose coefficients are functions of $\mathcal{F}^{p,s}_{\delta}$. We further denote by $\mathcal{S}^{p,s}_{\delta}(e)$ the space of symmetric, covariant, second-order tensors g' such that the difference $g' - e \in \mathcal{S}^{p,s}_{\delta}(\mathbb{R}^3)$. We topologize the set $\mathcal{S}^{p,s}_{\delta}(e)$ by imposing the condition that the mapping $g' \to h = g' - e$ from $\mathcal{S}^{p,s}_{\delta}(e)$ to $\mathcal{S}^{p,s}_{\delta}(\mathbb{R}^3)$ is a homeomorphism. We also denote by $\Omega^{p,s}_{\delta}(\mathbb{R}^3)$ the space of those 1-forms on $\mathbb{R}^3$ whose coefficients are functions of $\mathcal{F}^{p,s}_{\delta}(\mathbb{R}^3)$. Then consider the map

$$
\begin{array}{cccc}
 & \mathcal{S}^{p,s}_{\delta}(e) \times \mathcal{S}^{p,s-1}_{\delta+1}(\mathbb{R}^3) & & \\
 & \cup & & \\
\Phi : & \mathcal{U} & \longrightarrow & \mathcal{F}^{p,s-2}_{\delta+2}(\mathbb{R}^3) \times \Omega^{p,s-2}_{\delta+2}(\mathbb{R}^3) \\
 & (g', k') & \to & (\mathcal{H}(g',k'), \gamma(g',k'))\ ,
\end{array}
$$

where $\mathcal{U}$ is a convenient neighborhood from $(e, 0)$ in the space $\mathcal{S}^{p,s}_{\delta}(e) \times \mathcal{S}^{p,s-1}_{\delta+1}(\mathbb{R}^3)$. With this notation, we are able to state the following result:

Theorem VII.7. *If $-\frac{3}{p} < \delta < \frac{3}{q} - 2$ (where p and q are conjugate exponents; that is, $\frac{1}{p} + \frac{1}{q} = 1$), then the previous mapping Φ is stable by linearization at the point $(e, 0)$ in the sense of Definitions* V.3 *and* V.7.

This theorem, due originally to Y. Choquet-Bruhat and S. Deser, [22], and which we have adapted to the new definition of stability by linearization introduced in Definition V.3, means that Minkowski's metric of $\mathbb{R}^4$ is stable by linearization of Einstein's equation, *always providing that the deformations taken of the initial metric belong to Sobolev spaces with weights, introduced in Section* VII.3, *and with* $-\frac{3}{p} < \delta < \frac{3}{q} - 2$.

This justifies the calculations made by Einstein himself and which appear in detail in Chapter III. As regards the intuitive meaning of this theorem and the role played by the Sobolev spaces employed, the reader is referred once again to the closing paragraphs in Section V.3.

VII.5 The Euclidean asymptotic case

The aim of this section is to generalize the foregoing results to the case of any initial metric in $\mathbb{R}^4$ (instead of Minkowski's metric), but which tends to Minkowski's metric when $x \in \mathbb{R}^4$ tends to infinity in any space-like direction. Recall that in Section VII.2 of the present chapter the Lorentzian manifold V was assumed to be $\mathbb{R}^4$; that the submanifold M of V was $M = \{0\} \times \mathbb{R}^3$; that the stress-energy tensor was $T = 0$ (the vacuum), and that the initial pair (g, k) was the pair $(e, 0)$, where e was the Euclidean metric of $\mathbb{R}^3$. With these hypotheses we arrive at Theorem VII.7. Now we situate ourselves in the case of an initial metric $\tilde{g}$ in $\mathbb{R}^4$, which is no longer Minkowski's metric, and of an initial stress-energy tensor T of a perfect fluid,

$$T = (\rho + p)u^* \otimes u^* + p\tilde{g} \ ,$$

where u^* is the 1-form associated (by the metric $\tilde{g}$) to a time-like vector field u (of velocities) such that $\tilde{g}(u, u) = -1$. This case encompasses the previous case of the vacuum when ρ and p are identically null. Let M be a hypersurface of $\mathbb{R}^4$, chosen in such a way that M is perpendicular to the field of velocities u of the perfect fluid. Assume that M with the induced topology is homeomorphic to $\mathbb{R}^3$. In the particular case where $T = 0$ (without matter) we take M as in the previous section, $M = \{0\} \times \mathbb{R}^3$.

As we saw in Section V.3, the right-hand sides of the two constraint equations (V.2) are

$$F = -T(N, N) = -(\rho + p)(1 + g(v, v)) + p \ ,$$
$$X = i(N)T = -(\rho + p)\sqrt{1 + g(v, v)}v^* \ ,$$

where v denotes the tangential component of the field u of velocities of the fluid. In other words, $u = v + zN$, where v is a vector tangent to M at each point. Since we have taken the hypersurface M perpendicular to the field u at each point, the field of velocities u coincides at the points of M with the normal field N. Therefore, $v = 0$, whereby $X = 0$ and $F = -\rho$. With these conditions we are able to state the following theorem, which generalizes a result by Y. Choquet-Bruhat, A. Fischer and J. E. Marsden, [23].

Theorem VII.8. *Assume a real number $p > 3$ and a natural number $s \geq 3/p + 2$ to be fixed. Let us also assume the following conditions to be fulfilled:*

1) *g (restriction of $\tilde{g}$ to $M \simeq \mathbb{R}^3$) is asymptotically Euclidean, in the following sense:*

   a)
   $$\lim_{|x| \to \infty} |(g(x) - e(x)| = 0 \ ,$$
   where e denotes the Euclidean metric. This in fact means that the components g_{ij} of g fulfill
   $$\lim_{|x| \to \infty} |(g_{ij}(x) - \delta_{ij}| = 0 \ .$$

   b)
   $$\lim_{|x| \to \infty} |x|^{|\beta|} |\partial_\beta g| = 0 \, , \, 1 < |\beta| \leq s + 1 \ .$$

2) *The second fundamental form* $2k$ *of* M *is such that*

$$\lim_{|x|\to\infty} |x|^{|\beta|+2}|\partial_\beta(k\cdot k)| = 0\,,\ |\beta|\le s\,.$$

3) *The density* ρ *of the perfect fluid fulfills on* M *the condition*

$$\lim_{|x|\to\infty} |x|^{|\beta|+2}|\partial_\beta \rho| = 0\,,\ |\beta|\le s\,.$$

4) $\rho \le \frac{1}{4\chi} k\cdot k$ *at every point of* M.

5) $\operatorname{tr} k$ *is constant on* M.

Then the map

$$\begin{array}{cccc} & \mathcal{S}^{p,s}_{\delta}(g)\times\mathcal{S}^{p,s-1}_{\delta+1}(k) & & \\ & \cup & & \\ \Phi: & \mathcal{U} & \longrightarrow & \mathcal{F}^{p,s-2}_{\delta+2}(F)\times\Omega^{p,s-2}_{\delta+2}(X) \\ & (g',k') & \to & (\mathcal{H}(g',k'),\gamma(g',k'))\,, \end{array}$$

is stable by linearization at the initial point (g,k) *if* $-\frac{3}{p}<\delta<\frac{3}{q}-2$, *where* q *denotes the conjugate exponent of p; that is,* $\frac{1}{p}+\frac{1}{q}=1$.

Remark VII.9. This theorem intuitively means that a Lorentzian metric in $\mathbb{R}^4$ that fulfills Einstein's equation with matter for a stress-energy tensor of a perfect fluid with small ρ, and which tends asymptotically to Minkowski's metric in any space-like direction, is stable by linearization of Einstein's equation, always *providing that the deformations taken from the initial metric belong to the Sobolev spaces with weights introduced in Section* VII.3, *and with* $-\frac{3}{p}<\delta<\frac{3}{q}-2$. Y. Choquet-Bruhat, A. Fischer and J. E. Marsden, [23], obtained a similar result in the case of the vacuum. Since in the vacuum we have $\rho=0$, condition 4) is therefore automatically fulfilled. Furthermore, condition $-\frac{3}{p}<\delta<\frac{3}{q}-2$ from the previous theorem is more general than that obtained by Y. Choquet-Bruhat, A. Fischer and J. E. Marsden. Thus, this previous theorem generalizes the result established by these authors (both for the possible presence of matter and for the weight values δ allowed in the Sobolev spaces).

In order to prove Theorem VII.8, the following two results are required, which we prove subsequently below:

Theorem VII.10. *Assume* $n\ge 3$, *and also a natural number* $s>2$ *and a real number* $p>n/(n-2)$ *to be fixed. Let* g *be a Riemannian metric in* $\mathbb{R}^n$, *asymptotically Euclidean in the sense of property* 1) *of the previous theorem. Let* $a(x)$ *be a* C^∞ *function on* $\mathbb{R}^n$ *that fulfills* $a(x)\ge 0$ *for any* x, *and which satisfies*

$$\lim_{|x|\to\infty} |x|^{|\beta|+2}|\partial_\beta a(x)| = 0\,,\ |\beta|\le s\,.$$

Then the operator $u \to \Delta_g u + au$ *yields an isomorphism from* $\mathcal{F}^{p,s}_{\delta}(\mathbb{R}^n)$ *onto* $\mathcal{F}^{p,s-2}_{\delta+2}(\mathbb{R}^n)$ *if* $-\frac{n}{p} < \delta < \frac{n}{q} - 2$.

Theorem VII.11. *Let us assume* $n \geq 3$, *and also a real* $p > n/(n-2)$ *and natural number* $s > n/p + 2$ *as fixed. Let* g *be a Riemannian metric in* $\mathbb{R}^n$ *that fulfills the conditions of Theorem* VII.10. *Then the operator that associates the* 1*-form* $\nabla^* L_Y g$ *to every vector field* Y *on* $\mathbb{R}^n$ *yields an isomorphism from* $\mathcal{X}^{p,s}_{\delta}(\mathbb{R}^n)$ *onto* $\Omega^{p,s-2}_{\delta+2}(\mathbb{R}^n)$ *if* $-\frac{n}{p} < \delta < \frac{n}{q} - 2$.

Proof of Theorem VII.8. Since we have assumed that tr k is constant on M, $X = 0$ and $F = -\rho$, then the formulas (VII.2) become

$$(D_{(g,k)}\mathcal{H})(Y,\tau) = \frac{1}{8}\left[-2\tau k \cdot k - 4\, k \cdot L_Y g + 8\chi\rho\tau - 8\Delta\tau\right],$$
$$(D_{(g,k)}\gamma)(Y,\tau) = \nabla^* L_Y g\,.$$

The proof consists in checking (using Theorems VII.10 and VII.11) that the mapping $D_{(g,k)}\Phi$ is surjective and that its kernel has a topological supplement. Given that $(f,Z) \in \mathcal{F}^{p,s-2}_{\delta+2}(\mathbb{R}^n) \times \Omega^{p,s-2}_{\delta+2}(\mathbb{R}^n)$, in accordance with Theorem VII.11 there exists a unique $Y \in \mathcal{X}^{p,s}_{\delta}(\mathbb{R}^n)$ such that $\nabla^* L_Y g = Z$. We now apply Theorem VII.10, to which effect we denote by a the function $a = \frac{1}{4}k \cdot k - \chi\rho$. Note that since we have assumed that $\rho \leq \frac{1}{4\chi} k \cdot k$, function a will be > 0 at every point, as required by the hypothesis of Theorem VII.10. Thus by virtue of this theorem we know that there exists a unique function $\tau \in \mathcal{F}^{p,s}_{\delta}(\mathbb{R}^n)$ such that

$$-\frac{1}{2}k \cdot L_Y g = (\Delta_g + \frac{1}{4}k \cdot k - \chi\rho)\tau\ .$$

Therefore, we have $(D_{(g,k)}\mathcal{H})(Y,\tau) = f$, $(D_{(g,k)}\gamma)(Y,\tau) = Z$, and the surjectivity of $D_{(g,k)}\Phi$ is proven. The existence of a topological supplement to the kernel can then be seen by means of a standard argument already employed above. □

For the proofs of Theorems VII.10 and VII.11, we require some prior development.

VII.5.1 $\mathcal{W}^{p,s}_{\delta}(\mathbb{R}^n)$ Sobolev spaces and their duals

Denote by $\mathcal{W}^{p,s}_{\delta}(\mathbb{R}^n)$ the Sobolev space of functions on $\mathbb{R}^n$ which take values in $\mathbb{R}^k$, associated to the L^p space, with weight δ and degree of derivation s (note that k is not made explicit in the notation $\mathcal{W}^{p,s}_{\delta}(\mathbb{R}^n)$ of the space). We preserve the notation $\mathcal{F}^{p,s}_{\delta}$ employed thus far for the case $k = 1$ (that is, for the case of ordinary functions which take values in $\mathbb{R}$).

Recall that this type of space was defined in Section VII.3 only for $s \in \mathbb{Z}^+ \cup \{0\}$, $p \in \mathbb{R}^+$ and $\delta \in \mathbb{R}$. Now we denote by $\mathcal{W}^{q,-s}_{-\delta}$ the topological dual space of $\mathcal{W}^{p,s}_{\delta}$, where q is the conjugate exponent of p ($\frac{1}{p} + \frac{1}{q} = 1$). Note that if $s > 0$, then

$-s < 0$, and the meaning of $\mathcal{W}^{q,-s}_{-\delta}$ is not given by the definition in Section VII.3. Henceforth (as regards the requirements for this chapter), its meaning will be only $(\mathcal{W}^{p,s}_{\delta})'$, the dual of $\mathcal{W}^{p,s}_{\delta}$. To justify this notation, we note that when $s = 0$ it is well known that the dual of $\mathcal{W}^{p,0}_{\delta}$ is just $\mathcal{W}^{q,0}_{-\delta}$ (in this case, the meaning of the two spaces $\mathcal{W}^{p,0}_{\delta}$ and $\mathcal{W}^{q,0}_{-\delta}$ is given by the definition in Section VII.3). Let us pause for a moment in the proof of this fact. Let i be the mapping $i : \mathcal{W}^{q,0}_{-\delta} \to (\mathcal{W}^{p,0}_{\delta})'$ given by saying that the action of $i(f)$ on any $g \in \mathcal{W}^{p,0}_{\delta}$ is the number

$$\langle i(f), g \rangle = \int_{\mathbb{R}^n} f \cdot g \, dx \ .$$

This integral is finite as a consequence of Hölder's inequality. Indeed,

$$\begin{aligned} \left| \int_{\mathbb{R}^n} fg\,dx \right| &= \left| \int_{\mathbb{R}^n} (f(x)(1+|x|^2)^{\frac{-\delta}{2}}) \cdot (g(x)(1+|x|^2)^{\frac{\delta}{2}})\,dx \right| \\ &\le \left(\int_{\mathbb{R}^n} |f(x)|^q (1+|x|^2)^{\frac{-q\delta}{2}}\,dx \right)^{\frac{1}{q}} \left(\int_{\mathbb{R}^n} |g(x)|^p (1+|x|^2)^{\frac{p\delta}{2}}\,dx \right)^{\frac{1}{p}} < +\infty \end{aligned}$$

Of course $i(f) = 0$ implies $f = 0$. Therefore, $\mathcal{W}^{q,0}_{-\delta} \subset (\mathcal{W}^{p,0}_{\delta})'$. Then the equality $\mathcal{W}^{q,0}_{-\delta} = (\mathcal{W}^{p,0}_{\delta})'$ is proven in a standard way by using the Radon-Nikodym theorem (in the same way that the proof of the fact that the dual of L^p is L^q).

Now let us see an inclusion relation for these spaces. We prove that if $\delta_1 \le \delta_2$, then we have $\mathcal{W}^{p,s}_{\delta_2} \subset \mathcal{W}^{p,s}_{\delta_1}$ for any s (positive, negative or zero). Indeed, we begin by assuming that $s \ge 0$. Thus, the meaning of the spaces $\mathcal{W}^{p,s}_{\delta_1}$ and $\mathcal{W}^{p,s}_{\delta_2}$ is given by the definition introduced in Section VII.3. If $\delta_1 \le \delta_2$, then for any multi-index $k = (k_1, \ldots, k_n)$ we have that

$$(1+|x|^2)^{\frac{p(|k|+\delta_1)}{2}} \le (1+|x|^2)^{\frac{p(|k|+\delta_2)}{2}} ,$$

which implies the inclusion $\mathcal{W}^{p,s}_{\delta_2} \subset \mathcal{W}^{p,s}_{\delta_1}$. We now see that on passing to the duals, the direction of this inclusion changes. Let i be the canonical inclusion of $\mathcal{W}^{p,s}_{\delta_2}$ in $\mathcal{W}^{p,s}_{\delta_1}$. Let us prove that the map

$$\begin{array}{ccc} (\mathcal{W}^{p,s}_{\delta_1})' & \longrightarrow & (\mathcal{W}^{p,s}_{\delta_2})' \\ f & \to & f \circ i \end{array}$$

is an injection of $(\mathcal{W}^{p,s}_{\delta_1})'$ in $(\mathcal{W}^{p,s}_{\delta_2})'$ (which is not true in finite dimensional spaces). Let $f \in (\mathcal{W}^{p,s}_{\delta_1})'$ such that $f \circ i = 0$. We wish to check that $f = 0$. Let $\mathcal{D}$ be the space of C^∞ functions with compact support (which take values in $\mathbb{R}^k$). Since $\mathcal{D} \subset \mathcal{W}^{p,s}_{\delta_2} \subset \mathcal{W}^{p,s}_{\delta_1}$, if f vanishes on $\mathcal{W}^{p,s}_{\delta_2}$, then f vanishes on $\mathcal{D}$, and by density f vanishes on $\mathcal{W}^{p,s}_{\delta_1}$.

We now prove the inclusion $\mathcal{W}^{p,s}_{\delta_2} \subset \mathcal{W}^{p,s}_{\delta_1}$ when $s < 0$ (if $\delta_1 \le \delta_2$). We write $s = -s'$ with $s' > 0$. Since $\delta_1 \le \delta_2$, we have $-\delta_2 \le -\delta_1$, thus, $\mathcal{W}^{q,s'}_{-\delta_1} \subset \mathcal{W}^{q,s'}_{-\delta_2}$.

On passing to the duals, the direction of this inclusion changes. We therefore have that

$$\mathcal{W}^{p,-s'}_{\delta_2} = (\mathcal{W}^{q,s'}_{-\delta_2})' \subset (\mathcal{W}^{q,s'}_{-\delta_1})' = \mathcal{W}^{p,-s'}_{\delta_1} ,$$

as we wished to prove.

VII.5.2 Some results on elliptic operators and Fredholm operators in $\mathbb{R}^n$

Henceforth, it is necessary to work with second-order, linear differential operators in $\mathbb{R}^n$ that act on functions which take values in $\mathbb{R}^k$ and give functions which take values in $\mathbb{R}^k$. More precisely, if $u(x) = (u_1(x), \ldots, u_k(x))$ is one such function (which takes values in $\mathbb{R}^k$) and $A(u)(x) = (A(u)(x)_1, \ldots, A(u)(x)_k)$, we assume that the components $A(u)_i$ of A are obtained from u in the following way:

$$A(u)_i = \sum_{i=1}^{k}\sum_{j=1}^{k}\sum_{|\alpha|\le 2} (a_\alpha)_{ij}\partial_\alpha u_j ,$$

where $\alpha = (\alpha_1, \ldots, \alpha_n)$ is a multi-index and the functions $(a_\alpha)_{ij}(x)$ are assumed to be of class C^∞. The foregoing expression can be written in vector form, thus:

$$A(u) = \sum_{|\alpha|\le 2} a_\alpha \partial_\alpha u , \tag{VII.5}$$

where a_α are now $k \times k$ function matrices which act on vector function u. Recall that if we denote the set of real number $k \times k$ matrices by $\mathcal{M}(\mathbb{R})_k$, in Chapter IV the leading symbol of such an operator was defined as the map

$$\begin{array}{rccc} \sigma_L A: & \mathbb{R}^n \times \mathbb{R}^n & \longrightarrow & \mathcal{M}(\mathbb{R})_k \\ & (x,\xi) & \to & \sum_{|\alpha|=2} a_\alpha(x)\xi^\alpha \end{array} .$$

Let $\det_A(x,\xi)$ be the determinant of the matrix $(\sigma_L A)(x,\xi)$. An operator A is elliptic (by definition) if $\det_A(x,\xi) \neq 0$ for any $x \in \mathbb{R}^n$ and for any $\xi \neq 0$; or, equivalently, if $\det_A(x,\xi) \neq 0$ for any $x \in \mathbb{R}^n$ and for any $\xi \in S^{n-1}$. Operator A is said to be *uniformly elliptic* if there exists a positive constant C, independent of x and ξ, such that $|\det_A(x,\xi)| > C$ for any $\xi \in S^{n-1}$.

We now address Fredholm operators. A linear and continuous operator $T : X \to Y$ between two Banach spaces is said to be a *Fredholm operator* if the image $T(X)$ is closed in Y, the kernel $N(T)$ has finite dimension and the codimension of $T(X)$ within Y is finite. For such an operator, its index is defined as the number $\operatorname{ind}(T) = \dim N(T) - \operatorname{codim} T(X)$.

Henceforth, let us assume that the Banach spaces X and Y are reflexive, in the sense that the dual of the dual coincides with the original space.

Let $T : X \to Y$ be a Fredholm operator and $T^* : Y' \to X'$ its adjoint. Let m be the codimension of $T(X)$. Let $y_1, \dots, y_m \in Y$ be linearly independent such that

$$Y = \langle y_1, \dots, y_m \rangle \oplus T(X) ,$$

where $\langle y_1, \dots, y_m \rangle$ denotes the vector space generated by $y_1, \dots, y_m$. For every i such that $1 \le i \le m$, let $y_i^* \in Y'$ such that $y_i^*(y_j) = \delta_{ij}$ for $1 \le j \le m$ and such that y_i^* vanishes on $T(X)$. Then the $y_i^* \in N(T^*)$, because $(T^*(y_i^*))(x) = (y_i^*)(T(x)) = 0\ \forall x \in X$. It is immediately deduced that $N(T^*) = \langle y_1^*, \dots y_m^* \rangle$. Indeed, if $y^* \in N(T^*)$, we have $y^* = \sum_i \lambda_i y_i^*$ with $\lambda_i = y^*(y_i)$. It thereby follows that the codimension m of $T(X)$ coincides with the dimension of the kernel of T^*. Henceforth, we denote the dimension of the kernel of T by $\nu(T)$. We therefore have that $\mathrm{ind}(T) = \nu(T) - \nu(T^*)$.

A differential operator of the type (VII.5) gives rise to a continuous operator

$$A : \mathcal{W}_\delta^{p,s}(\mathbb{R}^n) \longrightarrow \mathcal{W}_{\delta+2}^{p,s-2}(\mathbb{R}^n) .$$

Wherever we are interested in making it clear that the initial space of this operator A is $\mathcal{W}_\delta^{p,s}(\mathbb{R}^n)$, we write $A_\delta^{p,s}$ instead of A.

Proposition VII.12. *Let A be a second-order differential operator of the type* (VII.5) *(it is not required to be elliptic). Assume that for two values of δ, $\delta = \delta_1$ and $\delta = \delta_2$, the operators $A_{\delta_1}^{p,s}$ and $A_{\delta_2}^{p,s}$ are Fredholm operators. Let $\nu_{\delta_1}^{p,s}$ and $\nu_{\delta_2}^{p,s}$ be the respective dimensions of the kernels of $A_{\delta_1}^{p,s}$ and $A_{\delta_2}^{p,s}$. If $\delta_1 \le \delta_2$, we have that $\nu_{\delta_2}^{p,s} \le \nu_{\delta_1}^{p,s}$ and $\mathrm{ind}(A_{\delta_2}^{p,s}) \le \mathrm{ind}(A_{\delta_1}^{p,s})$. If the equality $\mathrm{ind}(A_{\delta_2}^{p,s}) = \mathrm{ind}(A_{\delta_1}^{p,s})$ is fulfilled, then the equality $\nu_{\delta_2}^{p,s} = \nu_{\delta_1}^{p,s}$ is also fulfilled.*

Remark VII.13. Note that, in accordance with this proposition, for those δ such that $A_\delta^{p,s}$ is a Fredholm operator, both $\nu_\delta^{p,s}$ and $\mathrm{ind}(A_\delta^{p,s})$ are decreasing functions of δ.

Proof. In Subsection VII.5.1, we saw that if $\delta_1 \le \delta_2$, then $\mathcal{W}_{\delta_2}^{p,s} \subset \mathcal{W}_{\delta_1}^{p,s}$. Therefore, we have that the Kernel of $A_{\delta_2}^{p,s} \subset$ Kernel of $A_{\delta_1}^{p,s}$. Thus, $\nu_{\delta_2}^{p,s} \le \nu_{\delta_1}^{p,s}$. If we consider the operator A acting on $\mathcal{W}_\delta^{p,s}$ (for any δ), we have

$$A : \mathcal{W}_\delta^{p,s} \longrightarrow \mathcal{W}_{\delta+2}^{p,s-2} ,$$

and its adjoint A^* (by definition of the adjoint) will be an operator

$$A^* : (\mathcal{W}_{\delta+2}^{p,s-2})' \longrightarrow (\mathcal{W}_\delta^{p,s})' .$$

Bearing in mind what was previously stated about the duals of these Sobolev spaces, the adjoint of $A_\delta^{p,s}$ will act thus:

$$(A_\delta^{p,s})^* : \mathcal{W}_{-\delta-2}^{q,-s+2} \longrightarrow \mathcal{W}_{-\delta}^{q,-s} ,$$

where q is the conjugate exponent of p. If $\delta_1 \le \delta_2$, then $-\delta_2 - 2 \le -\delta_1 - 2$, and by applying the same reasoning as before to the adjoints, we have $\nu^{*q,-s+2}_{-\delta_1-2} \le$

${\nu^*}^{q,-s+2}_{-\delta_2-2}$, where the dimension of the kernel of the adjoint A^* is denoted by ν^*. Therefore, $-{\nu^*}^{q,-s+2}_{-\delta_2-2} \leq -{\nu^*}^{q,-s+2}_{-\delta_1-2}$. Thus, $\operatorname{ind}(A^{p,s}_{\delta_2}) = \nu^{p,s}_{\delta_2} - {\nu^*}^{q,-s+2}_{-\delta_2-2} \leq \nu^{p,s}_{\delta_1} - {\nu^*}^{q,-s+2}_{-\delta_1-2} = \operatorname{ind}(A^{p,s}_{\delta_1})$.

Let us suppose that $\operatorname{ind}(A^{p,s}_{\delta_2}) = \operatorname{ind}(A^{p,s}_{\delta_1})$. We wish to prove that $\nu^{p,s}_{\delta_2} = \nu^{p,s}_{\delta_1}$. Let us assume that this is not fulfilled, and we arrive at a contradiction. Indeed, if this equality is not fulfilled then we have $\operatorname{ind}(A^{p,s}_{\delta_2}) = \nu^{p,s}_{\delta_2} - {\nu^*}^{q,-s+2}_{-\delta_2-2} < \nu^{p,s}_{\delta_1} - {\nu^*}^{q,-s+2}_{-\delta_1-2} = \operatorname{ind}(A^{p,s}_{\delta_1})$. □

In the 1970s and 1980s, L. A. Bagirov, V. A. Kondrat'ev, R. B. Lockhart and R. C. McOwen each studied independently when an elliptic differential operator in $\mathbb{R}^n$ was a Fredholm operator. The results obtained by these authors (applied to our situation) may be summarized in the following theorem:

Theorem VII.14. *Let $s \in \mathbb{Z}^+$. Let A_∞ be a homogeneous, second-order, linear differential operator of the type* (VII.5), *where the coefficients are constant $k \times k$ matrices. Assume that A_∞ is elliptic. Let Q be a second-order, linear differential operator of the type* (VII.5), *$Q = \sum_{|\alpha|\leq 2} q_\alpha(x)\partial_\alpha$, where the coefficients are $k \times k$ function matrices. Consider the operator $A = A_\infty + Q$, and assume that A is elliptic. Let us also assume that the functions $q_\alpha(x)$ are C^∞ and fulfill*

$$\lim_{|x|\to\infty} |(1+|x|^2)^{\frac{2-|\alpha|+|\beta|}{2}} \partial^\beta q_\alpha(x)| = 0 \tag{VII.6}$$

for $|\beta| \leq s$. Then,

1) *$A^{p,s}_\delta : \mathcal{W}^{p,s}_\delta \to \mathcal{W}^{p,s-2}_{\delta+2}$ is a Fredholm operator if and only if one of the following two conditions is fulfilled:*
   a) *$\delta > -n/p$ and $\delta + 2 - n/q \notin \mathbb{N}$,*
   b) *$\delta \leq -n/p$ and $-\delta - n/p \notin \mathbb{N}$.*
2) *$(A_\infty)^{p,s}_\delta$ is an isomorphism for any δ such that $-n/p < \delta < n/q - 2$.*
3) *We denote by $\mathcal{D}_A$ the set of those $\delta \in \mathbb{R}$ such that $A^{p,s}_\delta$ is not a Fredholm operator. If I is an interval of $\mathbb{R}$ containing no $\delta \in \mathcal{D}_A$, then the indices of all $A^{p,s}_\delta$ for $\delta \in I$ are the same (the index of $A^{p,s}_\delta$ does not depend on $\delta \in I$).*

Remark VII.15. Statement 1) of the theorem was obtained in the particular case of $k = 1$ by L. A. Bagirov and V. A. Kondrat'ev, [6], with hypotheses slightly different from those employed here, although their work was unknown to specialists in the West until much later. Also in the case of $k = 1$, R. B. Lockhart, [45], obtained results from which 1), 2) and 3) are immediately deduced. Indeed, statement 3) of the theorem in the case of $k = 1$ is a consequence of Corollary 5.7. and of Theorem 4.3. in [45], since the index of A is equal to the index of A_∞. The case $k > 1$ was studied by R. C. McOwen, [51], and subsequently by R. B. Lockhart and R. C. McOwen together, [46], [47], [48]. Statements 1) and 2) of the theorem are a consequence of Theorem 1 in [46]. For $k > 1$, it is no longer true that the index of A is equal to the index of A_∞ for those values of δ for which both are Fredholm operators, [47]. Nevertheless, statement 3) of Theorem VII.14 is true as a consequence of Corollary 9.2 in [48]. Many of the works cited require that the operator A be *uniformly elliptic*, which is not explicitly required by Theorem VII.14, but the fact that A has constant coefficients at ∞ (a condition that is indeed required by Theorem VII.14) means that the ellipticity of A implies uniform ellipticity. Of course, all the cited works study more general situations, and here we adapt their results according to our own requirements.

VII.5.3 Proof of Theorems VII.10 and VII.11

For the proof of Theorem VII.10, the following lemma is required, which is an adaptation to our case of an argument by S. T. Yau (employed in the proof of Proposition 1 in [58]).

Lemma VII.16. *Let g be a C^∞ Riemannian metric in $\mathbb{R}^n$ fulfilling the condition $\lim_{|x|\to\infty} |(g(x) - e(x))| = 0$, where e denotes the Euclidean metric. Let $a(x)$ be a C^∞ function on $\mathbb{R}^n$, bounded, such that $a(x) \geq 0$ at every point. Let A be the operator on functions given by $A(u) = \Delta_g u + au$. Let u be a function of $L^p(\mathbb{R}^n)$. If $p > 1$ and $A(u) = 0$, then u vanishes identically.*

Proof. Let us first point out that the fact that the metric g tends to the Euclidean metric at infinity implies that the spaces $L^p(dx)$ and $L^p(d\mu)$ are the same, where dx denotes the usual measure of $\mathbb{R}^n$ and $d\mu$ the measure associated to the volume element of the metric g. This follows from the fact that the volume element of g is $\sqrt{\det(g_{ij}(x))}dx_1 \wedge \ldots \wedge dx_n$, and that $\det(g_{ij}(x))$ tends to 1 when x tends to infinity. Let φ be a C^∞ function on $\mathbb{R}$, with $0 \leq \varphi(t) \leq 1$ for every $t \in \mathbb{R}$, with $\varphi(t) = 1$ for every $t \leq 1$ and with $\varphi(t) = 0$ for $t \geq 2$. Given R_1 and R_2 positive numbers, with $R_1 < R_2$, let ω be the function on $\mathbb{R}^n$ defined by

$$\omega(x) = \varphi\left(\frac{\sqrt{\sum_j x_j^2} + R_2 - 2R_1}{R_2 - R_1}\right) .$$

For every $R > 0$, we denote by $B(R)$ the closed Euclidean ball of radius R centered at the origin. In other words, $B(R) = \{x \in \mathbb{R}^n \text{ such that } \sum_j x_j^2 \leq R^2\}$. The function ω, which depends on R_1 and R_2, has as support the ball $B(R_2)$ and equals 1 on the ball $B(R_1)$. At each point $x \in \mathbb{R}^n$, the norm $|d\omega|_g$ of the exterior derivative $d\omega$ with respect to the metric g will be

$$|d\omega|_g = \sqrt{\sum_{ij} g^{ij}(x)\partial_i\omega(x)\partial_j\omega(x)} .$$

The condition that g tends to the Euclidean metric at infinity implies that the functions $|g_{ij}(x)|$ and $|g^{ij}(x)|$ are bounded by a positive constant, whereby, and from the definition of $\omega(x)$, it is easily deduced that

$$|d\omega|_g < \frac{C}{R_2 - R_1}\,, \tag{VII.7}$$

for a positive constant C independent of R_2 and R_1.

Since A is elliptic, the condition $A(u) = 0$ implies that u is a C^∞ function. Let us fix two positive numbers R_1 and R_2 with $R_1 < R_2$. Having fixed these numbers, let us consider the previous function ω (which depends on these two numbers); denoting, as usual, by $\langle\ ,\ \rangle$ the global inner product of differential forms with respect to the metric g, we have

$$\begin{aligned} 0 &= \langle \omega^2 (u^2)^{\frac{p}{2}-1} u \,,\, A(u)\rangle = \langle \omega^2 (u^2)^{\frac{p}{2}-1} u \,,\, \delta du + au\rangle \\ &= 2\langle \omega (u^2)^{\frac{p}{2}-1} u d\omega \,,\, du\rangle + \langle \omega^2 d(u^2)^{\frac{p}{2}-1} u \,,\, du\rangle \\ &\quad + \langle \omega^2 (u^2)^{\frac{p}{2}-1} du \,,\, du\rangle + \langle \omega^2 (u^2)^{\frac{p}{2}-1} u \,,\, au\rangle\ . \end{aligned}$$

Note that, since ω has compact support, it is possible to apply the divergence theorem without difficulty in order to pass δ from one factor to another, transforming it into d. Since $a(x) \geq 0$ at every point, the last term of the previous equality is ≥ 0. Therefore, the previous equality implies

$$\langle \omega^2 (u^2)^{\frac{p}{2}-1} du \,,\, du\rangle + \langle \omega^2 d(u^2)^{\frac{p}{2}-1} u \,,\, du\rangle \leq -2\langle \omega (u^2)^{\frac{p}{2}-1} u d\omega \,,\, du\rangle\ .$$

Now, $d(u^2)^{\frac{p}{2}-1} = (p-2)(u^2)^{\frac{p}{2}-2} u du$; therefore, the previous inequality becomes

$$\begin{aligned} &(p-1)\langle \omega^2 (u^2)^{\frac{p}{2}-1} du \,,\, du\rangle \leq -2\langle \omega (u^2)^{\frac{p}{2}-1} u d\omega \,,\, du\rangle \\ &\leq 2|\langle \omega (u^2)^{\frac{p}{2}-1} u d\omega \,,\, du\rangle| = 2|\langle (u^2)^{\frac{p}{4}-1} \omega u du \,,\, (u^2)^{\frac{p}{4}} d\omega\rangle|\ . \end{aligned}$$

Consequently, in accordance with Schwarz's inequality, we have

$$(p-1)\langle \omega^2 (u^2)^{\frac{p}{2}-1} du \,,\, du\rangle \leq 2\left(\langle \omega^2 (u^2)^{\frac{p}{2}-1} du \,,\, du\rangle\right)^{\frac{1}{2}} \left(\langle (u^2)^{\frac{p}{2}} d\omega, d\omega\rangle\right)^{\frac{1}{2}}\,,$$

which can be written as follows:

$$(p-1)\langle \omega^2 |u|^{p-2} du \,,\, du\rangle \leq 2\left(\langle \omega^2 |u|^{p-2} du \,,\, du\rangle\right)^{\frac{1}{2}} \left(\int_{\mathbb{R}^n} |u|^p |d\omega|^2 d\mu\right)^{\frac{1}{2}}\ . \tag{VII.8}$$

Let

$$P = \langle \omega^2 |u|^{p-2} du \,,\, du\rangle = \int_{B(R_2)} \omega^2 |u|^{p-2} |du|^2 d\mu\ .$$

We now wish to check that the expression $|u|^{p-2}|du|^2$ vanishes identically on the ball $B(R_1)$. Notice that the definition of P implies that P is ≥ 0. Suppose that $P = 0$. Then the integrand $\omega^2|u|^{p-2}|du|^2$ vanishes identically on $B(R_2)$, and hence $|u|^{p-2}|du|^2$ would vanish identically on $B(R_1)$ because ω equals 1 in this ball.

Suppose now that $P > 0$ (the other possibility). Then squaring the left-hand and right-hand sides of (VII.8) and dividing by P, we obtain

$$(p-1)^2 P \leq 4 \int_{\mathbb{R}^n} |u|^p |d\omega|^2 d\mu \; . \tag{VII.9}$$

Bearing in mind that

$$\int_{B(R_1)} |u|^{p-2} |du|^2 d\mu \leq \int_{B(R_2)} |u|^{p-2} \omega^2 |du|^2 d\mu = P \; ,$$

and taking into account (VII.7), from the inequality (VII.9) we obtain

$$(p-1)^2 \int_{B(R_1)} |u|^{p-2} |du|^2 d\mu \leq 4 \int_{\mathbb{R}^n} |u|^p |d\omega|^2 d\mu \leq \; 4 \left(\frac{C}{R_2 - R_1} \right)^2 \int_{\mathbb{R}^n} |u|^p d\mu .$$

Now, by making R_2 tend to ∞ (taking into account that $u \in L^p(d\mu)$ and that $p > 1$), we obtain

$$\int_{B(R_1)} |u|^{p-2} |du|^2 d\mu = 0 \; .$$

Hence, we deduce also that $|u|^{p-2}|du|^2$ must vanish identically on $B(R_1)$.

The fact that $|u|^{p-2}|du|^2$ vanishes identically on $B(R_1)$ implies that $du = 0$ at every point of $B(R_1)$, since if $du \neq 0$ at a point x_0, then $|du|^2 \neq 0$ in a neighborhood of x_0. Therefore, u should vanish in this neighborhood (because $|u|^{p-2}|du|^2 \equiv 0$), and we would have $(du)_{x_0} = 0$, which is contrary to the hypothesis $(du)_{x_0} \neq 0$. In short, u must be constant on $B(R_1)$. However, as R_1 is any positive number, u is constant on $\mathbb{R}^n$. Since $u \in L^p(\mathbb{R}^n)$, this constant must be zero. □

For the proof of Theorems VII.10 and VII.11, the following result on Banach spaces is also required, a proof of which can be found in [18], page 8.

Lemma VII.17. *Let A_0 and A_1 be two linear and continuous operators between two Banach spaces X and Y. Assume that A_0 is an isomorphism. Assume also that there exists a continuous curve c mapping the interval $[0,1]$ into the space of bounded operators from X to Y such that $c(0) = A_0$ and $c(1) = A_1$, and that for every $t \in [0,1]$, $c(t)$ is an injection from X into Y with closed image. Then A_1 is an isomorphism.*

We are now able to address the proofs of Theorems VII.10 i VII.11.

Proof of Theorem VII.10. Bearing in mind expression (I.3), we have

$$\begin{aligned} \Delta_g f &= -\nabla^* \nabla f = -g^{ij} \{ \partial_i \partial_j f - \Gamma^k_{ij} \partial_k f \} \\ &= \Delta_e f + (\delta^{ij} - g^{ij}) \partial_i \partial_j f + g^{ij} \Gamma^k_{ij} \partial_k f \; , \end{aligned}$$

where the Euclidean Laplacian is denoted by Δ_e. From this expression and the condition required of a, it is deduced that the operator A defined on functions by $A(u) = (\Delta_g + a)u$ fulfills the conditions of Theorem VII.14, with $A_\infty = \Delta_e$ (the Euclidean Laplacian).

For every t such that $0 \leq t \leq 1$, let g_t be the Riemannian metric defined by $g_t = e + t(g - e)$, and a_t be the function ta. Let A_t be the operator on functions defined by $A_t(u) = (\Delta_{g_t} + a_t)u$. Note that for $t = 0$, A_0 is the Euclidean Laplacian, and that for $t = 1$, A_1 is the operator A. All the operators A_t also fulfill the hypotheses of Theorem VII.14, with $A_\infty = \Delta_e$. Therefore, the A_t and A, if we consider them to be acting on the space $\mathcal{F}_\delta^{p,s}$, will be Fredholm operators when $\delta \in I$, where I denotes the interval of those δ hat fulfill $-\frac{n}{p} < \delta < \frac{n}{q} - 2$.

We now prove the injectivity of the A_t as operators from $\mathcal{F}_\delta^{p,s}$ to $\mathcal{F}_{\delta+2}^{p,s-2}$, when $\delta \in I$. By statement 3) of Theorem VII.14, the index of $(A_t)_\delta^{p,s}$ does not depend on δ when $\delta \in I$. Then, by Proposition VII.12, the dimension of the kernel of $(A_t)_\delta^{p,s}$ does not depend either on δ when $\delta \in I$. Since we have assumed that $p > n/(n-2)$, we have that $-\frac{n}{p} < 0 < \frac{n}{q} - 2$ (indeed, by adding $\frac{n}{p}$ to this last inequality, we obtain the equivalent inequality $0 < \frac{n}{p} < n - 2$, which is equivalent to $p > n/(n-2)$). Therefore, in order to prove the injectivity of $(A_t)_\delta^{p,s}$ it suffices to prove the injectivity of $(A_t)_0^{p,s}$. However, for $\delta = 0$, the space where $(A_t)_0^{p,s}$ acts is $\mathcal{F}_0^{p,s} \subset L^p$. Therefore, Lemma VII.16 can be applied to those u of the kernel of $(A_t)_0^{p,s}$ to conclude that these u must be zero. This proves the injectivity of A_t.

Furthermore, the Euclidean Laplacian Δ_e (which is A_∞) is an isomorphism for every $\delta \in I$ by virtue of Corollary VII.4. We have a continuous curve of operators $t \mapsto A_t$ that fulfill the conditions of Lemma VII.17. This lemma ensures that $A_\delta^{p,s}$ is an isomorphism for $\delta \in I$, which is what we sought. □

Proof of Theorem VII.11. The differential operator A that it is now necessary to consider is $A(Y) = \nabla^* L_Y g$ (which acts on vector fields Y). If Y has components $Y^1 \dots Y^n$, recall that $L_Y g$ is the covariant 2-tensor given (in component-wise form) by $(L_Y g)_{ij} = \nabla_i Y_j + \nabla_j Y_i$ (Corollary I.8). Recall also that $\nabla^* L_Y g$ is the 1-form given (in component-wise form) by

$$(\nabla^* L_Y g)_j = -\nabla^i (L_Y g)_{ij} = -\nabla^i (\nabla_i Y_j + \nabla_j Y_i) \, .$$

As was done in the proof of Theorem VII.10, for every t such that $0 \leq t \leq 1$, consider the Riemannian metric g_t defined by $g_t = e + t(g - e)$, where e is the Euclidean metric, and also consider the operator A_t defined on vector fields Y by $A_t(Y) = (\nabla^*)_{g_t} L_Y g_t$, where $(\nabla^*)_{g_t}$ denotes the operator ∇^* associated to the metric g_t. Note that for $t = 0$, we have the operator $Y \to (\nabla^*)_e L_Y g_e$ associated to the Euclidean metric, and for $t = 1$ the operator A.

It is easy to prove the ellipticity of the operators A_t. These operators fulfill the hypotheses of Theorem VII.14 with $A_\infty(Y) = (\nabla^*)_e L_Y e$. Therefore, if they are considered to be acting on the space $\mathcal{W}^{p,s}_{\delta}$, these A_t will be Fedholm operators when $\delta \in I$, where I denotes the interval of those δ that fulfill $-\frac{n}{p} < \delta < \frac{n}{q} - 2$. Moreover, A_∞ is an isomorphism when $\delta \in I$. To repeat the proof for Theorem VII.10, it simply suffices to prove that the operators A_t give rise to injective maps from $\mathcal{W}^{p,s}_{\delta}$ to $\mathcal{W}^{p,s-2}_{\delta+2}$ when $\delta = 0$. This is already provided in [18]. We now set out the fundamental features of this proof in [18].

For every $R > 1$, let ω_R be a C^∞ function on $\mathbb{R}^n$, with $0 \le \omega_R(x) \le 1$ for every $x \in \mathbb{R}^n$, with support contained in the ball $B(2R)$, which has value 1 on the ball $B(R)$, and with $|D^\alpha \omega_R(x)| \le C$ for every $|\alpha| \le 2$ and $x \in \mathbb{R}^n$, where C is a positive constant that does not depend on $R > 1$. Such a function ω_R can be constructed by taking the function ω from the proof of Lemma VII.16, with $R_1 = R$ and $R_2 = 2R$. Let us say that the hypothesis $R > 1$ (or, in general, R greater than a certain positive constant) is essential for the constant C of the inequality $|D^\alpha \omega_R(x)| \le C$ for every $|\alpha| \le 2$ not to depend on R. What follows would not hold if we only assumed $R > 0$. For every $R > 1$, let $(g_t)_R$ be the Riemannian metric $(g_t)_R = e + \omega_R(g_t - e)$. Note that outside the ball $B(2R)$, the metric $(g_t)_R$ coincides with the Euclidean metric, and that when R tends to ∞ (since ω_R tends to 1), $(g_t)_R$ tends to g_t. Let $(A_t)_R$ be the operator defined on vector fields Y by $(A_t)_R(Y) = \nabla^*_{(g_t)_R} L_Y (g_t)_R$. For these operators, M. Cantor proves an inequality of the type

$$\|Y\|_{p,s,\delta} \le C \|(A_t)_R(Y)\|_{p,s-2,\delta+2} \tag{VII.10}$$

for $\delta \ge 0$. We require it only for $\delta = 0$ (M. Cantor derives this inequality from his Lemma 5.5.1, which works for more general operators). An explicit proof could be given of the inequality for the operator $(A_t)_R$ by transferring the calculations performed on pages 22 and 23 of [18] to our situation.

Inequality (VII.10) proves the injectivity of the operators $(A_t)_R$. The proof for the injectivity of A_t (which is also provided in [18]) works in the following way: since $(A_t)_R = A_t + ((A_t)_R - A_t)$, from (VII.10) we obtain

$$\|Y\|_{p,s,\delta} \le C \|A_t(Y)\|_{p,s-2,\delta+2} + C \|(A_t)_R - A_t\| \, \|Y\|_{p,s,\delta} \, .$$

It can be proven that the operators $(A_t)_R$ tend to A_t (with the norm of the space of operators) when R tends to ∞. This means that for R sufficiently large, we have $C\|(A_t)_R - A_t\| < 1/2$, and the foregoing inequality becomes

$$(1/2)\|Y\|_{p,s,\delta} \le C \|A_t(Y)\|_{p,s-2,\delta+2} \, ,$$

which proves the injectivity of A_t. □

Chapter VIII

Stability by Linearization of Einstein's Equation in Robertson-Walker Cosmological Models

This chapter will deal with a study of stability by linearization of Einstein's equation when the initial metric and the initial stress-energy tensor are those from a Robertson-Walker cosmological model. If we interpret the universe as a Robertson-Walker model with metric $\tilde{g}$ and stress-energy tensor T, the explosion of a supernova in a distant galaxy will produce a perturbation of the initial stress-energy tensor T which will give rise to a perturbation of the initial metric $\tilde{g}$. The perturbed universe will no longer be of Robertson-Walker type. In order to study the effects of this explosion, is it correct or not to work with a linearized Einstein equation? An important role in answering this question is played by the new concept of stability by linealization introduced in Section V.2, since we do not deal here with the case of linearization of Einstein's equation in the vacuum: we have to linearize it in presence of the matter. The authors of this monograph have just studied this case in [14], [15] and [16].

Cosmological Robertson-Walker models have already been introduced in Section II.5. Reading of this chapter only requires us to recall that in a Robertson-Walker model it is assumed that:

- The space-time V can be taken as a product $I \times S$ of an open interval I of $\mathbb{R}$ (which can coincide with the whole of $\mathbb{R}$) and a Riemannian manifold $(S, \hat{g})$ of dimension 3 with constant curvature $\kappa = 0$, $\kappa = 1$, or $\kappa = -1$.
- The metric $\tilde{g}$ of the space-time has the form $\tilde{g} = -dt^2 + \zeta(t)^2\hat{g}$. If we put

$g_t = \zeta(t)^2\hat{g}$, we can interpret g_t as the Riemannian metric of each manifold $M_t = \{t\} \times S$.

- The distribution of matter is given by a stress-energy tensor of perfect fluid type, $T = (\rho + p)dt \otimes dt + p\tilde{g}$. By homogeneity and isotropy, the pressure p and the density ρ only depend on the coordinate t.

Normally, an extra assumption is made that the manifold $(S, \hat{g})$ is complete and simply connected. Then $(S, \hat{g})$ has to be necessarily the Euclidean space $\mathbb{R}^3$, if $\kappa = 0$, the unit sphere $\mathbb{S}^3$, if $\kappa = 1$, or the hyperbolic space $\mathbb{H}^3$, if $\kappa = -1$. Taking into account the techniques employed in two previous chapters, it will be of utmost importance to recognize that the sphere is compact while the Euclidean space and the hyperbolic space are not.

It is clear that other cases are possible if we do not require S to be simply connected, as is seen from the following result: *any Riemannian manifold S of dimension n which is complete and has constant curvature κ, is of the type $\tilde{S}/\Gamma$, where $\tilde{S} = \mathbb{R}^n$ if $\kappa = 0$, $\tilde{S}$ is the sphere $\mathbb{S}^n$ if $\kappa = 1$, $\tilde{S}$ is the hyperbolic space $\mathbb{H}^n$ if $\kappa = -1$, and Γ is a discrete subgroup of the group of isometries of $\tilde{S}$ acting freely and properly discontinuous.* The first two cases, $\mathbb{R}^n/\Gamma$ and $\mathbb{H}^n/\Gamma$, can be compact or not, orientable or not, depending on the subgroup Γ. Here, linearization stability of the cases $\mathbb{R}^3/\Gamma$ and $\mathbb{H}^3/\Gamma$ will be studied when they are compact and orientable. The results we will obtain are as follows:

- If $S = \mathbb{R}^3$, Einstein's equation is **stable** by linearization (using weighted Sobolev spaces).
- If $S = \mathbb{H}^3$, Einstein's equation is **stable** by linearization (using non-weighted Sobolev spaces).
- If $S = \mathbb{S}^3$, Einstein's equation is **non-stable** by linearization.
- If S is a compact flat torus, $S = \mathbb{R}^3/\Gamma$, Einstein's equation is **non-stable** by linearization.
- If S is a **compact** hyperbolic manifold, $S = \mathbb{H}^3/\Gamma$, Einstein's equation is **stable** by linearizarion.

These results make clear that stability of Einstein's equation in Robertson-Walker models does not depend on whether or not the model is compact, neither on the sign of the curvature, since stability or non-stability can occur in compact models, as well as in the case of models having the same curvature.

We have said before that for any $t \in I$ we denote by M_t the hypersurface $M_t = \{t\} \times S$ endowed with the Riemannian metric $g_t = \zeta(t)^2\hat{g}$. From now on, we denote as usual by k_t twice the second fundamental form of the hypersurface M_t in V. Since $\tilde{g} = -dt^2 + \zeta(t)^2\hat{g}$, we have

$$k_t = \frac{\partial g_t}{\partial t} = 2\zeta(t)\dot{\zeta}(t)\hat{g} = 2\frac{\dot{\zeta}(t)}{\zeta(t)}g_t \, .$$

Take now some $t_0 \in I$ (initial time). With this t_0 fixed, denote by M the hypersurface $M_{t_0} = \{t_0\} \times S$, by g the metric g_{t_0}, and let $k = k_{t_0}$. As usual we also denote by F and X, respectively, the function and the vector field on M given by $F = T_0^0|_M$ and $X^i = T_0^i|_M$.

For brevity, we put $C = \dot{\zeta}(t_0)/\zeta(t_0)$. For further calculations, it is useful to mention the following identities which can easily be deduced from the above definitions and the expression for k_t:

$$\begin{aligned} &k \times_g k = 4\,C^2 g, && k \cdot_g k = 12\,C^2, && \mathrm{tr}_g k = 6\,C, \\ &F = T_0^0|_M = -\rho(t_0), && X^i = T_0^i = 0, && k \cdot L_Y g = 4\,C\,\mathrm{div} Y\,. \end{aligned} \tag{VIII.1}$$

In what concerns the constraint equations, $\gamma(g,k) = \chi X$ gives the identity $0 = 0$ while $\mathcal{H}(g,k) = \chi F$ relates, at the instant t_0 (and therefore at any instant), the functions ζ and ρ:

$$\left(\frac{\dot{\zeta}}{\zeta}\right)^2 + \frac{\kappa}{\zeta^2} = \frac{\chi\rho}{3}\,.$$

This gives $C^2 + \kappa/\zeta(t_0)^2 = \chi\rho(t_0)/3$. By substitution of all these expressions into expressions (VII.2) of the linear tangent map, we obtain (taking into account that now $\mathrm{tr}(k)$ is constant),

$$\begin{aligned} D_{(g,k)}\mathcal{H}(Y,\tau) &= \frac{3\,\kappa}{\zeta^2(t_0)}\tau - 2C\;\mathrm{div}\, Y - \Delta\tau, \\ D_{(g,k)}\gamma(Y,\tau)_i &= (\nabla^* L_Y g)_i\;. \end{aligned} \tag{VIII.2}$$

VIII.1 Euclidean model

Equations (VIII.2) now become

$$\begin{aligned} D_{(g,k)}\mathcal{H}(Y,\tau) &= 2C\,\nabla^* Y - \Delta\tau \\ D_{(g,k)}\gamma(Y,\tau)_i &= (\nabla^* L_Y g)_i\;, \end{aligned}$$

and the Riemannian manifold is $\mathbb{R}^3$, endowed by the metric $\zeta(t_0)^2 e$ proportional to the Euclidean metric e. The results in Section VII.3 on the isomorphism of the Laplacian in $\mathbb{R}^n$ enable us to adapt the proof of Theorem VII.7 to the case we consider now using the same weighted Sobolev spaces that we have used there. Indeed, we can prove along similar lines that $D_{(g,k)}\Phi$ is surjective by using the isomorphism $Y \to \nabla^* L_Y e$ (established in that chapter). Indeed, given a pair (f, Z), there exists a unique vector field Y such that $D_{(g,k)}\gamma(Y,\tau) = \nabla^* L_Y e = Z$. Once this field Y is known, the isomorphism of the Laplacian gives the function τ such that $D_{(g,k)}\mathcal{H}(Y,\tau) = f$. The pair (Y,τ) determines through transform (VII.1) the pair (h, K) which enables us to prove that the tangent linear map is surjective and that a topological supplement of its kernel exists.

This argument proves the following theorem.

Theorem VIII.1. *In a Robertson-Walker model of curvature zero, $V = I \times \mathbb{R}^3$, let M be a hypersurface of the form $\{t_0\} \times \mathbb{R}^3$ and let g, k, F, X be the 2-tensors, the function and the vector field on M defined above corresponding to this model: $g = \zeta(t_0)^2 e$, where e is the Euclidean metric, $k = 2(\dot{\zeta}(t_0)/\zeta(t_0))g$, $F = -\rho(t_0)$ and $X = 0$. Assume that Φ is the map*

$$\begin{array}{cccc} & \mathcal{S}^{p,s}_{\delta}(g) \times \mathcal{S}^{p,s-1}_{\delta+1}(k) & & \\ & \cup & & \\ \Phi : & \mathcal{U} & \longrightarrow & \mathcal{F}^{p,s-2}_{\delta+2}(F) \times \Omega^{p,s-2}_{\delta+2}(X) \\ & (g', k') & \to & (\mathcal{H}(g', k'), \gamma(g', k')) \end{array}$$

where $\mathcal{U}$ is an appropriate neighborhood of (g, k) in the space $\mathcal{S}^{p,s}_{\delta}(g) \times \mathcal{S}^{p,s-1}_{\delta+1}(k)$.

If $-\frac{3}{p} < \delta < \frac{3}{q} - 2$ (where p and q are conjugate exponents, that is $\frac{1}{p} + \frac{1}{q} = 1$), then the map Φ is stable by linearization at point (g, k), in the sense of Definitions V.3 *and* V.7.

This theorem means that complete and simply connected Robertson-Walker models of curvature zero are stable by linearization of Einstein's equation *whenever the perturbations taken from the initial model belong to weighted Sobolev spaces introduced in Section VII.3, with* $-\frac{3}{p} < \delta < \frac{3}{q} - 2$. We refer the reader interested in an intuitive meaning of this theorem, and the role played by the Sobolev spaces we have used, to read again the last paragraphs in Section V.3.

VIII.2 Hyperbolic model

In this case, we will also prove stability by linearization of Einstein's equation. However, in the same way we have used weighted Sobolev spaces in the previous section, we will need here the Sobolev spaces introduced in Section I.4. The map Φ we employ will be exactly the same as that introduced in Section V.3 (between the same Sobolev spaces as those appearing in that section).

To be concrete, we will prove the following theorem.

Theorem VIII.2. *In a Robertson-Walker model $V = I \times \mathbb{H}^3$ of curvature $\kappa = -1$, let M be a hypersurface of the form $\{t_0\} \times \mathbb{R}^3$ and let g, k, F, X be 2-tensors, the function and the vector field on M defined above and corresponding to this model. The map Φ in Section I.4,*

$$\begin{array}{cccc} & \mathcal{S}^{s}(g) \times \mathcal{S}^{s-1}(k) & & \\ & \cup & & \\ \Phi : & \mathcal{U} & \longrightarrow & \mathcal{F}^{s-2}(F) \times \Omega^{s-2}(X) \\ & (g', k') & \to & (\mathcal{H}(g', k'), \gamma(g', k')) \end{array}$$

is stable by linearization at point (g, k) in the sense of Definitions V.3 *and* V.7.

The proof of this theorem consists in checking that $D_{(g,k)}\Phi$ is surjective. In this case, equations (VIII.2) become

$$D_{(g,k)}\mathcal{H}(Y,\tau) = \frac{-3}{\zeta^2(t_0)}\tau + 2C\,\nabla^*Y - \Delta\tau,$$
$$D_{(g,k)}\gamma(Y,\tau)_i = (\nabla^* L_Y g)_i\,.$$

Checking the fact that $D_{(g,k)}\Phi$ is surjective is almost immediate using the following theorem in [16]:

Theorem VIII.3. *For any real number $q \geq 0$, the two maps*

$$\begin{aligned} \mathcal{F}^s(\mathbb{H}^3) &\longrightarrow \mathcal{F}^{s-2}(\mathbb{H}^3) \\ \tau &\longrightarrow \Delta\tau + q\tau \end{aligned} \qquad\qquad \begin{aligned} \Omega^s(\mathbb{H}^3) &\longrightarrow \Omega^{s-2}(\mathbb{H}^3) \\ Y &\longrightarrow \nabla^* L_Y g \end{aligned}$$

are topological isomorphisms, where $Y \in \Omega^s(\mathbb{H}^3)$ is interpreted as a vector field identifying fields and 1-forms by means of the metric.

We will devote Section VIII.3 to giving a hint to the proof of this theorem, as well as to explaining the reasons for using ordinary Sobolev spaces in the statement of this theorem, in contrast to similar theorems in $\mathbb{R}^3$ where we have used weighted Sobolev spaces.

Proof of Theorem VIII.2. Let (f, Z) be an element belonging to the space $\mathcal{F}^{s-2}(\mathbb{H}^3) \times \Omega^{s-2}(\mathbb{H}^3)$. By Theorem VIII.3, we can find Y satisfying $\nabla^* L_Y g = Z$. Once this Y is found, the same theorem assures that there exists τ such that $\Delta\tau + \frac{3}{\zeta^2(t_0)}\tau = -f + 2C\,\nabla^*Y$. The pair (τ, Y) constructed in this way is mapped by $D_{(g,k)}\Phi$ into the given pair (f, Z). □

VIII.3 Sobolev spaces and hyperbolic Laplacian

In this section, we will explain the results in [13] and [16] which will be required for proving Theorem VIII.3. We will also try to show why this theorem works with ordinary Sobolev spaces in contrast to similar results in $\mathbb{R}^3$ requiring weighted Sobolev spaces.

We will work with the model of the unit ball, $\mathbb{B}^3 = \{x \in \mathbb{R}^3,\ \|x\| < 1\}$, endowed with the metric

$$g = \frac{4}{(1-r^2)^2}(dx^1 \otimes dx^1 + dx^2 \otimes dx^2 + dx^3 \otimes dx^3)$$

where $r^2 = (x^1)^2 + (x^2)^2 + (x^3)^2$.

At the moment, we will be interested in the Laplace operator Δ applied to functions. In the sequel, we will also need to consider this operator applied to m-forms. We begin by checking that Δ gives an isomorphism between the spaces of functions $\mathcal{F}^s(\mathbb{H}^3)$ and $\mathcal{F}^{s-2}(\mathbb{H}^3)$ (a particular case of the first part of Theorem VIII.3 with $q=0$). We will give a draft of the proof of this result different from that contained in [16] and [13], similar to our proof of Corollary VII.4 in $\mathbb{R}^3$. We think that the proof we are going to outline now shows better than any other proof the reason for using ordinary (non-weighted) Sobolev spaces in this context.

VIII.3.1 Δ gives an isomorphism between $\mathcal{F}^s(\mathbb{H}^3)$ and $\mathcal{F}^{s-2}(\mathbb{H}^3)$

In order to adapt the proof of Corollary VII.4 we have given for the case of $\mathbb{R}^3$ to the hyperbolic situation, we begin with finding the coordinate-wise expression of the Laplacian. After that we will apply the Green's identity to identify the inverse operator. This last step will require us to know something about hyperbolic translations, an equivalent in $\mathbb{H}^3$ of the Euclidean translations.

Hyperbolic Laplacian on functions

By the expression for the Laplacian found in Section I.3, the Laplacian of a function f is given by $\Delta f = \nabla^*\nabla f = -\nabla^i\nabla_i f = -g^{ji}\nabla_j\nabla_i f$. The 1-form ∇f has components $\nabla_i f = \partial_i f$. For each j, the components $\nabla_j\nabla_i f$ of the covariant derivative $\nabla_{\partial_j}\nabla f$ will be (according to expression I.3):

$$\nabla_j\nabla_i f = \partial_j\partial_i f - \Gamma^k_{ji}\partial_k f \ .$$

One can check that Christoffel symbols of the hyperbolic metric

$$g_{ij} = \frac{4}{(1-r^2)^2}\delta_{ij}$$

are given as

$$\Gamma^k_{ij} = \frac{2}{1-r^2}(x_i\delta^k_j + x_j\delta^k_i - x^k\delta_{ij}) \ ,$$

so that

$$\begin{aligned}
-\Delta f = g^{ij}\nabla_j\nabla_i f &= \frac{(1-r^2)^2}{4}\delta^{ij}\left[\partial_j\partial_i f - \frac{2}{1-r^2}(x_i\delta^k_j + x_j\delta^k_i - x^k\delta_{ij})\partial_k f\right] \\
&= \frac{(1-r^2)^2}{4}\left[\Delta_e f + \frac{2}{1-r^2}x^k\partial_k f\right] \\
&= \frac{1}{4}(1-r^2)^2\Delta_e f + \frac{1}{2}(1-r^2)(x\partial_x + y\partial_y + z\partial_z)f \ ,
\end{aligned}$$

where Δ_e denotes the common Euclidean Laplacian $\partial_x^2 + \partial_y^2 + \partial_z^2$.

A radial solution of $\Delta v = 0$ in $\mathbb{B}^3 - \{0\}$, having a pole at zero, is

$$v(r) = A(r + 1/r) + B \ ,$$

with A and B constants. This can readily be checked in view of

$$\Delta_e r = \frac{2}{r} \quad ; \quad \Delta_e \frac{1}{r} = 0 \quad ; \quad (x\partial_x + y\partial_y + z\partial_z)(r + \frac{1}{r}) = r - \frac{1}{r} \ .$$

Green's formula for functions

If U is an open set of $\mathbb{B}^3$ with $\overline{U} \subset \mathbb{B}^3$, we have

$$\int_U (u\Delta v - v\Delta u)d\mu = -\int_{\partial U} \left(u\frac{\partial v}{\partial n} - v\frac{\partial u}{\partial n} \right) \omega \ ,$$

where $d\mu$ denotes the hyperbolic volume element, ω the element of area of the surface ∂U with respect to the hyperbolic metric, n is the unit exterior normal to ∂U with respect to the hyperbolic metric, and $\partial/\partial n$ stands for the operator of directional derivation with respect to n.

We have

$$d\mu = \sqrt{\det(g_{ij})}\, dx = \frac{8}{(1-r^2)^3}\, dx \ ,$$

where $dx = dx_1 dx_2 dx_3$ is the Euclidean volume element. If we denote by n_e the unit exterior normal to ∂U with respect to the Euclidean metric, we obtain

$$n = \frac{1}{2}(1-r^2)n_e \ ,$$

because $g(n,n) = \frac{1}{4}(1-r^2)^2 g(n_e,n_e)$ and $g(n_e,n_e) = \frac{4}{(1-r^2)^2}$.

The hyperbolic element of area will be

$$\omega = i(n)d\mu = \frac{4}{(1-r^2)^2}\, i(n_e)dx = \frac{4}{(1-r^2)^2}\, d\sigma$$

where $i(n)d\mu$ denotes interior contraction of $d\mu$ by n, $i(n_e)dx$ the interior contraction of dx by n_e, and $d\sigma$ stands for the Euclidean element of area of ∂U. Substitute all this into the above Green's formula and take into account that

$$\frac{\partial}{\partial n} = \frac{1}{2}\left(1-r^2\right)\frac{\partial}{\partial n_e}$$

to obtain

$$8\int_U (u\Delta v - v\Delta u)(1-r^2)^{-3}dx = -2\int_{\partial U} \left(u\frac{\partial v}{\partial n_e} - v\frac{\partial u}{\partial n_e} \right)(1-r^2)^{-1}d\sigma \ .$$

Hyperbolic translations

For every $x \in \mathbb{B}^n$ there exists an isometry φ_x of the metric given by

$$\varphi_x(y) = \frac{(|x|^2-1)y + (|y|^2 - 2xy + 1)x}{|x|^2|y|^2 - 2xy + 1} \ ,$$

with $\varphi_x(0) = x$ and $\varphi_x(x) = 0$, such that $\varphi_x^{-1} = \varphi_x$; this isometry φ_x is unique up to a rigid motion fixing zero. Therefore it is $|\varphi_x(y)|$ which is completely determined. To be concrete, it is well known that

$$1 - |\varphi_x(y)|^2 = \frac{(1-|x|^2)(1-|y|^2)}{|x-y|^2 + (1-|x|^2)(1-|y|^2)} = 1 - |\varphi_y(x)|^2 . \tag{VIII.3}$$

Since φ_x is an isometry and Δ is a metric operator, we have

$$\Delta(u \circ \varphi_x) = (\Delta u) \circ \varphi_x .$$

We see that the Euclidean transforms $y \to x - y$ should be substituted by φ_x in the hyperbolic context.

Inverse operator of the Laplacian on functions

Let us apply Green's identity to $u \in C_c^2(\mathbb{B}^3)$ and to the radial solution $v(r)$ given before on the region $U = B_R - B_\epsilon$, with R close to 1 in order for the support of u to be included into it:

$$4\int_{B_R - B_\epsilon} (-v\Delta u)(1-r^2)^{-3} dx = -\int_{\partial B_\epsilon} \left(u\frac{\partial v}{\partial n_e} - v\frac{\partial u}{\partial n_e} \right) (1-r^2)^{-1} d\sigma .$$

In $r = \epsilon$, we have $\frac{\partial v}{\partial n_e} = \frac{\partial v}{\partial r} = A(1 - \frac{1}{r^2})$ and

$$\int_{r=\epsilon} u\frac{\partial v}{\partial n_e}(1-r^2)^{-1} d\sigma = A(1 - \frac{1}{\epsilon^2})(1-\epsilon^2)^{-1} \int_{r=\epsilon} u d\sigma = -\frac{A}{\epsilon^2}\int_{r=\epsilon} u d\sigma .$$

By choosing $A = 1/4\pi$ and making $\epsilon \to 0$, the above expression tends to $-u(0)$. The other term in the surface integral tends to zero under the same assumptions since v has order $1/\epsilon$ at $x = \epsilon$ and the measure of ∂B_ϵ is proportional to ϵ^2,

$$\int_{x=\epsilon} v\frac{\partial u}{\partial r}(1-r^2)^{-1} d\sigma = (1-\epsilon^2)^{-1} O(1/\epsilon) \frac{4\pi\epsilon^2}{\sigma(\partial B_\epsilon)} \int_{x=\epsilon} \frac{\partial u}{\partial r} d\sigma \underset{\varepsilon \to 0}{\longrightarrow} 0 .$$

By making $R \to 1$, we obtain

$$u(0) = -4\int_{\mathbb{B}^3} [A(1 + \frac{1}{r}) + B](1-r^2)^{-3} \Delta u \, dx .$$

The term B is irrelevant since taking $v = 1$ in the Green's formula gives $\int_{\mathbb{B}^3} (1-r^2)^{-3} \Delta u \, dx = 0$ if $u \in C_c^2(\mathbb{B}^3)$. By putting $B = -2A$ and in view of

$$1 + \frac{1}{r} - 2 = \frac{1 + r^2 - 2r}{r} = \frac{(1-r)^2}{r} ,$$

we obtain:

$$u(0) = \frac{1}{\pi}\int_{r<1} \frac{(1-r)^2}{r}(1-r^2)^{-3} \Delta u \, dx = \frac{1}{\pi}\int_{|x|<1} \frac{(1-|x|)^2}{|x|} \Delta u(x) \, d\mu(x) .$$

Apply this formula to $u \circ \varphi_x$,

$$(u \circ \varphi_x)(0) = \frac{1}{\pi} \int_{|y|<1} \frac{(1-|y|)^2}{|y|} \Delta(u \circ \varphi_x)(y)\, d\mu(y)\ ,$$

that is

$$u(x) = \frac{1}{\pi} \int_{|y|<1} \frac{(1-|y|)^2}{|y|} \Delta u(\varphi_x(y))\, d\mu(y)\ .$$

By making the change $\varphi_x(y) = z$ and bearing in mind that $\varphi_x^{-1} = \varphi_x$, that $d\mu$ is invariant with respect to isometries and $|\varphi_x(y)| < 1$ when $x, y \in \mathbb{B}^3$, we have

$$u(x) = \frac{1}{\pi} \int_{|z|<1} \frac{(1-|\varphi_x(z)|)^2}{|\varphi_x(z)|} \Delta u(z)\, d\mu(z)\ .$$

The function $G(x,z) = \frac{1}{\pi}\frac{(1-|\varphi_x(z)|)^2}{|\varphi_x(z)|}$ will be the Green's function for the hyperbolic Laplacian and will replace $G(x,z) = |x-z|^{-3}$ appearing in the Euclidean case. In operator formalism, the inverse of $u \to \Delta u$ is $f \to G[f]$, with

$$G[f](x) = \int_{|y|<1} G(x,y) f(y) d\mu(y)\ .$$

If we define $a(z) = \frac{1}{\pi}\frac{(1-|z|)^2}{|z|}$, we obtain $G(x,y) = a(\varphi_x(y))$. Then we can interpret $G[f]$ as a *hyperbolic convolution* of a with f,

$$C[f] = a *_h f = \int_{|y|<1} a(\varphi_x(y)) f(y) d\mu(y)\ .$$

Once the inverse operator to the Laplacian is found, we are able to ask whether Sobolev spaces exist such that Δ and its inverse are topological isomorphisms, which basically means that we would determine whether the operator G is bounded. Let us determine that there is no need to add any weight to the L^p-norm.

Lemma VIII.4. *G is bounded in $L^p(d\mu)$, $1 < p < +\infty$.*

We are not going to give a detailed proof of this lemma, rather a draft. Observe first that in the region $|z| < 1/2$ the function a belongs to $L^1(d\mu)$. Therefore Schur's lemma implies that in this region $G[f] = a *_h f$ is bounded in $L^p(d\mu)$. Indeed, this lemma claims that if

$$\int |K(x,y)| d\mu(x) \le C \quad , \quad \int |K(x,y)| d\mu(y) \le C\,,$$

the operator

$$T_K f(x) = \int K(x,y) f(y) d\mu(y)$$

is bounded in $L^p(d\mu)$, $1 < p < +\infty$, having norm less than or equal to C. In our case, $K(x,y)$ is the function $a(\varphi_x(y))$ in the region $|z| < 1/2$ that is symmetric since a is radial and, by (VIII.3), $|\varphi_x(y)| = |\varphi_y(x)|$. The assumption of the theorem holds since

$$\int_{|z|<1/2} a(z)\, d\mu(z) = \frac{1}{\pi}\int_{|z|<1/2} \frac{(1-|z|)^2}{|z|}\, d\mu(z)$$
$$= \frac{1}{\pi}\int_{|z|<1/2} \frac{(1-|z|)^2}{|z|} \frac{1}{(1-|z|^2)^3} |z|^2\, d|z|$$

is finite in a neighborhood of $z = 0$.

As for the region $1/2 < |z| < 1$, we observe from the very beginning that $G[f]$ is defined here for almost all $f \in L^p$, $1 < p < +\infty$ with respect to the measure $d\mu$. Indeed, by the Hölder inequality,

$$\left| \int fg d\mu \right| \le \|f\|_p \|g\|_q \,,$$

meaning that $G[f]$ will be finite for $f \in L^p$ if $a \in L^q$ with $q > 1$, where p and q are conjugate exponents. One has

$$\int_{|z|>1/2} a(z)^q\, d\mu(z) = \frac{1}{\pi}\int_{|z|>1/2} \frac{(1-|z|)^{2q}}{|z|^q}\, d\mu(z)$$
$$= \frac{1}{\pi}\int_{|z|>1/2} \frac{(1-|z|)^{2q}}{|z|^q} \frac{1}{(1-|z|^2)^3}\, dv(z) = \frac{1}{\pi}\int_{|z|>1/2} \frac{(1-|z|)^{2q-3}|z|^2}{|z|^q(1+|z|)^3}\, d|z| \,.$$

Around $z = 1$ (representing infinity) this latter integral is finite if $2q - 3 > -1$, that is $q > 1$. In order to prove that G is moreover bounded in this region, we would need to see, using Lemma VII.3, that there exists a function $h \ge 0$ such that

$$\int G(x,y)h(y)^{-q} d\mu(y) = O(h(x)^{-q}) \,;\, \int G(x,y)h(x)^{-p} d\mu(x) = O(h(y)^{-p}) \,.$$

We are not continuing with this argument since the only reason we have referred to it has been to make it evident that appropriate Banach spaces are usual Sobolev spaces, non-weighted. The most general result concerning the isomorphism Δ is as follows [12]: the Laplacian $\Delta : \mathcal{W}^{p,s+2}(\mathbb{H}^n)_m \to \mathcal{W}^{p,s}(\mathbb{H}^n)_m$ is a topological isomorphism for $p \in (p_1, p_2)$, where

$$p_1 = \frac{2(n-1)}{n-2+|n-2m|} \,,\quad \frac{1}{p_1} + \frac{1}{p_2} = 1 \,,$$

when $m \ne \frac{n\pm1}{2}, \frac{n}{2}$. In this theorem, $\mathcal{W}^{p,s}(\mathbb{H}^n)_m$ denotes the (non-weighted) Sobolev space of m-forms on $\mathbb{H}^n$ whose regularity order is s, based on the L^p

space. Observe that, for functions ($m = 0$) one obtains $p_1 = 1$ and $p_2 = +\infty$, that is, the hypothesis of Lemma VIII.4. We should also cite [1], [38] and [42] for analogous results on asymptotically hyperbolic manifolds involving weighted Sobolev spaces (while the results in [13], [16] and [12] concern usual Sobolev spaces).

VIII.3.2 A draft of the proof of Theorem VIII.3

Recall that we have introduced the Laplacians Δ and Δ_R in Section I.3. The second Laplacian (rough Laplacian) has been defined on any tensor α of order r by $\Delta_R\alpha = \nabla^*\nabla\alpha$. In component-wise form, we have

$$(\Delta_R\alpha)_{i_1\dots i_r} = -\nabla^j\nabla_j\alpha_{i_1\dots i_r} \,.$$

If $r = 0$ (that is, for functions) both Laplacians Δ and Δ_R coincide.

Let F be the operator $Y \to \nabla^* L_Y g$ appearing in the statement of Theorem VIII.3. Since $(L_Y g)_{ij} = \nabla_i Y_j + \nabla_j Y_i$, we have

$$\begin{aligned} F(Y)_j = (\nabla^* L_Y g)_j &= -\nabla^i\nabla_i Y_j - \nabla^i\nabla_j Y_i \\ &= (\Delta_R Y)_j - (\nabla^i\nabla_j - \nabla_j\nabla^i)Y_i - \nabla_j\nabla^i Y_i \,. \end{aligned}$$

In the case of manifolds having constant curvature -1, the well-known Ricci identity looks like $(\nabla^i\nabla_j - \nabla_j\nabla^i)Y_i = -(n-1)Y_j$, where n is the dimension of the manifold. In the case of $\mathbb{H}^3$, we have $(\nabla^i\nabla_j - \nabla_j\nabla^i)Y_i = -2Y_j$. On the other hand, the term $-\nabla_j\nabla^i Y_i$ is $(d\delta Y)_j$. Therefore

$$F = \Delta_R + 2I + d\delta \;,$$

where I denotes the identity operator.

In order to prove Theorem VIII.3, we will base our argument on two lemmas we are going to state now which, however, will not be proved here.

Lemma VIII.5 (Proposition 2.3 and Lemma 3.3 in [13]). *If in a complete Riemannian manifold M having constant curvature -1 an r-form α and its Laplacian $\Delta_R\alpha$ belongs to $\mathcal{W}^{s-2}(M)_r$, $0 < r < n$, then $\alpha \in \mathcal{W}^s(M)_r$ and for all r-forms satisfying this condition we have $\|\alpha\|_s \leq \text{const}\ \|\Delta_R\alpha\|_{s-2}$. The result also holds for $r = 0, n$ if $M = \mathbb{H}^n$.*

Lemma VIII.6 (Proposition 1 in [16]). *Let (M, g) be a complete Riemannian manifold having curvature -1 and let r satisfy $0 < r < n$. Assume that G is the operator $G = \Delta_R + d\delta$. If $\alpha, G(\alpha) \in \mathcal{W}^{s-2}(M)_r$, then $\alpha \in \mathcal{W}^s(M)_r$ and for all r-forms satisfying this condition we have $\|\alpha\|_s \leq \text{const}\ \|G(\alpha)\|_{s-2}$. The result also holds for $r = 0, n$ if $M = \mathbb{H}^n$.*

The Sobolev spaces involved in these two lemmas are constituted by elements of L^2. The advantage of working with L^2 is that these Sobolev spaces become Hilbert spaces endowed with the inner product $\langle\alpha, \beta\rangle_s = \sum_{k\leq s}\langle\nabla^k\alpha, \nabla^k\beta\rangle$.

The following proofs will refer to a well-known result claiming that (as in finite-dimensional spaces) a Hilbert space is canonically isomorphic to its dual space through the inner product: *Let H be a Hilbert space and $\langle\ ,\ \rangle$ the inner product in it. Then for any linear and continuous $\omega : H \to \mathbb{R}$ there exists a unique $y \in H$ such that $\omega(x) = \langle x, y\rangle$.* In all proofs we will make use of the fact that any linear differential operator T of order 2 defined from $\mathcal{W}^s(M)$ to $\mathcal{W}^{s-2}(M)$ is continuous $\forall s \geq 2$ because $\|Tx\|_{s-2} \leq C\|x\|_s$. Recall that $\mathcal{F}$ and Ω denote the Sobolev space of functions and that of 1-forms, respectively. Finally, as we have already mentioned in Section I.4, if $\alpha, \beta \in \mathcal{W}^s(M)_p$, the following formulas hold:

$$\langle \nabla^{(k)}\alpha, \beta\rangle = \langle \alpha, \nabla^{*(k)}\beta\rangle ,$$

and if T is a self-adjoint operator (as Δ_R or δd are), then $\langle T\alpha, \beta\rangle = \langle \alpha, T\beta\rangle$.

In order to prove the second part of Theorem VIII.3, we have to show that for any integer $s \geq 2$ the operator F is a topological isomorphism between $\Omega^s(M)$ and $\Omega^{s-2}(M)$.

Let us first see that F is injective. We have

$$\langle F(\alpha), \alpha\rangle = \langle(\Delta_R + 2I + d\delta)(\alpha), \alpha\rangle = \langle \nabla\alpha, \nabla\alpha\rangle + 2\langle\alpha,\alpha\rangle + \langle\delta\alpha,\delta\alpha\rangle , \quad \text{(VIII.4)}$$

and $F(\alpha) = 0$ implies $\langle\alpha,\alpha\rangle = 0$ meaning at the same time that $\alpha = 0$.

We will now use induction to prove that it is surjective and has continuous inverse, that is $\|\alpha\|_s \leq c\|F(\alpha)\|_{s-2}$. Begin with $s = 2$. Consider the following inner product in $\Omega^1(M)$:

$$B[\alpha, \beta] = \langle\nabla\alpha, \nabla\beta\rangle + 2\langle\alpha,\beta\rangle + \langle\delta\alpha,\delta\beta\rangle$$

where $\langle\ ,\ \rangle$ is the usual L^2-product, that is, the norm associated with $\|\ \|$ is exactly $\|\ \|_0$. The norm $\|\ \|_B$ induced by the inner product B is equivalent to the Sobolev norm $\|\ \|_1$ and therefore $\|\ \| < c\|\ \|_B$. Let us first show that $F : \Omega^2(M) \to \Omega^0(M)$ is surjective. Given an element $f \in \Omega^0(M) \subset L^2(M)$, we construct the element belonging to the dual space

$$\begin{aligned} L : \Omega^1(M) &\longrightarrow \mathbb{R} \\ \beta &\longrightarrow \langle f, \beta\rangle . \end{aligned}$$

Since $|L(\beta)| = |\langle f,\beta\rangle| \leq \|f\|\ \|\beta\| \leq c\|f\|\ \|\beta\|_B$, L is a linear functional which is continuous in the topology induced on $\Omega^1(M)$ by $\|\ \|_B$. Hence there exists a unique $\alpha \in \Omega^1(M)$ such that $B[\alpha,\beta] = L(\beta)$ for all $\beta \in \Omega^1(M)$, that is

$$\langle\nabla\alpha, \nabla\beta\rangle + 2\langle\alpha,\beta\rangle + \langle\delta\alpha,\delta\beta\rangle = \langle f, \beta\rangle .$$

In particular, for β of class C^∞ and having compact support we have

$$\langle \alpha, (\Delta_R + 2I + d\delta)\beta\rangle = \langle f, \beta\rangle$$

which can be written in view of the definition of F as

$$\langle \alpha, F(\beta) \rangle = \langle f, \beta \rangle \ .$$

This means that $F(\alpha) = f$ (F acts in the sense of weak derivatives) and therefore $\alpha \in \Omega^1(M)$ is the pre-image of $f \in \Omega^0(M)$. This α also belongs to $\Omega^2(M)$ since $\alpha, F(\alpha) \in \Omega^0(M)$ also implies $G(\alpha) = F(\alpha) - 2\alpha \in \Omega^0(M)$. By Lemma VIII.6, $\alpha \in \Omega^2(M)$ with $\|\alpha\|_2 \leq c\|G(\alpha)\|$. The latter inequality enables us to check that the inverse is continuous in view of

$$\|\alpha\|_2 \leq c\|G(\alpha)\| \leq c\|F(\alpha) - 2\alpha\| \leq c(\|F(\alpha)\| + \|\alpha\|) \ .$$

By (VIII.4), $\|\alpha\|^2 \leq c\langle F(\alpha), \alpha\rangle \leq c\|F(\alpha)\| \, \|\alpha\|$, that is $\|\alpha\| \leq c\|F(\alpha)\|$. Substitute this into the above inequality to obtain $\|\alpha\|_2 \leq c\|F(\alpha)\|$.

Assume that F is surjective for some $s > 2$. Given $f \in \Omega^s(M) \subset \Omega^{s-1}(M)$, by the induction hypothesis there exists $\alpha \in \Omega^{s+1}(M)$ such that $F(\alpha) = f$. Since $G(\alpha) = F(\alpha) - 2\alpha \in \Omega^s(M)$, another application of Lemma VIII.6 implies $\alpha \in \Omega^{s+2}(M)$.

As far as continuity of the inverse is concerned, it suffices to show that $\|G(\alpha)\|_s \leq c\|F(\alpha)\|_s$ since Lemma VIII.6 implies that $\|\alpha\|_{s+2} \leq c\|G(\alpha)\|_s$. Suppose that this holds for $s-1$, then

$$\|\alpha\|_s \leq \|\alpha\|_{s+1} \leq \text{const } \|G(\alpha)\|_{s-1} \leq \text{const } \|F(\alpha)\|_{s-1} \leq \text{const } \|F(\alpha)\|_s \ .$$

Hence

$$\|G(\alpha)\|_s \leq \|F(\alpha)\|_s + 2\|\alpha\|_s \leq \text{const } \|F(\alpha)\|_s \ .$$

The proof of the first part of Theorem VIII.3 referred to the operator $T = \Delta_R + qI$, $q \geq 0$, is very similar. The formula

$$\langle Tf, f \rangle = \langle \Delta_R f, f \rangle + q\langle f, f \rangle$$

implies that it is injective and $Tf = 0$ implies $f = 0$.

The fact that it is surjective and the inverse is continuous are also proved by induction. We have to consider the inner product defined on $\mathcal{F}^1(M)$,

$$B[f, g] = \langle \nabla f, \nabla g \rangle + q\langle f, g \rangle \ ,$$

to conclude that for every $h \in \mathcal{F}^0(M)$ there exists $f \in \mathcal{F}^1(M)$ such that $Tf = h$. Since f and $\Delta_R f = Tf - qf$ belong to $\mathcal{F}^0(M)$, Lemma VIII.5 also implies that $f \in \mathcal{F}^2(M)$, with $\|f\|_2 \leq \|\Delta_R f\|$. The last inequality implies that the inverse is continuous since

$$\|\Delta_R f\| = \|Tf - qf\| \leq \|Tf\| + q\|f\|$$

and $\|f\| \leq \text{const } \|Tf\|$ in view of

$$\langle Tf, f \rangle = \langle \Delta_R f, f \rangle + q\|f\|^2 \geq q\|f\|^2 \ .$$

The induction will be completed if we prove that $\|\Delta_R f\|_s \leq \text{const}\ \|Tf\|_s$ for all $s > 2$. By Lemma VIII.5, we know that $\|f\|_s \leq \text{const}\ \|\Delta_R f\|_{s-2}$. Then

$$\begin{aligned}\|\Delta_R f\|_s &\leq \|Tf\|_s + \|f\|_s \leq \|Tf\|_s + c\|\Delta_R f\|_{s-2} \\ &\leq \|Tf\|_s + c\|Tf\|_{s-2} \leq c\|Tf\|_s\,.\end{aligned}$$

VIII.4 Spherical model

Consider the model $V = I \times \mathbb{S}^3$, where $\mathbb{S}^3$ is the unit sphere of $\mathbb{R}^4$ endowed with the metric $\hat{g}$ induced by $\mathbb{R}^4$. Take a hypersurface M in V of the form $\{t_0\} \times \mathbb{S}^3$ and consider, as usual, the covariant 2-tensors g and k on M, the function F and the vector field X on M defined as before and corresponding to this model.

In order to study stability by linearization of Einstein's equation in this model, we have to study (as we have already seen in Section V.3, stability by linearization of the map

$$\begin{array}{cccc} & \mathcal{S}^s(g) \times \mathcal{S}^{s-1}(k) & & \\ & \cup & & \\ \Phi: & \mathcal{U} & \longrightarrow & \mathcal{F}^{s-2}(F) \times \Omega^{s-2}(X) \\ & (g', k') & \rightarrow & (\mathcal{H}(g', k'), \gamma(g', k'))\,. \end{array}$$

First, in order to simplify calculations, we introduce the variable $p' = k' - \operatorname{tr}_{g'}(k')g'$. Since $\operatorname{tr}(p') = -2\operatorname{tr}(k')$, we have $k' = p' - \frac{1}{2}\operatorname{tr}(p')g'$. Then

$$\begin{aligned} k'\!\cdot\! k' &= p'\!\cdot\! p' - \operatorname{tr}(p')p'\!\cdot\! g' + \frac{1}{4}\operatorname{tr}^2(p')g'\!\cdot\! g' \\ &= p'\!\cdot\! p' - \operatorname{tr}^2(p') + \frac{3}{4}\operatorname{tr}^2(p') = p'\!\cdot\! p' - \frac{1}{4}\operatorname{tr}^2(p')\,. \end{aligned}$$

Substitute $\mathcal{H}(g',k') = \frac{1}{8}(k' \cdot k' - \operatorname{tr}^2(k') - 4R(g'))$ and $\gamma(g',k')_i = \frac{1}{2}\nabla^j(k'_{ij} - \operatorname{tr}(k')g'_{ij})$ to obtain

$$\begin{aligned} \mathcal{H}(g',p') &= \frac{1}{8}\left(p'\cdot p' - 4R(g') - \frac{1}{2}\operatorname{tr}^2(p')\right) \\ \gamma(g',p')_i &= \frac{1}{2}\nabla^j p'_{ij}\,. \end{aligned}$$

VIII.4.1 First and second derivatives of Φ

The operators $D_{(g,p)}\Phi$, $(D_{(g,p)}\Phi)^*$ and $D^2_{(g,p)}\gamma$ are necessary for proving that this model is not stable. The calculation of the component $D_{(g,p)}\mathcal{H}$ of course is very similar to that of $D_{(g,k)}\mathcal{H}$ in Section V.6. Substitute $g' = g + h$ and $p' = p + P$

into the last expression for $\mathcal{H}(g',p')$ to obtain, for the first term,

$$\begin{aligned} p' \cdot_{g'} p' &= g'^{ir} g'^{js} p'_{ij} p'_{rs} \approx (g^{ir} - h^{ir})(g^{js} - h^{js})(p_{ij} + P_{ij})(p_{rs} + P_{rs}) \\ &\approx g^{ir} g^{js} p_{ij} p_{rs} + g^{ir} g^{js} P_{ij} p_{rs} + g^{ir} g^{js} p_{ij} P_{rs} - h^{ir} g^{js} p_{ij} p_{rs} - g^{ir} h^{js} p_{ij} p_{rs} \\ &= p \cdot p - 2p \cdot P - 2h \cdot (p \times p) \, . \end{aligned}$$

The first-order approximation of the term $\mathrm{tr}^2_{g'}(p')$ at the initial pair (g,p) is two times the derivative of $\mathrm{tr}_{g'}(p')$ at (g,p) multiplied by $\mathrm{tr}_g(p)$:

$$\begin{aligned} &2\, \mathrm{tr}_g(p)(\,\mathrm{tr}_{g+h}(p+P) - \,\mathrm{tr}_g(p)) \approx 2\, \mathrm{tr}_g(p)\{(g^{ij} - h^{ij})(p+P)_{ij} - g^{ij} p_{ij}\} \\ &= 2\, \mathrm{tr}_g(p)(g^{ij} P_{ij} - h^{ij} p_{ij}) = 2\, \mathrm{tr}_g(p)\, \mathrm{tr}_g(P) - 2\, \mathrm{tr}_g(p) h \cdot p \, . \end{aligned}$$

Concerning the term $R(g')$, its derivative is given by (V.17). Summarizing, we have

$$\begin{aligned} D_{(g,p)}\mathcal{H}(h,P) =& \frac{1}{8}[\,2p \cdot P - 2h \cdot (p \times p) + 4h \cdot Ric(g) - 4\nabla^i \nabla^j h_{ij} \\ &+ 4\nabla^i \nabla_i \mathrm{tr}(h) - \mathrm{tr}(p)\mathrm{tr}(P) + \mathrm{tr}(p) h \cdot p\,]. \end{aligned}$$

Now we calculate the operators $D_{(g,p)}\gamma(h,P)$ and $D^2_{(g,p)}\gamma(h,P)$ at the same time since they are, respectively, the terms of order 1 and of order 2 with respect to h and P of $\gamma(g',p')_i = g'^{jk} \nabla'_k p'_{ij}$ when $g' = g + h$ and $p' = p + P$. Before beginning our calculation, we recall (Section III.1) that $Q(X,Y) = \nabla'_X Y - \nabla_X Y$. Hence

$$\begin{aligned} \nabla'_k p'_{ij} &= (\nabla'_{\partial_k} p')(\partial_i, \partial_j) = \partial_k p'(\partial_i, \partial_j) - p'(\nabla'_{\partial_k} \partial_i, \partial_j) - p'(\partial_i, \nabla'_{\partial_k} \partial_j) \\ &= \partial_k p'_{ij} - p'(\nabla_{\partial_k} \partial_i + Q(\partial_k, \partial_i), \partial_j) - p'(\partial_i, \nabla_{\partial_k} \partial_j + Q(\partial_k, \partial_j)) \\ &= \nabla_k p'_{ij} - Q^l_{ki} p'_{lj} - Q^l_{kj} p'_{li} = \nabla_k (p_{ij} + P_{ij}) \\ &\quad - Q^l_{ki}(p_{lj} + P_{lj}) - Q^l_{kj}(p_{il} + P_{il}) \end{aligned}$$

where, by equality (V.9), we have $Q^k_{ij} = \frac{1}{2} g'^{kl}(\nabla_i h_{jl} + \nabla_j h_{il} - \nabla_l h_{ij})$. Since $\gamma(g',p')_i = (1/2) g'^{jk} \nabla'_k p'_{ij}$, it will be necessary to know the terms of order 2 with respect to h of the matrix $g'^{-1} = (g+h)^{-1}$. These terms are given by equation (III.3), $g'^{jk} \cong g^{jk} - h^{jk} + (h \times h)^{jk}$. The terms of order 1 with respect to h and P of $\gamma(g',p')_i = 1/2 g'^{jk} \nabla'_k p'_{ij}$ will come either from multiplication of $-h^{jk}$ by terms of order zero of $\nabla'_k p'_{ij}$, that is $-h^{jk} \nabla_k p_{ij}$, or from $g^{jk} \nabla_k P_{ij}$, or from multiplication of g^{jk} by terms of order 1 of $\nabla'_k p'_{ij}$. The latter ones appear from multiplication of g^{jk} by the terms $Q^l_{ki} p_{lj}$ and $Q^l_{kj} p_{il}$ replacing g' by g in the expression of Q. The first of these terms is

$$g^{jk} Q^l_{ki} p_{lj} \cong g^{jk} \frac{1}{2} g^{lm} (\nabla_k h_{im} + \nabla_i h_{km} - \nabla_m h_{ki}) p_{lj} = \frac{1}{2} p^{km} \nabla_i h_{km} \, .$$

The first and the third terms are opposite since the index i, which is the only one that is not mute, is placed at the same place. The second product is

$$g^{jk} Q^l_{kj} p_{il} \cong g^{jk} \frac{1}{2} g^{lm} (\nabla_k h_{jm} + \nabla_j h_{km} - \nabla_m h_{kj}) p_{il} = p_{il} \nabla^j h^l_j - \frac{1}{2} p_{il} \nabla^l \,\mathrm{tr}(h) \, .$$

Hence the operator $D_{(g,p)}\Phi(h,P)$ is

$$\begin{aligned} D_{(g,p)}\mathcal{H}(h,P) =& \frac{1}{8}[2p\cdot P - 2h\cdot(p\times p) + 4h\cdot Ric(g) - 4\nabla^i\nabla^j h_{ij} \\ &+ 4\nabla^i\nabla_i \mathrm{tr}(h) - \mathrm{tr}(p)\mathrm{tr}(P) + \mathrm{tr}(p) h\cdot p]\,, \\ D_{(g,p)}\gamma(h,P)_i =& \frac{1}{2}\left[\nabla^s P_{is} - h^{rs}\nabla_r p_{is} - \frac{1}{2}p^{rm}\nabla_i h_{rm}\right. \\ &\left. - p_i^m \nabla^s h_{sm} + \frac{1}{2}p_i^m\nabla_m\, \mathrm{tr}(h)\right]\,. \end{aligned} \tag{VIII.5}$$

The terms $(h\times h)^{jk}\nabla_k p_{ij}$ are of order 2 in h and P. The other terms of order 2 in h and P come from products of g^{jk} by terms of order 2 in h and from products of g^{jk} by terms of first order in h and P. So we have

$$\begin{aligned} (D^2_{(g,p)}\gamma)(h,P)_i =& \frac{1}{2}\,[\nabla_r((h\times h)^{rs}p_{is}) + h^{rs}p_s^m\nabla_i h_{rm} \\ &- \frac{1}{2}h^{rs}p_i^m\nabla_m h_{rs} - \frac{1}{2}h^{lm}p_{il}\nabla_m\,\mathrm{tr}(h) \\ &- \frac{1}{2}P^{mr}\nabla_i h_{rm} + \frac{1}{2}P_i^m\nabla_m\,\mathrm{tr}(h) \\ &- \nabla_r(h^{rs}P_{is})]\,. \end{aligned} \tag{VIII.6}$$

VIII.4.2 Adjoint map of $D\Phi$

The adjoint $(D_{(g,p)}\Phi)^*$ applied to an element $(f,Y)\in\mathcal{F}(M)\times\Omega(M)$ belonging to the class C^∞ and having a compact support will be a pair $(A,B)\in\mathcal{S}_2(M)\times\mathcal{S}_2(M)$, both related by

$$\langle D_{(g,p)}\Phi(h,P),(f,Y)\rangle = \langle (h,P),(D_{(g,p)}\Phi)^*(f,Y)\rangle\,.$$

Using the inner products introduced in (VI.1) the last definition of the adjoint operator is translated into

$$\begin{aligned} &4\int_M D_{(g,p)}\mathcal{H}(h,P)\cdot f(x)dx + 4\int_M g(D_{(g,p)}\gamma(h,P),Y)dx \\ &= \int_M h\cdot A dx + \int_M P\cdot B dx \end{aligned} \tag{VIII.7}$$

for any pair (h,P).

In order to identify $A(f,Y)$ and $B(f,Y)$ as that accompanying h and P, the integrands on the left-hand side of (VIII.7) should be appropriately handled. For example, in the expression of $8D_{(g,p)}\mathcal{H}(h,P)f$, the term $4f\nabla^i\nabla^j h_{ij}$ should be transformed in order for a divergence to appear:

$$4f\nabla^i\nabla^j h_{ij} = \mathrm{div} - 4(\nabla^j h_{ij})\nabla^i f = \mathrm{div} + 4h_{ij}\nabla^j\nabla^i f\,.$$

As far as the term $4f\nabla^i\nabla_i \mathrm{tr}(h)$ is concerned, we have

$$4f\nabla^i\nabla_i \mathrm{tr}(h) = \mathrm{div} + 4g^{jk}h_{jk}\nabla^i\nabla_i f = -4(g\cdot h)\Delta f\,.$$

Hence we arrive at

$$\begin{aligned}8D_{(g,p)}\mathcal{H}(h,P)f =& 2fp\cdot P - 2fh\cdot(p\times p) + 4fh\cdot Ric(g) - 4h_{ij}\nabla^j\nabla^i f\\ &- 4(g\cdot h)\Delta f - f\ \mathrm{tr}(p)\ \mathrm{tr}(P) + f\ \mathrm{tr}(p)h\cdot p + \ \mathrm{div}\ .\end{aligned}$$

The mechanics is the same for the other integrand $g(D_{(g,p)}\gamma(h,P),Y)$. For example the first term $(\nabla^s P_{is})Y^i$ is handled as follows: $(\nabla^s P_{is})Y^i = \nabla^s(P_{is}Y^i) - P_{is}\nabla^s Y^i = -P_{is}\nabla^s Y^i + \mathrm{div}$. By doing this with all terms containing as factors derivatives of h and those of P, we obtain:

$$\begin{aligned}2g((D_{(g,p)}\gamma)(h,P),Y) &= -P_{is}\nabla^s Y^i - h^{rs}(\nabla_r p_{is})Y^i\\ &+ \frac{1}{2}h_{mr}(\nabla_i p^{mr})Y^i + \frac{1}{2}h_{mr}p^{mr}\nabla_i Y^i + h_{sm}(\nabla^s p_i^m)Y^i\\ &+ h_{sm}p_i^m\nabla^s Y^i - \frac{1}{2}\ \mathrm{tr}(h)(\nabla^l p_{il})Y^i - \frac{1}{2}\ \mathrm{tr}(h)p_{il}\nabla^l Y^i + \mathrm{div}\ .\end{aligned}$$

The second and the fifth term on the right-hand side of the above expression are cancelled out since they are equal. The third and the sixth term, in view of Proposition I.7, are grouped in the term $\frac{1}{2}h.L_Y p$. The first term (by Corollary I.8) is written as $-\frac{1}{2}P.L_Y q$. For the same reason, the eighth term is $-\frac{1}{4}\ \mathrm{tr}(h)p\cdot L_Y g$. The seventh term, since $\gamma(g,p)_i = 1/2\nabla^j p_{ij}$, equals $-\ \mathrm{tr}(h)g(\gamma(g,p),Y)$. Finally

$$\begin{aligned}2g(D_{(g,p)}\gamma(h,P),Y) =& -\frac{1}{2}P\cdot L_Y g + \frac{1}{2}(h\cdot p)\mathrm{div}(Y) - \ \mathrm{tr}(h)g(\gamma(g,p),Y)\\ &+ \frac{1}{2}(h\cdot L_Y p) - \frac{1}{4}\ \mathrm{tr}(h)p\cdot L_Y g + \ \mathrm{div}\ .\end{aligned}$$

Since the manifold M is compact and has no boundary, the integral over M of a divergence equals zero. By simple equating, we arrive at the following expressions for A and B:

$$\begin{aligned}A(f,Y) =& - fp\times p + 2fRic(g) - 2\mathrm{Hess}(f)\\ &- 2(\Delta f)g + \frac{1}{2}f\ \mathrm{tr}(p)p + L_Y p + \mathrm{div}(Y)p\\ &- \frac{1}{2}(p\cdot L_Y g)g - g(\gamma(g,p),Y)g,\\ B(f,Y) =& - L_Y g + fp - \frac{1}{2}f\ \mathrm{tr}(p)g\ .\end{aligned} \tag{VIII.8}$$

VIII.4.3 Proof of instability

As we have seen in (VIII.1), we have here $F = -\rho(t_0)$ and $X = 0$, and M is the submanifold $\{t_0\}\times\mathbb{S}^3$ of $V = I\times\mathbb{S}^3$ endowed by the Riemannian metric $g = \zeta(t_0)^2\hat{g}$, where $\hat{g}$ is the usual metric (of curvature 1) on $\mathbb{S}^3$.

The main result of this section is the following.

Theorem VIII.7. *There is no neighborhood $\mathcal{U}$ of (g,p) in $\mathcal{S}^s(g)\times\mathcal{S}^{s-1}(p)$ such that the set of pairs $(g',p')\in\mathcal{U}$ satisfying $\Phi(g',p')=\chi(-\rho(t_0),0)$ is a differentiable submanifold of $\mathcal{U}$.*

Before proving this theorem, we can already state the following.

Corollary VIII.8. *There cannot exist any closed subspace $\mathcal{K}$ of $\mathcal{F}^{s-2}(-\rho(t_0))\times\Omega^{s-2}(0)$ such that Φ is stable by linearization at point (g,p) in the direction of $\mathcal{K}$.*

Proof. Indeed, in Definition V.3 of stability by linearization with respect to any subspace $\mathcal{K}$ it is required that $\Phi^{-1}(\Phi(g,p))$ should be a differentiable manifold in a neighborhood of (g,p) which is claimed false by the theorem. □

Proof of Theorem VIII.7. Assume that there exists a neighborhood $\mathcal{U}$ of (g,p) in $\mathcal{S}^s(g)\times\mathcal{S}^{s-1}(p)$ such that the set of pairs $(g',p')\in\mathcal{U}$ satisfying $\Phi(g',p')=-\chi(\rho(t_0),0)$ is a differentiable submanifold of $\mathcal{U}$. Let $\lambda\to(g'(\lambda),p'(\lambda))$ be a small curve belonging to this manifold and passing through (g,p) for the value $\lambda=0$. Denote by $(h,P)\in\mathcal{S}^s(M)\times\mathcal{S}^{s-1}(M)$ the tangent vector to this curve at the point (g,p). Let (h',P') be the second derivative $d^2(g'(\lambda),p'(\lambda))/d\lambda^2$ at the point $\lambda=0$. Since $\Phi(g'(\lambda),p'(\lambda))=-\chi(\rho(t_0),0)$ for all λ, by deriving this equality at the point $\lambda=0$, we obtain $(D_{(g,p)}\Phi)(h,P)=0$. Another derivation at $\lambda=0$ gives

$$(D^2_{(g,p)}\Phi)(h,P)+(D_{(g,p)}\Phi)(h',P')=0\;. \tag{VIII.9}$$

Let now (f,Y) be an arbitrary element of $\mathcal{F}^\infty(M)\times\Omega^\infty(M)$. By scalar multiplication of (VIII.9) by (f,Y), we have

$$\langle(D^2_{(g,p)}\Phi)(h,P),(f,Y)\rangle+\langle(D_{(g,p)}\Phi)(h',P'),(f,Y)\rangle=0\;.$$

It can be deduced from the equality

$$\langle(D_{(g,p)}\Phi)(h,P),(f,Y)\rangle=\langle(h,P),(D_{(g,k)}\Phi)^*(f,Y)\rangle$$

that for all $(h,P)\in\mathcal{S}^s(M)\times\mathcal{S}^{s-1}(M)$ that belong to $\ker(D_{(g,p)}\Phi)$ and all $(f,Y)\in\mathcal{F}^\infty(M)\times\Omega^\infty(M)$ belonging to $\ker(D_{(g,p)}\Phi)^*$ we have

$$\langle(D^2_{(g,p)}\Phi)(h,P),(f,Y)\rangle=0\;. \tag{VIII.10}$$

The strategy for proving Theorem VIII.7 (using an idea from [2]) will consist in finding a pair (h,P) in the kernel of $D_{(g,p)}\Phi$ and a pair (f,Y) belonging to the kernel of $(D_{(g,p)}\Phi)^*$ for which (VIII.10) will fail to hold.

In a Robertson-Walker model $\gamma(g,p)=0$. Therefore the last term of the expression for A in (VIII.8) equals zero. Since $g=\zeta^2(t_0)\hat g$ and $k=2C\,g$, with $C=\dot\zeta(t_0)/\zeta(t_0)$, we obtain $p=k-\operatorname{tr}_g(k)g=-4Cg$. The new expressions (VIII.1) in the variables (g,p) are:

$$\begin{aligned}
&Ric(g)=\frac{2\kappa g}{\zeta^2(0)}, && R(g)=\frac{6\kappa}{\zeta^2(0)}, && p=-4Cg,\\
&p\times_g p=16C^2g, && p\cdot_g p=48C^2, && \operatorname{tr}_g(p)=-12C\,.
\end{aligned} \tag{VIII.11}$$

The components (A,B) of the adjoint operator $(D_{(g,p)}\Phi)^*$ given by (VIII.8) are now, putting $\kappa = 1$ in the above expressions,

$$A(f,Y) = \left(8C^2 + \frac{4}{\zeta^2(0)}\right) fg - 4CL_Y g - 2\mathrm{Hess}(f) - 2(\Delta f)g\ ,$$
$$B(f,Y) = 2Cfg - L_Y g\ .$$

The metric $g = \zeta^2(t_0)\hat{g}$ is the restriction of the Euclidean metric of $\mathbb{R}^4$ to the sphere $\mathbb{S}^3(\zeta(t_0))$ or radius $\zeta(t_0)$ centered at the origin. Choose $f = 0$ and Y to be a Killing vector field of g, $L_Y g = 0$, as elements of the kernel of $(D_{(g,p)}\Phi)^*$. For any pair $(h,P) \in \mathrm{Ker}(D_{(g,p)}\Phi)$ and a pair of the form $(0,Y) \in \mathrm{Ker}(D_{(g,p)}\Phi)^*$, we have

$$\langle (D^2_{(g,p)}\Phi)(h,P), (0,Y)\rangle = 4\int_M g((D^2_{(g,p)}\gamma)(h,P),Y)dv\ ,$$

where dv denotes the volume element of $M = \mathbb{S}^3(\zeta(t_0))$ by g. Expression (VIII.6) of $(D^2_{(g,p)}\Phi)(h,P)$ in a Robertson-Walker model, using $p_{ij} = -4C\, g_{ij}$, is

$$\begin{aligned}(D^2_{(g,p)}\gamma)(h,P)_i =& \frac{1}{2}[-4C\,\nabla^r(h\times h)_{ri} - 2C\,h^{rm}\nabla_i h_{rm} \\ &+ 2C\,h_{im}\nabla^m\ \mathrm{tr}(h) - \frac{1}{2}P_{mr}\nabla_i h^{rm} \\ &+ \frac{1}{2}P_{im}\nabla^m\ \mathrm{tr}(h) - \nabla_r(h^{rs}P_{is})]\ .\end{aligned} \tag{VIII.12}$$

Therefore

$$\begin{aligned}& g((D^2_{(g,p)}\gamma)(h,P),Y) \\ &= \frac{1}{2}\ [4C\,(\nabla^r Y^i)(h\times h)_{ri} - 2C\,h^{rm}(\nabla_Y h)_{rm} \\ &\quad - 2C\,(\nabla^m Y^i)h_{im}\ \mathrm{tr}(h) - 2C\,Y^i\ \mathrm{tr}(h)\nabla^m h_{im} \\ &\quad - \frac{1}{2}P^{mr}\nabla_Y h_{rm} - \frac{1}{2}Y^i\ \mathrm{tr}(h)\nabla^m P_{im} \\ &\quad - \frac{1}{2}\ \mathrm{tr}(h)P_{im}\nabla^m Y^i - Y^i\nabla_r(h^{rs}P_{is})] + \ \mathrm{div}\,,\end{aligned} \tag{VIII.13}$$

where div means divergence term. If (h,P) has to belong to the kernel of $D_{(g,p)}\Phi$, both components of (VIII.5) should be equal to zero. Since the second expression of (VIII.5) becomes zero and in view of $p_{ij} = -4C\,g_{ij}$, we deduce that $\nabla^m P_{im} = -4C\,\nabla^m h_{im}$. By substituting into (VIII.13), we obtain

$$g((D^2_{(g,p)}\gamma)(h,P),Y) = \frac{1}{2}\,\{-2C\,h\cdot\nabla_Y h - \frac{1}{2}(\nabla_Y h)\cdot P - Y^i\nabla_r(h^{rs}P_{is}\,\} + \mathrm{div}\,.$$

Upon integration the last identity over a compact manifold without boundary, the term $\int_M (h.\nabla_Y h)dv$ vanishes. Indeed,

$$h\cdot\nabla_Y h = h_{rm}Y^i\nabla_i h^{rm} = -h_{rm}(\nabla_i Y^i)h^{rm} - (\nabla_i h_{rm})Y^i h^{rm} + \mathrm{div}\,.$$

Since Y is a Killing vector field, it has divergence zero and therefore the term $h_{rm}(\nabla_i Y^i)h^{rm}$ in the above expression vanishes. By integrating on M, we obtain $\int_M (h \cdot \nabla_Y h)dv = -\int_M (h \cdot \nabla_Y h)dv$. Hence this integral equals zero. Finally

$$4\int_M g(D^2_{(g,p)}\gamma(h,P),Y)dv = -\int_M (\nabla_Y h \cdot P)dv - 2\int_M Y^i \nabla_r (h^{rs} P_{is})dv \,. \tag{VIII.14}$$

The problem consists then in searching for h and P that would vanish in (VIII.5) and a Killing vector field Y on the sphere M for which (VIII.14) would not vanish.

Consider the Euclidean space $\mathbb{R}^4$ where the sphere M is immersed. Denote by (x,y,z,t) the canonical coordinates of $\mathbb{R}^4$. The vector fields $e_1 = (y,-x,t,-z)$, $e_2 = (z,-t,-x,y)$, $e_3 = (t,z,-y,-x)$ are pairwise orthogonal; their Lie brackets are $[e_1\,,e_2] = 2e_3\,,\ [e_2\,,e_3] = 2e_1\,,\ [e_3\,,e_1] = 2e_2$. For example,

$$[e_1,e_2](f) = -2y\partial_z f - 2x\partial_t f + 2t\partial_x f + 2z\partial_y f = 2e_3(f)\,.$$

If $\varepsilon(i,j,k)$ is the sign of the permutation (i,j,k) if the three indices are different and zero otherwise, we have $[e_i,e_j] = 2\varepsilon(i,j,k)e_k$. The fields e_i $(i=1,2,3)$ are perpendicular to the radial field $N = (x,y,z,t)$ with respect to the usual Euclidean metric of $\mathbb{R}^4$. Therefore their restrictions to the sphere M are tangent fields which are pairwise perpendicular at each point. Hence they constitute at each point of the 3-sphere a basis of the tangent space. The Riemann formula (I.4) applied to the fields e_1, e_2, e_3 gives

$$\begin{aligned}
2g(\nabla_{e_i}e_j,e_k) &= -g(e_i,[e_j,e_k]) - g(e_j,[e_i,e_k]) - g(e_k,[e_j,e_i]) \\
&= -g(e_i,2\varepsilon(j,k,i)e_i) - g(e_j,2\varepsilon(i,k,j)e_j) - g(e_k,2\varepsilon(j,i,k)e_k) \\
&= -2\zeta^2(t_0)(\varepsilon(j,k,i) + 2\varepsilon(i,k,j) + 2\varepsilon(j,i,k)) \\
&= 2\varepsilon(i,j,k)\zeta^2(t_0)\,.
\end{aligned}$$

Thus, $\nabla_{e_i}e_j = \varepsilon(i,j,k)e_k$, that is the Christoffel symbols Γ^k_{ij} in the basis $\{e_1,e_2,e_3\}$ defined by $\nabla_{e_i}e_j = \Gamma^k_{ij}e_k$ are just $\varepsilon(i,j,k)$. A Killing vector field X of M satisfies $(L_X g)_{ij} = g(\nabla_{e_i}X,e_j) + g(\nabla_{e_j}X,e_i) = 0$. In particular, e_1, e_2 and e_3 are Killing fields of M. By the fact that the e_i's form a basis and taking into account that the tensors h,P should be symmetric, we choose $Y = e_1$, $h = e_1 \otimes e_2 + e_2 \otimes e_1$, $P = e_1 \otimes e_3 + e_3 \otimes e_1$. We would like to confirm that this triple (Y,h,P) satisfies all required assumptions. First, let us see whether these h and P make the right-hand side in (VIII.5) equal to zero. In a Robertson-Walker model with $\kappa = 1$, equations (VIII.5) take the form

$$\begin{aligned}
(D_{(g,p)}\mathcal{H})(h,P) =& \frac{1}{8}[4C\ \mathrm{tr}(P) + \left(16C^2 + \frac{8}{\zeta^2(0)}\right)\ \mathrm{tr}(h) \\
&- 4\nabla^i\nabla^j h_{ij} - 4\Delta\ \mathrm{tr}(h)],
\end{aligned} \tag{VIII.15}$$

$$(D_{(g,p)}\gamma)(h,P)_i = \frac{1}{2}\nabla^s(4C\,h + P)_{is}\,.$$

Both the trace of h and that of P vanish since, for example,

$$\mathrm{tr}(h) = (g)^{ij} h_{ij} = (1/\zeta^2(t_0))(\sum_i h_{ii}) = 0.$$

The divergences of h and P also vanish. Indeed, if $\{\omega^i\}$ is the dual basis of $\{e_i\}$ and A is an arbitrary symmetric 2-tensor, we have

$$\begin{aligned}\nabla_k A^{ij} &= \nabla_{e_k} A(\omega^i, \omega^j) = e_k(A^{ij}) - A(\nabla_{e_k}\omega^i, \omega^j) - A(\omega^i, \nabla_{e_k}\omega^j) \\ &= e_k(A^{ij}) + \Gamma^i_{kl} A^{lj} + \Gamma^j_{kl} A^{li} \ .\end{aligned}$$

Therefore $\nabla_i A^{ij} = e_i(A^{ij}) + \Gamma^i_{il} A^{lj} + \Gamma^j_{il} A^{il}$. If, as it happens with h and P, the components A^{ij} are constant, the first term in the last expression vanishes; the other two terms do the same: the second term equals zero since $\Gamma^k_{ij} = \varepsilon(i,j,k)$ and the third term equals zero since Γ^k_{ij} is antisymmetric with respect to i, j, while A^{ij} is symmetric with respect to the same indices.

Moreover the term $\nabla_i \nabla_j h^{ij}$ in (VIII.15) vanishes because we have proven that div h=0. Thus tensors h and P make the right-hand side in (VIII.15) equal to zero.

It remains to check that (VIII.14) is not zero. We have $\nabla_Y h = \nabla_{e_1}(e_1 \otimes e_2 + e_2 \otimes e_1) = e_1 \otimes e_3 + e_3 \otimes e_1 = P$ and $(\nabla_Y h) \cdot P = P \cdot P = P^{13} P_{31} + P^{31} P_{13} = 2/\zeta^2(t_0)$. On the other hand, since $Y = e_1$ and h has divergence zero,

$$\begin{aligned} Y^i \nabla_r (h^{rs} P_{is}) &= h^{rs} \nabla_r P_{1s} = h^{12} \nabla_1 P_{12} + h^{21} \nabla_2 P_{11} \\ &= (1/\zeta^2(t_0))(\nabla_1 P_{12} + \nabla_2 P_{11}) \ .\end{aligned}$$

But

$$\begin{aligned} \nabla_1 P_{12} &= \nabla_{e_1} P(e_1, e_2) = e_1(P(e_1, e_2)) - P(\nabla_{e_1} e_1, e_2) - P(e_1, \nabla_{e_1} e_2) \\ &= -P(e_1, e_3) = -1 \ , \\ \nabla_2 P_{11} &= \nabla_{e_2} P(e_1, e_1) = e_2(P(e_1, e_1)) - P(\nabla_{e_2} e_1, e_1) - P(e_1, \nabla_{e_2} e_1) \\ &= P(e_3, e_1) + P(e_1, e_3) = 1 + 1 = 2 \ . \end{aligned}$$

Therefore

$$Y^i \nabla_r (h^{rs} P_{is}) = 1/\zeta^2(t_0) \ .$$

By substituting these results into (VIII.14), we obtain

$$\begin{aligned} 4 \int_M g(D^2_{(g,p)} \gamma(h, P), Y) dv &= - \int_M (\nabla_Y h \cdot P) dv \\ - 2 \int_M Y^i \nabla_r (h^{rs} P_{is}) dv &= (1/\zeta^2(t_0))(-4 \,\mathrm{vol}\{M\}) \neq 0 \ . \end{aligned}$$

□

VIII.5 Universes that are not simply connected

In completion of this chapter, we will consider those Robertson-Walker models, $V = I \times S$, where the Riemannian manifold $(S, \hat{g})$ is **compact**, has constant sectional curvature, but it is **not simply connected**. Let us study the cases with constant curvature ≤ 0, since the compact case with positive curvature only contains $\mathbb{S}^3$ which is simply connected. If $\kappa = 0$, S is the flat torus $\mathbb{R}^3/\Gamma$, where Γ is the group of translations. If the curvature is negative, we have $S = \mathbb{H}^3/\Gamma$, where Γ is a discrete subgroup of isometries of $\mathbb{H}^3$.

Theorem VIII.9. *Einstein's equation is linearization unstable when the initial metric is that of a universe of Robertson-Walker, $I \times S$, where S is the flat torus $\mathbb{T}^3$. On the contrary, it is stable when S is a compact orientable Riemannian manifold having constant negative curvature.*

Proof. Following the same method as that applied in the case of the 3-sphere, we should find a pair (h, P) belonging to $\ker D_{(g,p)}\Phi$ and a Killing vector Y of the 3-torus $\mathbb{T}_3$ for which the integral

$$4 \int_{\mathbb{T}_3} g((D^2_{(g,p)}\gamma)(h,P), Y)dv = -\int_{\mathbb{T}_3} (\nabla_Y h \cdot P)dv - 2\int_{\mathbb{T}_3} Y^i \nabla_r (h^{rs} P_{is})dv$$

would be different from zero. Since the components $g_{ij} = \delta_{ij}$ are independent of the coordinates x^i, any vector field ∂_i is a Killing vector field. By transforms (VIII.11), expressions (VIII.5) of $D_{(g,p)}\Phi$ for $\kappa = 0$ are as follows:

$$(D_{(g,p)}\mathcal{H})(h,P) = \frac{1}{8}\left[4C \operatorname{tr}(P) + 16C^2 \operatorname{tr}(h) - 4\nabla^i\nabla^j h_{ij} - 4\Delta \operatorname{tr}(h)\right],$$
$$(D_{(g,p_o)}\gamma)(h,P)_i = \frac{1}{2}\nabla^s(4C\, h + P)_{is} .$$

Choose $Y = \partial_x$, $h = \partial_x \otimes \partial_y + \partial_y \otimes \partial_x$ and $P = x(\partial_x \otimes \partial_y + \partial_y \otimes \partial_x) - z(\partial_y \otimes \partial_z + \partial_z \otimes \partial_y)$. Then h and P have both the trace and the divergence equal to zero. Moreover, $\nabla_Y h = 0$ and the first integral are also equal to zero. The second term is

$$Y^i \nabla_r (h^{rs} P_{is}) = \nabla_r (h^{rs} P_{1s}) = h^{12}\nabla_1 P_{12} + h^{21}\nabla_2 P_{11} = \nabla_1 P_{12} = \partial_x(x) = 1 .$$

Hence

$$\int_{\mathbb{T}_3} Y^i \nabla_r (h^{rs} P_{is})dv = \operatorname{vol}(\mathbb{T}_3) \neq 0 .$$

Consider now the case where S is compact, orientable and has constant negative curvature. By the Berger-Ebin theorem [9] (see also Theorem VI.12), if M is a compact orientable manifold and if one of the operators $D_{(g,p)}\Phi : \mathcal{S}^s \times \mathcal{S}^{s-1} \to \mathcal{F}^{s-2} \times \Omega^{s-2}$ or its adjoint $(D_{(g,p)}\Phi)^*$ has injective symbol, the following splittings hold:

$$\mathcal{F}^{s-2} \times \Omega^{s-2} = \operatorname{Rang}(D\Phi) \oplus \ker(D\Phi)^* , \; \mathcal{S}^s \times \mathcal{S}^{s-1} = \ker(D\Phi) \oplus \operatorname{Rang}(D\Phi)^*$$

and the fact that $D_{(g,p)}\Phi$ is surjective is equivalent to the fact that $(D_{(g,p)}\Phi)^*$ is injective.

By equations (VIII.8) and (VIII.11), we obtain that the components A and B of $(D_{(g,p)}\Phi)^*$ are written, in a Robertson-Walker model with $\kappa = -1$, as follows:

$$\begin{aligned} A(f,Y) &= \left(8C^2 - \frac{4}{\zeta^2(t_0)}\right) fg - 4CL_Y g - 2\mathrm{Hess}(f) - 2(\Delta f)g, \\ B(f,Y) &= 2Cfg - L_Y g\ . \end{aligned} \tag{VIII.16}$$

The symbol of $(D_{(g,p)}\Phi)^*$ (for each $x \in M$ and $\xi \neq 0$, $\xi \in T_x(M)^*$) is the linear map

$$\begin{array}{rccc} \sigma_L(x,\xi): & T_x(M)\times\mathbb{R} & \longrightarrow & S_2\times S_2 \\ & (Y,f) & \rightarrow & (\alpha,\beta) \end{array}$$

for which we will now write down α and β. By recalling that $(L_Y g)_{ij} = \nabla_i Y_j + \nabla_j Y_i$, $(\mathrm{Hess}\ f)_{ij} = \nabla_i\nabla_j f$ and $\Delta f = -\nabla^i\nabla_i f$, the components of α and β are

$$\begin{aligned} \alpha_{rs} &= -2\xi_r\xi_s f + 2\xi\cdot\xi f g_{rs} - 4C(\xi_r Y_s + \xi_s Y_r) \\ \beta_{rs} &= -\xi_r Y_s - \xi_s Y_r\ . \end{aligned}$$

Let us prove that the symbol is injective. Given $x \in M$, take a system of coordinates in a neighborhood of x such that the matrix (g_{ij}) at x is the identity and $\xi = (\xi_1, 0, 0)$, with $\xi_1 \neq 0$. Let us assume that $\alpha = 0$ and $\beta = 0$. We wish to prove that $f = 0$ and $Y = 0$. From $\beta_{11} = 0$ we obtain $Y_1 = 0$. From $\beta_{21} = 0$ and $\beta_{31} = 0$ we deduce that $Y_2 = Y_3 = 0$. Finally $\alpha_{22} = 0$ implies $f = 0$.

To see that $(D\Phi)^*$ is injective, if we impose $B = 0$ on (VIII.16), we obtain $L_Y g = 2Cfg$. By substitution into the expression of $A(f,Y)$, we have

$$A(f,Y) = \frac{-4}{\zeta^2(t_0)} fg - 2\mathrm{Hess}(f) - 2(\Delta f)g\ .$$

If $A = 0$, then $\mathrm{tr}(A) = 0$ and we obtain from the previous equation

$$\frac{-12}{\zeta^2(t_0)} f - 4\Delta f = 0\ ,$$

that is

$$\Delta f + \frac{3}{\zeta^2(t_0)} f = 0\ .$$

Inner multiplication by f of this equation and the fact that $\langle\Delta f, f\rangle = \langle\delta df, f\rangle = \langle df, df\rangle$ give

$$\langle df, df\rangle + \frac{3}{\zeta^2(t_0)}\langle f, f\rangle = 0\ ,$$

implying that $f = 0$. Coming back to (VIII.16), we obtain $L_Y g = 0$. Therefore the kernel of $(D_{(g,p)}\Phi)^*$ is formed by those pairs that have the form $(0, Y)$, Y

being a Killing vector field of S. On the other hand, any compact differentiable Riemannian manifold whose Ricci tensor is negative definite has a finite number of isometries, that is, it does not have infinitesimal isometries [10]. In a Robertson-Walker model, $V = I \times S$, we have $Ric(g) = \frac{2\kappa g}{\zeta^2(t_0)}$. Therefore, since $\kappa = -1$, $Ric(g)$ is negative, $L_Y g = 0$ implies $Y = 0$ and the operator $(D_{(g,p)}\Phi)^*$ is injective. □

References

[1] **L. Anderson**, Elliptic systems on manifolds with asymptotically negative curvature, *Indiana Univ. math. J.*, 42 (1993), 1359–1388.

[2] **J. M. Arms, J. E. Marsden**, The Absence of Killing Fields is Necessary for Linearization Stability of Einstein's Equations, *Indiana University Math. J.*, 28 (1979), 119–125.

[3] **J. M. Arms, J. E. Marsden, V. Moncrief**, The structure of the space of solutions of Einstein's equations. II. Several Killing fields and the Einstein-Yang-Mills equations, *Ann. Physics*, 144 (1982), 81–106.

[4] **T. Aubin**, *Some Nonlinear Problems in Riemannian Geometry*, Springer Verlag, Berlin, 1998.

[5] **T. Aubin**, Espaces de Sobolev sur les variétés riemannianes, *Bull. Sci. Math.* 100 (1976), 149–173.

[6] **L. A. Bagirov, V. A. Kondrat'ev**, Elliptic equations in $\mathbb{R}^n$ (in Russian), *Differencial'nye Uravnenija*, 11 (1975), 498–504.

[7] **D. Bao, J. E. Marsden, R. Walton**, The Hamiltonian Structure of General Relativistic Perfect Fluids, *Commun. Math. Phys.*, 99 (1985), 319–345.

[8] **R. Bartnik**, Phase space for the Einstein equations, http://arXiv.org/gr-qc/0402070.

[9] **M. Berger, D. Ebin**, Some decompositions of the space of symmetric tensors on a Riemannian manifold, *J. of Diff. Geom.*, 3 (1969), 379–392.

[10] **S. Bochner**, Vector fields and Ricci curvature, *Bull. Am. Math. Soc.* 52 (1946), 776–797.

[11] **U. Brauer**, *Singularitäten in relativistischen Materiemodellen*, PhD Thesis, University of Postdam, 1995.

[12] **J. Bruna**, L^p-estimates for Riesz transforms on forms in the Poincaré space $\mathbb{H}^n$, *Indiana Univ. Math. J.*, 54 (2005), 153–186.

[13] **J. Bruna, J. Girbau**, Mapping properties of the Laplacian in Sobolev spaces of forms on complete hyperbolic manifolds, *Annals of Global Analysis and Geometry*, 25 (2004), 151–176.

[14] **L. Bruna, J. Girbau**, Linearization stability of the Einstein equation for Robertson-Walker models (I), *J. Math. Phys.*, 40 (1999), 5117–5130.

[15] **L. Bruna, J. Girbau**, Linearization stability of the Einstein equation for Robertson-Walker models (II), *J. Math. Phys.*, 40 (1999), 5131–5136.

[16] **L. Bruna, J. Girbau**, Laplacian in the hyperbolic space $\mathbb{H}_n$ and stability of the Einstein equation for Robertson-Walker models. *J. Math. Phys.*, 46 (2005), 072501-1 to 21.

[17] **M. Cantor**, Spaces of functions with asymptotic conditions on $\mathbb{R}^n$, *Indiana Univ. Math. J.*, 24 (1975), 897–902.

[18] **M. Cantor**, Some problems of global analysis on asymptotically simple manifolds, *Comp. Math.*, 38 (1979), 3–35.

[19] **M. Castrillón López, M. J. Gotay, J. E. Marsden**, Parametrization and stress-energy-momentum tensors in metric field theories, *J. Phys. A*, 41 (2008), 1–10.

[20] **M. Castrillón López, M. J. Gotay, J. E. Marsden**, Concatenating Variational Principles and Kinetic Stress-Energy-Momentum Tensor, in: *Variations, Geometry and Physics* (editors: O. Krupková, D. J. Saunders), 2008, pp. 117–128.

[21] **Y. Choquet-Bruhat**, Théorème d'existence pour certaines systèmes d'équations aus derivées partielles non linéaires, *Acta Mathematica*, 88 (1952), 141–225.

[22] **Y. Choquet-Bruhat, S. Deser**, On the Stability of Flat Space, *Ann. Phys.*, 81, 165–178 (1973).

[23] **Y. Choquet-Bruhat, A. Fischer, J. E. Marsden**, Équations des constraintes sur une variété non compacte, *C. R. Acad. Sci., Ser. I: Math.*, 284 (1977), A975–A978.

[24] **Y. Choquet-Bruhat, J. York**, The Cauchy problem, in: *Einstein Centenary Volume* (editor: Held), 1980, pp. 99–172.

[25] **Y. Choquet-Bruhat, J. W. York**, Geometrical well posed systems for Einstein equations, *C.R. Acad. Sci. Paris série I*, 231 (1995), 1089–1095.

[26] **P.T. Chruściel, E. Delay**, Manifold structures for sets of solutions of the general relativistic constraint equations, *Journal of Geometry and Physics*, 51 (2004) 442–472. Available online at http://arxiv.org/abs/gr-qc/0309001.

[27] **P.T. Chruściel, H. Friedrich** (Editors), *The Einstein Equations and the Large Scale Behavior of Gravitational Fields (50 Years of the Cauchy Problem in General Relativity)*, Birkhäuser Verlag, Basel, 2004.

[28] **M. Christ**, Weighted norm inequalities and Schur's lemma, *Studia Math.*, 78 (1984), 309–319.

[29] **J. Eells, J. H. Sampson**, Harmonic mappings of Riemannian manifolds, *Amer. J. of Math.*, 86 (1964), 109–160.

[30] **A. Einstein**, Näherungsweise Integration der Feldgleichungen der Gravitation, *Sitzungsberichte der Preussischen Akademie der Wissenschaften* (1916), 688–696.

[31] **A. Einstein**, Über Gravitationswellen, *Sitzungsberichte der Preussischen Akademie der Wissenschaften* (1918), 154–167.

[32] **A. Fischer, J. E. Marsden**, The Einstein Equations of Evolution – a Geometric Approach, *J. Math. Phys.*, 13 (1972), 546–568.

[33] **A. Fischer, J. E. Marsden**, The Einstein evolution equations as a first-order quasi-linear symmetric hyperbolic system, *Comm. Math. Phys.*, 28 (1972), 1–38.

[34] **A. Fischer, J. E. Marsden**, Linearization Stability of the Einstein Equations, *Bull. Am. Math. Soc.*, 79 (1973), 997–1003.

[35] **A. Fischer, J. E. Marsden**, The initial value problem and the dynamical formulation of general relativity, in: *Einstein Centenary Volume* (editors: Hawking and Israel), Cambridge University Press, Cambridge, 1979, pp. 138–211.

[36] **A. Fischer, J. E. Marsen, V. Moncrief**, The structure of solutions of Einstein's equations. I. One Killing field, *Ann. Inst. H. Poincaré, Sec. A (N.S.)*, 33 (1980), 147–194.

[37] **H. Friedrich, A. Rendall**, The Cauchy problem for the Einstein equations, in: *Einstein's Field Equations and Their Physical Implications*, Lecture Notes in Physics, 540, 2000, pp. 27–223.

[38] **C. R. Graham, J. M. Lee**, Einstein metrics with prescribed infinity on the ball, *Adv. in Math.*, 87 (1991), 186–225.

[39] **M. J. Gotay, J. E. Marsden**, Stress-energy-momentum tensors and the Belinfante-Rosenfeld formula, *Contemporary Mathematics*, 132 (1992), 367–392.

[40] **S. W. Hawking, G. F. R. Ellis**, *The Large Scale Structure of Space-time*, Cambridge University Press, 1973.

[41] **T. J. R. Hughes, T. Kato, J. E. Marsden**, Well-posed quasi-linear second-order hyperbolic systems with applications to nonlinear elastodynamics and general relativity, *Arch. Ration. Mech. Anal.*, 63 (1977), 273–294.

[42] **J. M. Lee**, *Fredholm operators and Einstein metrics on conformally compact manifolds*, Memoirs AMS, 183, 2006.

[43] **J. Leray**, *Hyperbolic differential equations*, Institute for Advanced Studies, 1953.

[44] **A. Lichnerowicz**, *Théories relativistes de la gravitation et de l'électromagnétisme*, Masson et Cie, éditeurs, Paris, 1955.

[45] **R. B. Lockhart**, Fredholm properties of a class of elliptic operators on non-compact manifolds, *Duke Math. J.*, 48 (1981), 289–311.

[46] **R. B. Lockhart, R. C. McOwen**, On Elliptic Systems in $\mathbb{R}^n$, *Acta Math.*, 150 (1983), 125–135.

[47] **R. B. Lockhart, R. C. McOwen**, Correction for On Elliptic Systems in $\mathbb{R}^n$, *Acta Math.*, 153 (1984), 303–304.

[48] **R. B. Lockhart, R. C. McOwen**, Elliptic Differential Operators on Noncompact Manifolds, *Ann. Scuola Norm. Sup. Pisa Cl. Sci.* (4), 12 (1985), 409–447.

[49] **J. E. Marsden**, The Initial Value Problem and Dynamics of Gravitational Fields, in: *Proceedings of the ninth International Conference on General Relativity and Gravitation* (editor: E. Schmutzer), Cambridge University Press, 1983, pp. 115–126.

[50] **J. E. Marsden**, *Lectures on Geometric Methods in Mathematical Physics*, CBMS-NSF Regional Conference Series in Applied Mathematics, 37, Society of Industrial and Applied Mathematics (SIAM), Philadelphia, Pa., 1981.

[51] **R. C. McOwen**, On Elliptic Operators in $\mathbb{R}^n$. *Comm. in Partial Diff. Eq.*, 5 (1980), 913–933.

[52] **V. Moncrief**, Space-Time Symmetries and Linearization Stability of the Einstein Equations (I), *J. Math. Phys., 16* (1975), 493–498.

[53] **B. O'Neill**, *Semi-Riemannian geometry*, Academic Press, New York, 1983.

[54] **G. de Rham**, *Variétés différentiables*, Hermann, Paris, 1960.

[55] **B. F. Schutz**, *A first course in general relativity*, Cambridge University Press, 1985.

[56] **J. Stewart**, *Advanced general relativity*, Cambridge University Press, 1990.

[57] **H. Weyl**, *Space, Time, Matter*, Dover Publications, Inc., New York, 1952.

[58] **S. T. Yau**, Some Function-Theoretic Properties of Complete Riemannian Manifolds and Their Applications to Geometry, *Indiana University Math. J.*, 25, 659–670 (1976).

Index

MIX
Papier aus verantwortungsvollen Quellen
Paper from responsible sources
FSC® C105338

If you have any concerns about our products,
you can contact us on
ProductSafety@springernature.com

In case Publisher is established outside the EU,
the EU authorized representative is:
Springer Nature Customer Service Center GmbH
Europaplatz 3, 69115 Heidelberg, Germany

Printed by Libri Plureos GmbH
in Hamburg, Germany